ENVIRONMENTAL
AND SOCIAL
IMPACT ASSESSMENT

ENVIRONMENTAL AND SOCIAL IMPACT ASSESSMENT

Edited by

FRANK VANCLAY
Charles Sturt University, Australia

DANIEL A. BRONSTEIN
Michigan State University, USA

JOHN WILEY & SONS
Chichester · New York · Brisbane · Toronto · Singapore

Copyright © 1995 by International Association for Impact Assessment

Published in 1995 by John Wiley & Sons Ltd,
Baffins Lane, Chichester,
West Sussex PO19 1UD, England

Telephone National (01243) 779777
International (+44) 1243 779777

Reprinted July 1996

Other Wiley Editorial Offices

John Wiley & Sons, Inc., 605 Third Avenue,
New York, NY 10158-0012, USA

Jacaranda Wiley Ltd, 33 Park Road, Milton,
Queensland 4064, Australia

John Wiley & Sons (Canada) Ltd, 22 Worcester Road,
Rexdale, Ontario M9W 1L1, Canada

John Wiley & Sons (SEA) Pte Ltd, 37 Jalan Pemimpin #05-04,
Block B, Union Industrial Building, Singapore 2057

British Library Cataloguing in Publication Data

A catalogue record for this book is available from the British Library

ISBN 0-471-95764-X

Camera-ready copy supplied by the editors
Printed and bound in Great Britain by Biddles Ltd, Guildford and King's Lynn

This book is printed on acid-free paper responsibly manufactured from sustainable forestation,
for which at least two trees are planted for each one used for paper production.

Contents

Part III: Tools for the Future

List of Contributors
Environmental and Social Impact Assessment

Editors: Frank Vanclay and Daniel A. Bronstein
Frank Vanclay is a key researcher with the Centre for Rural Social Research and teaches
sociology at Charles Sturt University, Wagga Wagga, Australia. Daniel Bronstein is
professor in the Departments of Resource Development and Psychiatry at Michigan State
University, and is managing editor of *Impact Assessment*, IAIA quarterly journal.

Frank Vanclay	phone number:	+61-69-332680
Centre for Rural Social Research	fax number:	+61-69-332792
Charles Sturt University	e-mail:	FVANCLAY@CSU.EDU.AU
Locked Bag 678		
Wagga Wagga NSW 2678 AUSTRALIA		
Daniel A. Bronstein	phone number:	+1-517-353-5326
Dept of Resource Development	fax number:	+1-517-353-8994
Michigan State University	e-mail:	IMPACTS@MSU.EDU
312 Natural Resources Bldg		
East Lansing MI 48824-1222 USA		

Chapter 1:   **Environmental Impact Assessment**
   Leonard Ortolano and Anne Shepherd
Leonard Ortolano is the UPS Foundation Professor of civil engineering at Stanford
University, USA. Anne Shepherd is a postdoctoral researcher in the Department of Civil
Engineering at Stanford University.

Leonard Ortolano	phone number:	+1-415-723-2937
Dept of Civil Engineering	fax number:	+1-415-725-8662
Stanford University	e-mail:	ORTOLANO@CIVE.STANFORD.EDU
Stanford CA 94305-4020 USA		
Anne Shepherd	phone number:	+1-415-328-4650
Dept of Civil Engineering	fax number:	+1-415-725-8662
Stanford University	e-mail:	ANNE@LELAND.STANFORD.EDU
Stanford CA 94305-4020 USA		

Chapter 2:   **Social Impact Assessment**
   Rabel J. Burdge and Frank Vanclay
Rabel Burdge is professor of rural sociology in the Institute of Environmental Studies at
the University of Illinois at Urbana/Champaign, USA. He is a former president of IAIA.
(For information on Frank Vanclay, see above.)

Rabel J. Burdge	phone number:	+1-217-333-2916
Institute for Environmental Studies	fax number:	+1-217-333-8046
University of Illinois	e-mail:	BURDGE@UX1.CSO.UIUC.EDU
1101 W Peabody Dr		
Urbana IL 61801 USA		

Chapter 3:   **Technology Assessment**
                 Alan L. Porter
Alan Porter is director of the Technology Policy and Assessment Center and professor of
Industrial and Systems Engineering, and Public Policy, Georgia Tech, Atlanta, Georgia
USA. He will be president of IAIA in 1995–96.

Alan L. Porter   phone number: +1-404-894-2330
Industrial & Systems Engineering   fax number: +1-404-894-2301
Georgia Tech   e-mail: ALAN.PORTER@ISYE.GATECH.EDU
Atlanta GA 30332-0205 USA

Chapter 4:   **Policy Assessment**
                 Peter Boothroyd
Peter Boothroyd teaches in the School of Community and Regional Planning at the
University of British Columbia in Vancouver, Canada.

Peter Boothroyd   phone number: +1-604-822-4155
Centre for Human Settlements   fax number: +1-604-822-6164
University of British Columbia   e-mail: PETERB@UNIXG.UBC.CA
2206 East Mall
Vancouver BC V6T 1Z3 CANADA

Chapter 5:   **Economic and Fiscal Impact Assessment**
                 F. Larry Leistritz
Larry Leistritz is professor of agricultural economics at North Dakota State University,
USA. He is a former president of IAIA.

F. Larry Leistritz   phone number: +1-701-237-7455
Dept of Agricultural Economics   fax number: +1-701-237-7400
NDSU   e-mail: LLEISTRI@NDSUEXT.NODAK.EDU
Morrill Hall
Fargo ND 58105 USA

Chapter 6:   **Demographic Impact Assessment**
                 Henk A. Becker
Henk Becker is professor of sociology at Utrecht University in The Netherlands. He is a
former president of IAIA.

Henk A. Becker   phone number: +31-30-532101
Dept of Sociology   fax number: +31-30-534405
Utrecht University   e-mail: BECKER@FSW.RUU.NL
PO Box 80410
3508 TC Utrecht THE NETHERLANDS

Chapter 7:   **Health Impact Assessment of Development Projects**
                 Martin H. Birley and Genandrialine L. Peralta
Martin Birley is a senior lecturer with the Liverpool School of Tropical Medicine in the
United Kingdom. He is manager of the British Overseas Development Agency-funded
Liverpool Health Impact Programme and of the Joint WHO/FAO/UNEP/UNCHS Collabo-
rating Centre for environmental management for vector control. Gene Peralta is in charge
of the Environmental Engineering Program at the University of the Philippines and, until
December 1996, is on research leave at the University of Toronto.

Martin H. Birley phone number: +44-51-708-9393
Health Impact Programme fax number: +44-51-708-8733
Liverpool Sch of Tropical Medicine e-mail: MHB@LIVERPOOL.AC.UK
Pembroke Place
Liverpool L3 5QA UK
Genandrialine L. Peralta (until 12/96) phone number: +1-416-978-8654
Dept of Chemical Engineering fax number: +1-416-978-8605
  and Applied Chemistry e-mail: PERALTA@ECF.TORONTO.EDU
University of Toronto
200 College St
Toronto ON M5S 1A4 CANADA
  or
National Engineering Center phone number: +63-2-982471 ext 5819
University of the Philippines fax number: +63-2-987190
Diliman Quezon City 1101 PHILIPPINES e-mail: GPERALTA@ENGG.UPD.EDU.PH

Chapter 8: **Ecological Impact Assessment**
Joanna Treweek

Jo Treweek is a research scientist with the Institute of Terrestrial Ecology at Monks Wood in Cambridgeshire, United Kingdom.

Joanna Treweek phone number: +44-4873-381
Institute of Terrestrial Ecology fax number: +44-4873-590
Monks Wood e-mail: MO_JRT@VAXA.NERC-MONKSWOOD.AC.UK
Abbots Ripton
Huntingdon, Cambridgeshire PE17 2LS UK

Chapter 9: **Environmental Risk Assessment**
Richard A. Carpenter

Richard Carpenter is an environmental risk assessment consultant in Virginia, USA. He was previously associated with the East West Centre in Hawaii.

Richard A. Carpenter phone number: +1-804-974-6010
Rte 5 Box 277 fax number (faxmodem) same as phone
Charlottesville VA 22901 USA e-mail: none available

Chapter 10: **Public Involvement: From Consultation to Participation**
Richard Roberts

Richard Roberts is president of Praxis, a Canadian consulting group, and adjunct professor in the Faculty of Environmental Design, University of Calgary, Canada. He is vice president of the Int'l Assoc for Public Participation Practitioners and former president of IAIA.

Richard Roberts phone number: +1-403-245-6404
Praxis Inc. fax number: +1-403-229-3037
2215 19th Street SW e-mail: PRAXIS1@NOVALINK.COM
Calgary AB T2T 4X1 CANADA

Chapter 11: **Climate Impact Assessment**
Roslyn Taplin and Rochelle Braaf

Ros Taplin is deputy director and Rochelle Braaf is a research fellow with the Climatic Impacts Centre, Macquarie University, Sydney, Australia.

Roslyn Taplin   phone number: +61-2-8508473
Climatic Impacts Centre   fax number: +61-2-8508428
Macquarie University   e-mail: ROS.TAPLIN@MQ.EDU.AU
Sydney NSW 2109 AUSTRALIA
Rochelle Braaf   phone number: +61-2-8509466
Climatic Impacts Centre   fax number: +61-2-8508428
Macquarie University   e-mail: ROCHELLE.BRAAF@MQ.EDU.AU
Sydney NSW 2109 AUSTRALIA

Chapter 12:   **Development Impact Assessment: Impact Assessment of Aid Projects in Nonwestern Countries**
Janice Jiggins

Janice Jiggins, a consultant based in The Netherlands, is currently a visiting professor at the University of Guelph, Ontario, Canada. She has served as president of the International Association of Farming Systems Research and Extension.

Janice Jiggins   phone number: +31-8880-51016
De Dellen 4   fax number: +31-8880-53710
6673 MD Andelst THE NETHERLANDS   e-mail: none available

Chapter 13:   **Environmental Auditing**
Ralf Buckley

Ralf Buckley is professor and director of the Centre for Environmental Management and the International Centre for Ecotourism Research at the Gold Coast campus of Griffith University, Australia.

Ralf Buckley   phone number: +61-75-948668
Engineering & Applied Science   fax number: +61-75-948679
PMB 50   e-mail: R.BUCKLEY@EAS.GU.EDU.AU
Gold Coast Mail Ctr QLD 4217 AUSTRALIA

Chapter 14:   **Environmental Sustainability**
Robert Goodland and Herman Daly

Robert Goodland, an ecologist, is the environmental assessment advisor to the World Bank, which he joined in 1978. He was the 1994–95 president of IAIA. Herman Daly was until recently a senior environmental economist with the World Bank. He is now a senior research scholar at the School of Public Affairs, University of Maryland, USA.

Robert Goodland   phone number: +1-202-473-3203
The World Bank   fax number: +1-202-477-0565
Environment Dept   e-mail: RGOODLAND@WORLDBANK.ORG
1818 H Street NW, S-5043
Washington DC 20433 USA
Herman Daly   phone number: +1-301-405-6360
School of Public Affairs   fax number: +1-301-403-4675
University of Maryland   e-mail: HDALY@PUAFMAIL.UMD.EDU
College Park MD 20742-1821 USA

Editors' Preface:
The State of the Art of Impact Assessment

FRANK VANCLAY
DANIEL A. BRONSTEIN

Impact assessment can be broadly defined as the prediction or estimation of the consequences of a current or proposed action (project, policy, technology). Impact assessment spans the full range of human intellectual endeavour, with a specific but growing number of subfields recognised. But there is considerable diversity of opinion about how many subfields of impact assessment there are and what they might be. For our purposes, we have identified four macro fields of impact assessment (part I) and a range of important subfields (part II) that act as inputs to the four major fields and are regarded as tools of impact assessment, although there may be situations when they can be termed fields in their own right. Finally, we have identified a number of areas of impact assessment that are likely to be increasingly important in the future (part III), some of which provide an evaluation of impact assessment.

Although there were many precursors, common understanding would date the emergence of impact assessment in the form of *environmental impact assessment (EIA)* to the passage of the National Environmental Policy Act of 1969 (NEPA) in the USA, which served as a model for similar legislation in jurisdictions around the world. Within a few years, a separate field of *social impact assessment (SIA)* emerged, largely because of the perceived deficiency of NEPA to respond fully to social impacts. *Technology assessment (TA)* developed in the 1960s as a general area of concern about the unintended consequences of technology on society, but was formalised only in 1972 with the establishment of the US Office of Technology Assessment, an agency that concentrated on policy concerns for the US government. With the emergence of other forms of IA, TA has probably declined in importance in other areas of the world. *Policy assessment (PA)*, which developed out of the policy science discipline, was concerned with impact assessment activities long before NEPA, although under a different rubric.

These four fields of impact assessment have different origins, theoretical bases, and methodological traditions, but all overlap in focus and interest, and to an extent, compete with each other. While EIA and SIA tend to be structured by the regulatory framework in which they are applied (and therefore limited or facilitated as the case may be), TA and PA are generally free of regulatory strictures. A common complaint with all forms of IA is that they tend to be too project based and too locally concerned, ignoring distant and cumulative impacts. Thus there is widespread agreement, at least among practitioners in these four major fields, on the need to expand conceptually and broaden the focus of IA. Such expansion, of course, would mean considerable overlap, complicating the process, perhaps, but also improving the product.

Because impact assessments could be applied to the full range of human activities, a comprehensive list of potential forms of impact assessment would be limitless, and not very helpful. It is more useful to list the fields where there is a discernible discourse and a professional identification. In this book, six subfields of IA are highlighted; four relate to specific forms of impacts (*economic and fiscal, demographic, health,* and *ecological*), and two are widely used tools of IA (*risk assessment* and *public consultation*). Our list probably covers the range or professional fields; although other fields are conceivable (for example, aesthetic, cultural, heritage), they could equally be included amongst the other fields. In short, we believe that we have selected the fields where there is an identifiable discourse and group of professionals, and which have made a major contribution to impact assessment.

In addition to our six tools of IA (part II), we have identified two relatively new areas of IA that are often ignored in other forms of IA but which will become increasingly important. *Climate impact assessment* refers both to the impacts of climate change on human society and the environment and to the impacts of human activities on creating climate change (e.g., global warming, greenhouse gases, and ozone layer depletion). *Development impact assessment* refers to the impacts of development assistance projects in nonwestern countries. Another chapter deals with *environmental auditing,* combining evaluation of past impact assessments and regulatory compliance issues. The final chapter deals with what all impact assessment is ultimately about, *environmental sustainability.*

This book is an initiative of the International Association for Impact Assessment (IAIA). With a growing number of subfields of impact assessment (IA) and much ignorance about IA generally, especially about some of the subfields and how they relate to each other, the IAIA thought it appropriate to commission a state-of-the-art series of papers. Initially these papers were to be published in the IAIA quarterly journal, *Impact Assessment*; Frank Vanclay was appointed to coordinate the project. Very quickly, however, the value of collecting this series of papers into one published volume that would be widely available became evident. Dan Bronstein, the managing editor of *Impact Assessment,* was invited to co-edit the collection. While some of the papers have been or will be published in *Impact Assessment,* they appear here in a more developed form. Other chapters appear only in this volume.

The book aims to outline the state of the art in each of the branches of impact assessment. It is not meant to be an introductory textbook covering basic issues, methodologies, or techniques; rather it addresses the major issues and controversies in the field. The book is aimed at those already involved with impact assessment as well as graduate and advanced students, rather than to novices or amateurs. It is designed to be a resource book to practitioners all over the world. The contributors, all leading figures in their respective subfields, were specifically selected for their professional contributions. All are members of the IAIA, and many have served in executive positions with IAIA. They were instructed to give for their subfield:

- A definition of the field
- A description of the field in general terms
- A brief history of the field
- A discussion about the importance of the field, and the contribution of the field to policymaking and environmental management
- A discussion about the relationship between their field and other fields of impact assessment
- An outline of the substantial issues, conflicts, paradoxes, etc., in the field
- An overview of the theoretical positions in the field
- A discussion of the practical problems in the field
- A discussion of the likely future of the field

Most have responded to this task directly, others have addressed these issues circuitously.

In total, the book presents a broad overview of the field of impact assessment, and its subfields, and how they relate to each other. We hope it will provide a general understanding of impact assessment, as well as a basis for the discussion of the future of impact assessment.

IAIA and the contributors to this volume are concerned about global environmental sustainability; indeed, social and environmental sustainability are what impact assessment is all about. In keeping with this professional commitment, then, IAIA will use the royalties from this book to assist the development of impact assessment in developing nations.

Acknowledgements

In some ways, this book is a tribute to technology. It is an international effort with contributors from all over the planet. Some chapters were co-written by authors who live on different continents and exchanged views and drafts over the electronic network. Many authors e-mailed their completed chapters to the editors for editing. The editors, again on different continents, discussed chapters by e-mail. And the final edited versions were e-mailed to Nancy Gendell at Michigan State University for formatting and copy preparation (electronic typesetting). Thus this book would not have been possible, at least in the time in which it was produced, without this technology. But the technology did not always work. There were frequent problems and frustrations, and high on the of list of obstacles were electronic incompatibilities. Computer experts and systems managers at many sites helped with varying degrees of success in the electronic transfers of the chapters of this book. But special thanks are due to Shawn Lock at Michigan State University who had to deal with all the chapters e-mailed for 'typesetting'.

The greatest acknowledgement possible is due to Nancy Gendell who undertook the massive task of converting the edited chapters into camera-ready copy for the publishers. Her meticulous attention to detail and steadfast work, together with a sense of humour to see her through, has ensured a punctual and professional presentation, for which we are all grateful.

Most of the chapters have been refereed by various individuals known to the editors, but not to the authors. While it would be inappropriate to name them, they are sincerely thanked for their efforts.

This book is a project of the International Association for Impact Assessment, commissioned by the IAIA executive board through the publications committee. As chair of that committee, Rita Hamm initially invited us to undertake this task, and she provided encouragement and advice throughout.

—Frank Vanclay and Daniel A. Bronstein

About IAIA

IAIA is the International Association for Impact Assessment, organized in 1980 to bring together researchers, practitioners, and users of various types of impact assessment from all parts of the world. IAIA involves people from many disciplines and professions. Our members, who represent more than 95 countries, include corporate planners and managers, public interest advocates, government planners and administrators, private consultants and policy analysts, university and college teachers and their students. Regional chapters are active in Brazil, Europe, Canada (Ontario and Québec), South Africa, and Cameroon. One of the unique features of IAIA is the mix of professions represented, which provides outstanding opportunities for interchange to—

▶ Advance the state of the art of impact assessment in applications ranging from local to global.

▶ Develop international and local capability to anticipate, plan, and manage the consequences of development to enhance the quality of life for all.

International conferences are held annually, and training programs are held regularly in conjunction with them. Our quarterly journal, *Impact Assessment*, contains a variety of peer-reviewed research articles, professional practice ideas, and book reviews of recent published titles. *Impact Assessment* provides a one-source link to the latest ideas in the wide-ranging field of impact assessment. The IAIA newsletter, published four times annually, provides members with current information concerning association activities and events.

IAIA has entered into strategic partnerships with a number of national and international organizations. Among these are the World Bank, various United Nations organizations, the Canadian International Development Agency (CIDA), the US Council on Environmental Quality (CEQ), and the Canadian Federal Environmental Assessment Review Office (FEARO). These partnerships have been mutually beneficial, enable our organization to accomplish jointly a number of projects that IAIA would probably not have been able to undertake alone.

IAIA seeks to develop approaches and practices for integrated and comprehensive impact assessment; improve assessment procedures and methods for practical application; promote training of impact assessment and public understanding of the field; provide professional quality assurance by peer review and other means; and share information networks, timely publications, and professional meetings.

For further information or a membership application, contact—

Rita Hamm, Executive Director, IAIA
IBID, North Dakota State University
Hastings Hall, PO Box 5256
Fargo ND 58105-5256 USA
phone: +1-701-231-1006
fax: +1-701-231-1007   e-mail:RHAMM@NDSUEXT.NODAK.EDU

Part I:

MACRO IMPACT ASSESSMENT

Chapter 1
Environmental Impact Assessment[1]

LEONARD ORTOLANO
Stanford University, USA
ANNE SHEPHERD
Stanford University, USA

Environmental impact assessment (EIA) is required, in one form or another, in more than half the nations of the world. This chapter examines how EIA requirements have influenced projects, programs, and organisations. EIAs have had far less influence than their original supporters had hoped they would. This chapter provides organisational and methodological reasons for this disparity and indicates ways in which EIA might be used more productively in the future.

ALTERNATIVE CONCEPTIONS OF EIA

Perhaps the most common conception of EIA is as a 'planning tool': assessments are done to forecast and evaluate the impacts of a proposed project and its alternatives. This perspective of EIA as a planning tool has been referred to as the 'technocratic paradigm', since it is a view widely held by engineers and scientists who conduct EIAs (Formby 1990:191).

According to this technocratic paradigm, EIA is an element of the 'rational model' of planning and decision making. In this model, objectives and criteria for evaluating alternative projects are identified at the outset. Engineers and planners then design alternative projects and do studies (including cost-benefit analyses and EIAs) to predict impacts and evaluate the alternatives. The information generated is then used to select one project from among the alternatives.

As a planning tool, EIA serves largely to inform interested parties of the likely environmental impacts of a proposed project and its alternatives. It illuminates environmental issues to be considered in making decisions. Generating and circulating information on impacts has salutary effects—it forces a 'hard look' at the environmental effects of projects, and it facilitates coordination among those affected by the proposed project.

This technocratic paradigm for EIA has been criticised because it ignores politics and models decision making in an unrealistic way (Formby 1990). As noted by

Environmental and Social Impact Assessment - Edited by F. Vanclay and D.A. Bronstein. Copyright © 1995 by the International Association of Impact Assessment. Published in 1995 by John Wiley & Sons Ltd. A version of this chapter appeared in *Impact Assessment* **13**(1), the quarterly journal of IAIA.

Culhane (1993:74), "the critical literature on EISs [environmental impact statements] has consistently documented the failure of EISs, written both in the United States and abroad, to meet basic tests of the rational-scientific model." Decisions on significant public or private development projects are not, in fact, made following the logic of the rational model. Instead, decisions are influenced by 'non-scientific' factors, such as agency and corporate power, and interest group politics. Courses of action are often determined more by the project sponsor's narrow goals, intra-organisational politics and interorganisational rivalries than by scientific studies of environmental impacts. Formby (1990:193) adds, "[T]he disadvantage of the techno-cratic view of EIA is that it can blind those concerned to the political realities of the EIA process and the need to take account of these. . . . [And] while EIA continues to be carried out, it becomes decreasingly related to actual decisions."

A more realistic conception of decision making embraces political realities and recognises that "the ultimate purpose of EIA is not just to assess impacts; it is to improve the quality of decisions" (Formby 1990:193). EIA is placed in a political context; requirements to conduct EIAs can influence the attitudes of top officials, the strategies of project opponents, and the standard operating procedures of organi-sations proposing projects. Using this broader conception of EIA, the focus is not just on scientific studies or EISs, it is on improving decisions.

The scope of the term 'environment' as used commonly in EIA work is clarified to show its links with other chapters in this book. The early literature on EIA (in the 1970s) sometimes was equivocal on whether 'environment' meant only the bio-physical (or natural) environment. However, by the 1990s, the normative literature on EIA generally used the term 'environment' in a broad sense, and EIAs were meant to include all nonmonetary impacts (i.e., impacts not included in a benefit-cost analysis). Thus, social impact evaluations, risk assessments, visual impact studies, cumulative impact analyses, etc., are all viewed herein as elements of an environmental impact assessment. In practice, EIA is often narrowly focused on bio-physical impacts, in part because social impacts and other nonbiophysical effects are not fully included in environmental impact assessment legislation.

ORIGINS AND EVOLUTION OF EIA

The origins of EIA have been so well told by others (e.g., Caldwell 1982) that only a brief note on the subject is offered here. Although predictions of how human actions affect the environment are as old as recorded history, the contemporary usage of 'environmental impact assessment' has its origins in the United States National Environmental Policy Act of 1969 (NEPA).[2] The impetus for that law was

---

[2] NEPA was passed by US Congress in 1969 and signed into law on January 1, 1970, by Pres. Richard Nixon. In the US, the date on a statute is typically the year of congressional passage, not presidential signature. The proper reference is the National Environmental Policy Act of 1969 or NEPA 1969. (NEPA is

the widespread recognition, in the late 1960s, that some significant environmental problems in the USA resulted from actions by the US government itself. The appetites of large infrastructure agencies in charge of water resources projects, highways, and energy facilities appeared to be unquenchable, and the mission statements of those agencies did not force them to account for the adverse environmental impacts of their actions.

NEPA changed that. By a single act of Congress, all federal agencies were required to consider the environmental impacts of their decisions. NEPA included 'action-forcing provisions' to ensure that agencies gave more than lip service to their new responsibilities. The most widely known of these provisions is in Section 102(2)(C), which states that "all agencies of the Federal Government shall include in every recommendation or report on proposals for legislation and other major Federal actions significantly affecting the quality of the human environment, a detailed statement by the responsible official" on the environmental impacts of the proposed action and its alternatives. This detailed statement came to be called an 'environmental impact statement' (EIS), and the activity of preparing and distributing the statement became the 'NEPA process'. The process, which was formalised by regulations (US Council on Environmental Quality 1986), includes preliminary assessments to determine if an EIS is necessary, a 'scoping process' to identify the main environmental issues to be examined, provisions for the public and agencies to comment on a draft EIS, and opportunities for citizens to sue federal agencies that fail to meet their responsibilities under NEPA.

As is summarised in a later section of this chapter, NEPA has influenced significantly both federal projects and federal agencies. It has also influenced, indirectly, the decision-making processes of hundreds of other political jurisdictions. For example, many of the 50 states of the USA have their own programs calling for EIAs. The laws setting up these programs are often referred to as 'little NEPAs', and they vary greatly in scope. One of the most far-reaching of the little NEPAs is the California Environmental Quality Act (CEQA) that governs the actions of local and state government agencies. Because it is practically impossible to implement a major private development project without some discretionary action by local government, CEQA applies to virtually all significant projects in California. In this sense, CEQA's reach is much broader than that of NEPA and most other little NEPAs (Mandelker 1993; Ferester 1992; Bass and Herson 1994).

The influence of NEPA has not been limited to the United States. By the early 1990s, over 40 countries had EIA programs (Robinson 1992). Some, like the EIA program in the Philippines under former President Ferdinand Marcos, included language similar to Section 102(2)(C) of NEPA. Others (e.g., the Chinese EIA program) were quite different and reflected well-thought-out efforts to tailor requirements for environmental impact statements to the local political context. Some countries set up their EIA programs using laws, while others (e.g., Taiwan)

---

sometimes erroneously referred to as NEPA 1970.) The official code citation is National Environmental Policy Act of 1969, §§2-207, 42 USC §§4321-4347 (1988).

relied on executive actions and administrative orders. In addition to national-level programs, the states (or provinces) in some countries (e.g., Brazil, Canada, and Australia) have established their own EIA requirements. Not surprisingly, there are enormous variations in the scope and quality of EIA among and within countries.

Requirements for EIAs are even imposed on countries that have no formal programs, because bilateral and multilateral aid agencies often call for EIAs on projects they fund. Although aid agencies have spotty records in implementing their own EIA requirements (see Jiggins' chapter), they have been under pressure to improve the ways EIAs are conducted for projects they fund. Some aid agencies have embraced EIA to ensure that their projects contribute to 'sustainable development'.

Environmental impact assessment activities throughout the world have provided extraordinary opportunities to share information at an international level. Indeed, each national, state, or local EIA program can be viewed as a policy experiment whose results can inform the EIA programs of others. Until recently, jurisdictions learned from each other primarily through articles in professional journals, meetings of professional associations (particularly IAIA, the International Association for Impact Assessment), and *ad hoc* communication among government agencies' staff. These avenues of EIA information exchange were augmented and made more systematic in 1993 with the initiation of "The International Study of the Effectiveness of Environmental Assessment." Launched to improve EIA performance by involving countries' environmental agency staff in formal appraisals of EIA practices, this study was a joint initiative of IAIA and Canada's Federal Environmental Assessment Review Office (Sadler 1994a).

IMPACT ASSESSMENT METHODS

The early 1970s witnessed much activity on the development of 'EIA methodologies' as government agencies and consultants struggled to figure out what constituted an EIA and what methods could be used to conduct one. Attempts to discover a single, applicable methodology were creative, but did not yield a widely agreed-upon algorithm. While these attempts generated information helpful in conducting EIAs, the results are not usefully termed a 'methodology'. (Many reviews of what are called 'EIA methodologies' exist; see, e.g., Jain et al. 1993.)

For this discussion of methods, the impact assessment exercise is viewed narrowly; it involves the identification, prediction, and evaluation of impacts. The identification of probable impacts worthy of study is aided by the scoping process—technical specialists, individuals from agencies and nongovernmental organisations (NGOs), and citizens potentially affected by a project give their views on types of impacts likely to be important. Scoping is enriched if the technical specialists know the impacts typically associated with the type of project being considered. To assist in impact identification, there are manuals and computer programs that characterise the impacts generally expected with certain types of projects (see, e.g., Fedra 1991). Some agencies require EIA issue checklists that guide the process of identifying impacts.

The second part of the impact assessment exercise involves prediction. An EIA for a proposed new highway in an urban area, for example, would routinely call for predicting changes in noise levels. Unless that forecasting exercise were simple, it would be conducted by an acoustics specialist. The same principle holds true for other predictions in an EIA. In fact, groups that conduct EIAs often consist of specialists from different disciplines. Methods used for impact prediction are not unique to the EIA process; they are based on engineering, natural science, and social science methods.

Surveys of the methods used to predict impacts in EIAs (see, e.g., Leon 1993; Culhane 1987) find that technical specialists often rely heavily on professional judgement to forecast environmental impacts, and predictions are often so vague they cannot be validated. Mathematical models are also used in making predictions; this practice is sometimes criticised because models are presented as 'black boxes', and the bases for predictions are not made clear. Indeed, because EIAs generally contain so little information about models and their assumptions, "errors that are inherent in this approach are not readily traceable, and the results are not subject to scrutiny" (Leon 1993:657).

Perhaps the most difficult aspect of environmental impact assessment involves evaluating predicted impacts. Attempts have been made to develop algorithms that combine predictions and the subjective values of affected parties to create an overall index to rate individual projects. While these algorithms are sometimes employed in preparing EIAs, they are not universally embraced and their use is controversial (Lawrence 1993). For example, Tu (1993) reports on a major political battle over the validity of applying Battelle's 'environmental evaluation system' (Dee et al. 1972) to appraise a proposed dam on the Liwu River in Taiwan. This project would have destroyed a treasured scenic resource in Taiwan, and the application of the Battelle approach attempted to evaluate the loss of visual resources using a numerical rating. Public criticism of this approach was intense and project opponents ridiculed the effort. After a second, independent EIA was conducted and the economic aspects of the dam were reevaluated, the project was halted and the project area was made into a national park.

In some contexts, consultants preparing EIAs deal with impact evaluation by presenting impact predictions without evaluative judgements or rankings. Regardless of whether an evaluation of impacts is attempted, the amount of information presented in EIA documents can sometimes overwhelm even the most persistent reader. This has prompted the search for clearer formats to display EIA results, e.g., network diagrams and matrices to compare the environmental impacts of project alternatives. Some formats display qualitative descriptions of impacts or ordinally scaled ratings, while others show results as quantitative, weighted impact scores (e.g., Smit and Spaling 1995; Jain et al. 1993).

In summary, EIA practitioners use a variety of methods for involving citizens and agencies in planning and for identifying, predicting, and evaluating impacts. Each project requires a set of methods tailored to the local situation and the time and budget available. There is no single, universally applicable EIA methodology.

INFLUENCE OF EIA ON PROJECTS AND ORGANISATIONS

What has all the effort in setting up policies and programs to implement EIA led to? Unfortunately, most of the resources devoted to EIA have gone into the day-to-day work of preparing environmental documents and administering programs, and relatively little has gone to investigating systematically what EIA has accomplished. This is true for both developing and industrialised countries. Consider, for example, the situation in France. According to Sánchez (1993:262), the French produce over 5,000 EIAs per year, yet "very little attention has been given to the follow-up study of EIA results. . . . [so] it is hard to evaluate the actual impact of EIA on project design and managerial practices" in France.

There have been few systematic studies of how a government's EIA program has affected the organisational structures and decision-making procedures of project proponents. However, much case study work has been done on how EIAs have been conducted for particular projects. Case studies suggest that, in many instances, EIA is not yielding all the benefits it could because the process is undertaken too late and project proponents are concerned primarily with meeting administrative requirements. (Of course, there are exceptions.) The discussion below provides a perspective on what are generally considered as positive effects of EIA. Subsequent sections highlight what many view as shortcomings of EIA.

Effects of EIA on Projects

While EIAs sometimes amount to nothing more than exercises in *pro forma* compliance with legal requirements, there are many cases where EIA has significantly influenced projects. For example, a candid evaluation of the EIA system in the Netherlands (conducted by BCR Consultants) revealed a range of positive influences of EIA on projects:

- *Withdrawal of unsound projects.* DuPont withdrew its proposal to construct a facility for producing an alternative to freon gas because the EIA process revealed 'better and more sensible alternatives' (van de Gronden 1994:12).

- *Legitimation of sound projects.* In planning for a proposed expansion of its refinery at Rotterdam, Esso knew that it would have to meet demanding environmental requirements. The EIS for the refinery legitimised Esso's proposal in that it provided a basis for the public and government authorities to understand the impacts involved and to become convinced of the project's merits. The refinery expansion was approved.

- *Selection of improved project location.* The EIA process of a proposed highway segment between Schoondijke and Sluis in the Netherlands was credited with yielding a corridor location that was "the most environmentally friendly and was ultimately accepted by the Minister" (van de Gronden 1994:17–18).

- *Reformulation of plans.* The EIA process of the industrial park Groot Bijsterhuizen has been characterised as a 'textbook case' demonstrating the potential contributions of EIA in land-use planning. The proposed project was a 350-acre industrial park in two contiguous municipalities. The EIA process structured the debate among interested parties, particularly the two municipalities involved. This process yielded a new zoning proposal, one that "gives considerable attention to 'green zones' and to conditions that businesses should meet in order to be allowed in that area" (van de Gronden 1994:15).

- *Redefinition of goals and responsibilities of project proponents.* The EIA process for a complex combination of private factories and public incineration facilities at Boeldershoek in the region of Twente led to a clarification of the participants' roles and responsibilities. Eventually, the project proposal was reformulated as one that did not involve the original public–private initiative, and a new EIS was prepared. The new proposal involved a collaboration of the Twente Cooperation Unit and the provincial electric company. While the EIA process was characterised as 'arbitrary and faltering', it led to a 'clarification of distribution of responsibilities' for the participants in the venture (van de Gronden 1994:14).

Another positive influence of EIA, one that is more difficult to document than those treated in the Dutch consultants' appraisal, is the role of EIA in discouraging project proponents from proposing an environmentally damaging project for fear that it will not survive a review of its environmental impacts.

The positive influences listed above are neither exhaustive nor typical. Based on case studies in the literature, the most common positive outcomes of EIA are suggestions for measures needed to mitigate (or offset) the adverse effects of a proposed plan. Although the term 'mitigation' is widely used in EIA, it is confusing and needs to be elaborated upon. Some analysts say that a mitigation occurs when environmentally damaging elements of a proposed plan are dropped. In this sense, 'to mitigate' is to avoid having the damage take place. It is more common to say that mitigation involves one or more of the following:

- Minimising adverse effects by scaling down or redesigning a project (e.g., adding fish ladders to allow anadromous fish to reach spawning grounds upstream of a proposed dam).

- Repairing, rehabilitating, or restoring those parts of the environment that are adversely affected by a project (e.g., replanting native vegetation in areas cleared for pipeline installation).

- Creating or acquiring environments similar to those adversely affected by an action (e.g., donation of wetlands to a public land trust to compensate for wetlands destroyed by a project).

These types of mitigations are presented here in the order of their desirability for attenuating adverse effects. Acquisition of environments to compensate for those destroyed by a project is generally considered the least desirable form of mitigation.

A result of a project EIA is usually suggestions for mitigation measures, rather than changes in fundamental decisions concerning the types of alternative actions considered or the size or location of a proposed project (see, e.g., Hill and Ortolano 1978). The reason mitigations (in contrast to changes in project scale or location) are the most common positive outcomes of EIAs is that EIAs are often conducted after important decisions are made and, in some cases, after proposed construction has started (Brown et al. 1991). While it is widely agreed that EIA should be done early enough to influence fundamental decisions, there are reasons why this is not often done; the reasons are reviewed in the section below on 'Perennial Problems in EIA Implementation'. Even though many EIAs suggest measures to mitigate adverse environmental impacts, few EIA programs require that mitigation measures be implemented, and few systematic studies have been done to determine if mitigations agreed to by project proponents were carried out. As elaborated later in this chapter, lack of follow-up on implementing mitigation measures is a common shortcoming of EIA programs.

EIA as an Impetus for Administrative Change

In some countries, EIA has reformed public decision making by giving information on project impacts to citizens, NGOs, and agencies interested in a proposed project. Indeed, the NEPA process in the USA is often hailed as a program of administrative reform because it opened the decision processes of federal agencies to public scrutiny. And in the USA, citizens and nongovernmental organisations have sued government agencies frequently to ensure the full disclosure of impacts as required by the NEPA process.

Administrative changes effected by EIA in the USA are somewhat unique for at least three reasons: (1) EIA implementation is heavily influenced by court actions brought by NGOs; (2) freedom-of-information laws make it relatively easy for citizens to obtain copies of documents in the files of government agencies; and (3) the NEPA process encourages citizen participation in agency decision making. In many countries, citizens are neither accustomed nor encouraged to participate in agency decision making. Thus, EIA may not always increase citizen involvement, as it does in the USA. Moreover, even in countries with strong democratic traditions and a highly informed citizenry, implementation of EIA does not necessarily translate into increased citizen participation in government decision making. This is demonstrated by EIA in France. The French EIA program requires an explicit consideration of environmental factors in decision making for a large number of projects and plans. However, the decision process is dominated by technical specialists and civil servants and is not heavily influenced by public participation in EIA (Sánchez 1993). This appears to be changing, however, as NGOs have made increased use of appeals to administrative tribunals to ensure that France's EIA requirements are implemented carefully.

EIA programs often influence administrative processes by enhancing interagency coordination. Many EIA programs require that environmental assessment documents be reviewed by an environmental protection agency (or ministry) and, possibly, other governmental bodies. These reviews help disseminate information about proposed actions and their impacts, which is generally viewed as an administrative improvement.

Another influence of EIA programs on administrative processes concerns power relations between ministries. These effects can be notable. For example, in the Philippines (under Marcos), the environmental agency in charge of EIA upset the traditional power balance among national ministries. When the environmental unit attempted to influence decisions normally made by economic development-oriented ministries, those ministries impeded implementation of the EIA program (Abracosa 1987). A similar outcome was reported in Kenya where concerns of development-oriented ministries were so great that the environmental unit was incapable of promoting an EIA program, except for private industrial projects unconnected to the ministries (Hirji 1990).

Sometimes the introduction of an EIA program enhances the influence of environmental protection agencies or environmental review boards or both. As an example, the Ministry of Environment for the province of Ontario, Canada, was granted significant authority when legislation establishing the provincial EIA program was passed. Gibson (1993:18) reports that in Ontario, "reviews and decision making on cases subject to full individual assessment requirements are the responsibility of the minister of the environment, and in controversial cases there is usually a referral to an independent administrative tribunal, the Environmental Assessment Board, which carries out public hearings and makes the final decision (subject to Cabinet revision or reversal)." In most cases, independent environmental review boards (e.g., the EIA Commission in the Netherlands) do not make final decisions; they offer advice to government decision-making authorities. However, in the cases reviewed by the Environmental Assessment Board in Ontario, the board *is* the decision maker.

Effects of EIA on Project Proponents

Few studies have investigated the influence of impact assessment programs on organisations that propose projects subject to EIA requirements. Organisation theory suggests that if an EIA program were to threaten the autonomy or survival of a project proponent, it would cause an organisational reaction. Sometimes the reaction involves efforts to avoid the EIA requirement entirely, or to comply with it only in a *pro forma* manner. These responses are demonstrated in Abracosa's (1987) study of EIA in the Philippines and Hirji's (1990) investigation of EIA in Kenya.

Sometimes project proponents react by changing how they do business. In these cases, EIA requirements are typically only one of several pressures for change. For example, the Taiwan Power Company (Taipower), compelled by strong citizen protests against Taipower projects that caused environmental degradation (Tu 1993), created a large environmental department to conduct EIAs. The environmental department, which employs environmental specialists in disciplines relatively new to Taipower, works both to satisfy environmental regulations and to influence the design of projects that Taipower proposes.

A recent study by Gariépy and Hénault (1994) details the influence of EIA on the organisational structure and behavior of Hydro-Québec in Canada. Hydro-Québec faced strong resistance to some of its enormous hydroelectric power schemes during the 1980s. It responded by making substantial organisational changes to accommodate the new forces opposing its projects and the new EIA requirements it faced. These organisational changes included co-opting project opponents by allowing them to participate in some decision processes; expanding programs of public consultation; restructuring the planning process to conduct EIA parallel with (not after) other planning activities; and elevating the status of environmental activities within the organisation.

For instance, during the late 1970s, when EIA was introduced as a project planning task at Hydro-Québec, an environmental unit produced EISs that were primarily 'add-ons' to projects that had already been planned. By the early 1990s, EIA's status had risen to include a full vice-presidency focused on environment and the integration of EIA into both corporate and project-level planning activities. All of these changes reflect Hydro-Québec's efforts to learn from its experiences and expand its EIA activities in the face of new environmental constraints.

An extensively documented case of organisational change precipitated, in part, by EIA requirements is that of the US Army Corps of Engineers. During the early 1970s, the Corps was coping with new EIA requirements, substantial public pressure to improve its environmental record, and other federal government requirements to include environmental quality in its water resource project evaluations. Many Corps projects were halted or delayed by court actions initiated by NGOs.

Several noteworthy changes in the Corps of Engineers took place in the 1970s (Mazmanian and Nienaber 1979; Taylor 1984). More than 35 Corps district and division offices were each augmented by creating new environmental units to meet EIA and related requirements. Several hundred environmental professionals were hired. For the first time, specialists with disciplinary training in environmental science and environmental engineering were integrated (at some level) into the Corps' project planning and decision-making processes.

While many of the new environmental specialists were hired specifically to produce EISs, some learned how to influence the engineers responsible for project design. For example, some environmental specialists were able to engage in internal politicking to derail or modify environmentally insensitive proposals. In addition to creating new units and hiring environmental specialists, the Corps rewrote its planning procedures to (1) enhance the importance of environmental quality as a planning objective; (2) accommodate requirements for EIAs under the NEPA process; and (3) facilitate the direct participation of citizens in some aspects of project planning.

The changes in the Corps of Engineers were extraordinary, given that its enormous bureaucracy was dominated by engineers with a tradition of building, and its Congressional allies were interested in having new projects constructed in their home districts. Changes in the Corps indicate how EIA programs, coupled with substantial societal pressures, can affect the organisational structure and behavior of project proponents.

PERENNIAL PROBLEMS IN EIA IMPLEMENTATION

This section details problems that have been associated with EIA since the early 1970s. It proposes that some of these problems are systemic and will persist because many project proponents do not view EIA as useful. Rather, project proponents often view EIA as a requirement to be completed, a hurdle to be jumped along the way to project implementation. This particular hurdle imposes risks on project proponents because EIA often forces a public disclosure of impacts, and the information on impacts can strengthen the hand of a project's opponents.

EIA Requirements Are Often Avoided

Some countries leave the decision on whether an EIA is required for a proposed project up to either the government unit responsible for deciding on the project or an environmental agency. When the former occurs, the exercise of administrative discretion by the responsible government unit can lead to situations in which EIAs are not conducted even though the environmental impacts of proposed projects are significant.

This is illustrated by EIA programs in Australia, where a relatively small fraction of development projects are subject to EIA procedures. Critics argue that this is a result of the discretionary nature of the various programs. For example, in analysing the Commonwealth of Australia EIA program between 1975 and 1985, Formby (1987) found that fewer than 10 EISs per year were called for, only 4 percent of the proposals considered significant. The small number of EISs resulted because the decision to initiate the Commonwealth EIS process was in the hands of the minister proposing an action, not the minister responsible for environmental affairs. In addition, citizens not directly harmed by the project were unable to gain standing to sue government agencies in the courts of the Commonwealth.

Sometimes pure politics leads to efforts to get around EIA requirements. An illustrative case is the Linha Vermelha, a highway in Brazil that connects the airport serving Rio de Janeiro with downtown Rio. As reported by Ortolano (1993:356), the agency responsible for environmental assessments in the State of Rio de Janeiro was put under considerable political pressure to exempt the project from EIA requirements. The exemption was granted. Ironically, political pressure to build the highway in time to serve the 1992 United Nations Conference on Environment and Development was partially responsible for the short-circuiting of EIA procedures. Other examples of political maneuvering to avoid EIA requirements are given by Abracosa (1987) and Hirji (1990).

The implementation of NEPA during the 1970s is filled with instances in which federal agencies tried to avoid preparing environmental impact statements as required by the act. During that decade, about half of the hundreds of the NEPA-related court challenges centered on cases in which agencies had to defend their decision not to prepare an EIS under NEPA. Frequently, the plaintiffs prevailed. Agencies eventually learned that they risked being sued and losing court challenges if they tried to avoid NEPA's requirement to disclose significant environmental impacts of their actions.

EIA Is Often Not Carefully Integrated into Planning

In the more than 20 years since NEPA's enactment, the law has been criticised for establishing "little more than a bureaucratic exercise that requires federal agencies to complete paperwork they subsequently file and ignore" (Fogleman 1993:79). Similar comments have been made about EIA programs in other countries (see, e.g., Abracosa 1987; Hirji 1990). The arguments are often the same—EIA is not well-integrated into decision making; and EIA occurs at the project level, but not generally at the policy or program level where decisions are made that foreclose some types of project alternatives. (For example, a program-level decision to build dams or enlarge channels to control floods rules out the consideration of flood-proofing structures or flood plain zoning as ways to reduce damages caused by floods.)

Even at the project level, EIA is typically done after planners and decision makers begin advocating a particular proposal, and EIA serves largely to suggest mitigations for a project already selected. As Ensminger and McLean (1993:48–49) have pointed out, "Major decisions, including the action to be carried out and its location, are often made before the EIS is prepared and. . . the EIS is then drawn up to support those decisions." This use of EIA as an *ex post facto* rationalisation for decisions reflects a failure to integrate EIA into project planning and is termed herein 'the integration problem'.

The integration problem persists because, in many contexts, a project proponent will not undertake an EIA until after a project is well-defined and there is a high likelihood that it will be funded and approved (Nelson 1993; Hirji 1990). Many project proponents feel it would be irrational to do otherwise. Why use resources to conduct an EIA if the proposed project is not likely to go forward? Another cause of the integration problem is that many project proponents do not give the same weight to environmental objectives as they give to economic performance measures such as the internal rate of return. If project proponents gave environmental impacts the attention they give to economic performance measures, the integration problem might not exist.

What explains the cases where EIA *is* integrated effectively into project planning and decision making? According to Ortolano, Jenkins, and Abracosa (1987), these cases involve a 'control mechanism' that causes project proponents to conduct an EIA. Table 1 provides a list of control mechanisms, and Ortolano (1993) uses EIA programs in several countries to illustrate relationships between control mechanisms and dimensions of EIA effectiveness. One illustration of a control mechanism is an EIA program that gives citizens acting through courts the ability to block projects with inadequate environmental impact statements (*judicial control* as exemplified by

EIS requirements under NEPA in USA). Another is where an independent environmental board sets the terms of reference for an EIS, reviews the final document, and influences the fate of the proposed project (*evaluative control* as illustrated by the EIA system established by the Canadian Environmental Assessment Act).

While control mechanisms can make project proponents take EIA requirements seriously, they do not necessarily force project proponents to consider environmental factors early or continually as the conception of a project evolves. Ridgway and Codner (1994:4) elaborate on this point: "One major criticism of EIA is that it occurs at only one point in time whilst a project changes over time—the process fails to recognise, or allow for, the iterative nature of engineering design."

Table 1. **Control mechanisms influencing EIA implementation**

Procedural control	Centralised administrative unit promulgates environmental impact assessment requirements, but does not have power to modify projects
Judicial control	Court has power to judge allegation of inadequate attention to EIA, but does not have direct control over the project proponent in relation to EIA compliance
Evaluative control	Centralised administrative unit issues recommendations to decision makers based on an appraisal of the proposed project and the EIA
Development aid agency control	Multilateral or bilateral lending institution requires an EIA before it makes a final decision to fund a project
Professional control	Project planners have professional standards and codes of ethical behavior that lead them to undertake environmental impact assessments for proposed projects
Direct public and agency control	Citizens or government agencies apply pressure to influence the environmental impact assessment process, but outside the context of the above-listed controls

Source: Adapted from Ortolano, Jenkins, and Abracosa (1987:287)

EIA Does Not Ensure Environmentally Sound Projects

An issue related to the integration problem is that EIA does not ensure that projects with significant adverse effects will be stopped. In many contexts, in fact, this point is moot—officials often promote environmentally damaging projects if the economic benefits outweigh their negative environmental impacts.

A manifestation of the problem is the debate over whether NEPA imposes substantive (as opposed to procedural) obligations on federal agencies. There is no question that NEPA imposes procedural obligations on agencies to conduct environmental assessments and, where impacts are significant, to prepare and circulate environmental impact statements for comment by governmental bodies, NGOs and individual citizens. Many legal scholars believe that NEPA imposes both procedural *and* substantive obligations (Yost 1990). They feel that substantive obligations are set out in Section 101 of NEPA, which declares that "it is the continuing responsibility of the Federal Government to use all practicable means, consistent with other considerations of national policy," to improve and coordinate its actions to fulfill substantive objectives of the act (e.g., "assure for all Americans safe, healthful, productive, and esthetically and culturally pleasing surroundings").

US Supreme Court rulings through the early 1990s held that NEPA imposes *only* procedural obligations. If that were true, an agency could carry out an environmentally damaging project as long as it met all the procedural requirements of the NEPA process (as articulated by the US Council on Environmental Quality 1986). Under these circumstances, courts could rule on whether an agency was supposed to prepare an environmental impact statement or whether the statement was legally acceptable, but not on whether the agency made an environmentally sound decision. Other federal legislation, specifically the Administrative Procedure Act, allows judges to review agency decisions. However, before such a review can take place, plaintiffs must demonstrate that the agency action was "arbitrary and capricious" or otherwise in violation of the law. The arbitrary and capricious standard places significant demands on plaintiffs seeking judicial review of agency decisions.

In sharp contrast to the situation in the USA (and other countries) where EIA has often been reduced to an exercise in producing legally adequate assessment documents, some jurisdictions have mandated EIA procedures to ensure environmentally sound projects. The Environmental Assessment Board in Ontario, Canada, provides a notable example. Except for the possibility of an intervention by the provincial cabinet, the Environmental Assessment Board's decisions on the acceptability of projects are final.

The issues of whether and how EIA can be used to yield environmentally sound decisions have taken on increasing significance as governments attempt to use EIA to foster sustainable development. Many analysts and some organisations (for example, the World Bank) have embraced EIA as a principal tool for ensuring the sustainability of development. However, there are many instances in which EIA has proven seriously deficient as a mechanism for attaining environmental policy goals. Those who see EIA as a linchpin in the quest for sustainable development may be disappointed.

The ability of EIA to contribute to sustainable development requires, at the outset, an unambiguous conceptual definition of sustainable development and a translation of that definition into operationally meaningful criteria for decision making. As the literature on sustainable development makes clear, this is no simple task (see, e.g., Lélé 1991). While the prospect of translating 'sustainable development' into meaningful decision criteria may be daunting, some nations have taken up the challenge. For example, the federal government in Canada recently acted on its commitment to sustainable development by authorising "the Minister of Environment to issue guidelines setting out criteria for judging the significance of adverse environmental effects and the acceptability of projects" (Gibson 1993:15).

A second condition required before EIA can contribute to sustainable development is the existence of a governmental body (or process) that ensures that criteria for defining sustainable development are used as a basis for decision making. As experience with NEPA demonstrates, if there is no governmental body responsible for stopping environmentally unsound projects, EIA alone may not prevent unsustainable development.

EIA Is Done Primarily for Projects, Not Programs or Policies

The influence of EIA could be far greater if it were applied at the level of programs; i.e., collections of individual projects, such as a coordinated series of dams, or an integrated set of research investigations. Some have even argued that EIAs should be done for proposed policies and legislation. The term 'strategic environmental assessment' (SEA) has been introduced to mean the application of EIA in strategic planning and policy making. (The terms 'programmatic EIA' and 'strategic environmental assessment' are sometimes difficult to distinguish, as EIA specialists have only recently started using the latter term. As Rosario Partidario (1993:37) states, "[The concept of strategic environmental assessment] still lacks a practical conceptualisation.")

Although a number of program and policy level EIAs have been completed (see, e.g., Sadler 1994b), programmatic EIAs (or SEAs) are not done as frequently as many feel they should be. This is problematic inasmuch as taking decisions one at a time makes it easy to miss accounting for the cumulative effects of a series of decisions. An EIA for programs or policies would provide an opportunity to mitigate or abandon environmentally unsound concepts before they were turned into projects. In addition, programmatic EIAs can enhance interagency coordination and yield efficiencies. If an EIA were done for a program (e.g., a future set of land development projects), then any future project consistent with the program could proceed without having to redo the analysis of environmental impacts already accounted for in the programmatic EIA. This approach is demonstrated by the Chinese practice of preparing EIAs for industrial development zones. The designation of these zones involves coordination between local environmental protection agencies and economic development officials. If a factory chooses to locate in an industrial development zone that has an EIA for the entire zone, the factory's EIA requirements are minimal. If the factory locates in the same city but outside the zone, it must generally do a complete EIA (Sinkule 1993).

If programmatic EIAs have such advantages, why are they not conducted more frequently? One reason is that program and policy decisions often evolve over time, making it difficult to identify what constitutes 'the program'. The scope of a program may be difficult to define, both spatially and temporally, and this makes assessing impacts even more uncertain than usual. Even when spatial boundaries can be delineated, the land areas involved may be huge and involve many decision-making authorities. In addition, agencies or private developers trying to promote an entire program may be wary of giving potential opponents a complete perspective on program impacts. Project proponents who view EIA as an administrative hurdle would resist a requirement that arms opponents with information on the adverse effects of an entire program. In many countries, the absence of top-level commitments to programmatic EIA—from either legislative bodies or agency leaders—has led to their underutilisation.

Commitment at the highest level of government is crucial in determining whether environmental assessments are conducted for policies and strategic plans. This point is clarified by experiences in Canada, which is often cited as being among the few nations that has extended EIA to high-level policy making. (A comparative analysis of the performance of leaders in EIA for policies and plans is given by Rosario Partidario (1993).) The province of Ontario has included explicit requirements for environmental assessments for "proposals, plans, or programs" in the provincial statute that sets out its EIA programs. These assessments have been conducted for waste management master plans, long-range electric power system plans, and timber management strategies. Gibson (1993:16) observes: "None of these initiatives has proceeded smoothly, but there is a general recognition in Ontario that assessment at the plan stage is appropriate both as a means of addressing overall effects and as a foundation for better focussed assessments of individual undertakings proposed under these plans."

In contrast to the provincial government of Ontario, the federal government of Canada appears less committed to assessing the environmental effects of programs. According to Gibson (1993:16), in passing a new federal EIA process in the Canadian Environmental Assessment Act, ". . .the government refused to include policy assessment, or even program and plan assessment, in the new law." Although the federal government has stated a commitment to assessing the environmental impacts of policy initiatives, it is not yet a legislated, enforceable requirement. More generally, governments have been reluctant to demand environmental assessments for new policies and legislation.

Cumulative Impacts Are Not Assessed Frequently

Cumulative impacts have been defined as the "result of additive and aggregative actions producing impacts that accumulate incrementally or synergistically over time and space" (Contant and Wiggins 1993:341). Using this definition, *additive actions* are repeated similar activities, such as a series of small dams to generate hydro-electric energy; and *aggregative actions* are groupings of dissimilar activities, such as a collection of demonstration projects to improve the commercial feasibility of using oil shale to produce energy. There are more complex characterisations of cumulative impacts that involve consideration of 'nonlinear functional attributes', 'growth induction', and related concepts, but the definition above suffices for this discussion. (For more definitions, see Contant and Wiggins 1993.)

One reason cumulative impacts continue to be (in the words of Ensminger and McLean 1993:53) "consistently underassessed" is that the programmatic EIA is one of the few workable approaches for dealing with them. For reasons mentioned above, programmatic EIAs are not performed frequently.

In addition to methodological difficulties in assessing cumulative impacts, there are also significant institutional impediments. For example, Contant (1984) developed a procedure for the US Army Corps of Engineers to account for the cumulative impacts of Corps decisions permitting land development projects affecting navigable waterways. Contant's approach required the Corps to influence land use explicitly by using a carrying capacity analysis to propose limits on waterfront development. The Corps chose not to implement Contant's procedure, in part because the agency was wary of encroaching on the prerogatives of local governments to control land use. Examples presented by Irving and Bain (1993) demonstrate that when there are appropriate institutional arrangements for addressing cumulative impacts, an analysis of those impacts can yield beneficial results.

Public Participation in EIA is Often Inadequate

Many, but certainly not all, countries with EIA programs have mandated some level of public participation in EIA. Typical goals of public involvement in EIA are set out in Table 2, which is abstracted from a public involvement manual prepared by the Federal Environmental Assessment Review Office (1988) in Canada. A continuing problem plaguing many EIA programs is that public involvement occurs too late to attain fully the goals summarised in Table 2.

Table 2. **Possible goals of public involvement in EIA**

- ► To identify public concerns and values

- ► To gather economic, environmental, and social information from the public

- ► To inform the public about potential actions or alternatives, and the potential consequences of these actions

- ► To develop and maintain credibility

- ► Ultimately, to improve the overall decision making of the agency

Source: Federal Environmental Assessment Review Office (1988)

Consider, for example, the NEPA process in the United States, which has had formal requirements for public involvement since its establishment, and is frequently lauded for involving citizens in decision making. Public involvement begins when a 'notice of intent' to prepare an environmental impact statement is published.

Citizens may respond to the notice of intent by participating in scoping, a process that encourages citizens, NGOs, and agencies to contribute their views on which impacts should be analysed. However, at this point, a proposed project has already been conceived, and thus key decisions on project size and location will have already been made. The only other required public involvement occurs when citizens and NGOs have opportunities to review and comment on a draft EIS. In addition, if citizens or NGOs feel that the final EIS does not satisfy NEPA's requirements, they can challenge an agency's implementation of NEPA in the federal courts.

The above-noted public involvement opportunities allow citizens to be informed and to influence the scope of an EIA. However, those public involvement opportunities are limited since, by the time they occur, agency decision makers have often become attached to a particular course of action. Public involvement is often reduced to public relations or defending a decision that has (with the possible exception of mitigation measures) already been made. In many cases, the influence of citizens opposed to a plan is limited to attempts at either halting a project or forcing the inclusion of mitigation measures.

During the past decade, even the modest opportunities for public participation noted above have often been short-circuited as US federal agencies can avoid preparing an EIS by proposing mitigation measures. In the NEPA process, an agency can legally proceed as follows: An agency conducts an 'environmental assessment' (EA) to determine if a full EIS, subject to public review, is required. An EIS is required if the proposed action would have significant effects on the quality of the human environment. If not, a 'finding of no significant impact' (FONSI) is issued. If the EA reveals significant impacts, the agency may include mitigations to reduce impacts enough to issue a FONSI. (When mitigations are used to justify the finding, the result is termed a 'mitigated FONSI'.) An agency notifies interested parties by listing its FONSI in the *Federal Register*, a government document that is not read by most citizens. By relying on the mitigated FONSI, an agency can essentially avoid involving citizens in EIA.

There is a widespread fear that this is precisely what US federal agencies are doing. In recent years, the number of EISs has fallen to less than 500 per year (down from well over 1,000 per year in the 1970s), and the number of mitigated FONSIs has skyrocketed. Indeed, a recent survey conducted by the Council on Environmental Quality (CEQ) indicates that well over 40,000 EAs are prepared each year and that "agencies consistently involve the public in fewer than half of their EA preparations" (Blaug 1993:57). The CEQ survey results, which are summarised by Blaug (1993), provide evidence that some agencies are relying heavily on the mitigated FONSI to avoid preparing EISs, thereby reducing greatly the opportunities citizens have to participate in EIA.

In many countries with democratic political traditions, opportunities for public involvement in EIA typically consist of making environmental assessment documents available to the public and, in some cases, conducting public hearings to discuss the EIS. Although some project proponents (e.g., the US Army Corps of Engineers and Hydro-Québec) sometimes conduct more elaborate citizen participation programs, those programs are not required by EIA regulations.

In countries where governments are not elected, public involvement in EIA, as would be expected, is not a priority. As MacDonald (1994:31–32) points out, "political situations such as single-party governments may strongly discourage any public opposition to development projects, making those who have concerns regarding a proposal afraid of the reprisals that may result from speaking out." She goes on to raise concerns about the degree to which public involvement in EIA leads to any 'real changes' in projects. In MacDonald's view, utilisation of information provided by citizens in the context of EIA is 'problematic' for several reasons:

> Among other factors, this may be the result of a reluctance on the part of some experts to accept and respect the information from those with less formal training, or it may stem from the subjective nature of some of the concerns of local citizens that are difficult to verify or mesh with the often technical and scientific focus of an [EIA] (MacDonald 1994:32).

In addition, as previously noted, public involvement often occurs so late in planning that project proponents are not receptive to ideas that require modifications in plans.

Proposed Mitigations May Not Be Implemented

It is common for an EIA to recommend actions to mitigate adverse impacts of a proposed project. What is far less common is to have assurances that a proposed mitigation will be implemented. Indeed, in some cases, the mitigations recommended in an EIA consist of actions that the project proponent has no authority to implement (e.g., a measure that calls on residents near a proposed road to install double-glazed windows to offset increased traffic noise). Moreover, there are many cases in which project proponents completely ignore mitigations (in an EIA) that they could implement (see, e.g., Hirji 1990).

The degree to which proposed mitigation measures are ignored is significant. For example, many EIA specialists surveyed by Ensminger and McLean (1993) felt that the "lack of guidelines and action-forcing mechanisms" to ensure implementation of impact mitigations was an important deficiency of the NEPA process in the USA. Several Congressional proposals to amend NEPA in the early 1990s (which were not passed) would have *required* that "environmental mitigation and monitoring measures and other conditions discussed [in the context of an agency's NEPA process for a project]. . . be implemented by the appropriate agency" (Bear, as cited by Smith 1993:83). The absence of follow-up to check on whether mitigation measures were implemented is part of a broader problem—few investigations are conducted to determine the impacts caused by projects after they are implemented.

In recent years, some government agencies have developed programs to ensure that mitigations specified in an EIA are actually undertaken. An example is the US Department of Energy's (DOE) mitigation action plan. The essence of the plan is that, "if a program promises in an EA or EIS to carry out a mitigation, the DOE will carry out a tracking program to ensure that the mitigation is, in fact, carried out" (Culhane 1993:73). As of 1993, four such post-project follow-ups had been conducted. Although experience with the DOE mitigation action plan has been limited, it represents an initiative that may provide precedents for others.

In some political jurisdictions, there are mechanisms in place to enforce the implementation of mitigations that government decision makers call for in the course of approving a project. In New South Wales, Australia, for example, the final EIS (which includes any required mitigations) becomes a legal document and citizens have the right to take a project proponent to court if mitigations agreed to by the proponent are not implemented (Ridgway, B.M., Monash University, Australia, private communication, July 12 1994).

Post-project Monitoring is Rarely Conducted

In commenting on the state of post-EIS project monitoring, Culhane (1993:67 ff) characterises the 'baseline approach' as one that seems to have agency managers—

> . . .(1) fight their way through the thickets of project planning and environ-
> mental review to obtain a favorable decision, (2) implement the decision by,
> for example, seeing the project through its construction phase, then (3) move
> onto planning and implementing their next major project, (4) leaving behind
> no official with any particular stake in monitoring the effectiveness or
> dysfunctions of the implemented project. . . . [Furthermore,] relatively few
> post-EIS audits have been conducted by anyone.

There have been calls for extensive post-project environmental impact monitoring since the 1970s. Two propositions in support of post-EIS audits are typically advanced: one concerns enhancing forecasting capabilities; the other is based on improving project outcomes.

The argument based on enhancing forecasting capabilities considers an EIA as containing a set of predictions of how the environment will change if a proposed action is implemented. Under the circumstances, the process of conducting an environmental impact study can be viewed as part of a scientific experiment in which predicted impacts constitute a hypothesis that can be tested by gathering data on impacts that occur after the proposed action is taken. In this way, the process of doing EIAs "provides an opportunity. . . to contribute to scientific advances" (Caldwell 1982, as interpreted and cited by Culhane 1993:69).

The second argument supporting post-project monitoring concerns opportunities to ameliorate adverse impacts and evaluate the effectiveness of mitigation measures. The state-of-the-art of impact prediction is such that unanticipated impacts occur often. Monitoring provides an opportunity to identify adverse impacts and intervene with mitigation measures if impacts are unacceptable. This is a strategy that was advocated during the 1970s by Holling (1978) and his colleagues as "adaptive environmental management." Holling's arguments are no less compelling today.

Although post-project monitoring to mitigate adverse impacts is not commonly undertaken, there are cases that demonstrate the value of monitoring. Canter (1993) summarises several such cases, including the comprehensive reservoir monitoring programs run by the Tennessee Valley Authority (TVA) in the United States. The TVA program, which focuses on 16 dams and reservoirs in the Tennessee River basin, monitors river flow and several water quality parameters. Data collected are used to, among other things, "improve water quality and aquatic habitat by increasing minimum flow rates and aerating releases from TVA dams to raise dissolved oxygen levels" (Canter 1993:80). Data collected by TVA are also used to study how changes in a reservoir's operations influence the dissolved oxygen levels of downstream water.

Assessments of Risk and Social Impacts Are Often Omitted from EIAs

Social impact assessment and risk assessment have long been considered an integral part of EIA in the normative literature on impact assessment. However, they are frequently left out of EIAs for projects in which either social impacts or risks to human health and the environment are significant.

Sometimes social impacts are left out of EIAs because the legislation establishing EIA requirements defines environment narrowly with an emphasis on the bio-physical environment. Beckwith's (1994) analysis of how environment is defined in the EIA program in Western Australia illustrates this point—the environmental agency interprets the Environmental Protection Act of Western Australia to restrict the range of social impacts to be included in an EIA. However, even when social impact assessments are required by law, they are often not conducted. An analysis of the institutional factors contributing to the underassessment of social impacts has been given by Rickson et al. (1990). (For a more general discussion of social impact assessment, see the chapter by Burdge and Vanclay in this book.)

Although risk assessment has advanced considerably as a field over the past few decades, these assessments continue to be left out or inadequately treated for many projects that pose major risks to human health and the environment (see, e.g., Arquiaga, Canter, and Nelson 1994). Examples include industrial projects where explosives and toxics are stored, and offshore oil drilling facilities.

The analysis by Carpenter (elsewhere in this book) details a practical approach to risk assessment and ways in which it can be integrated into EIA. Although there are challenges in integrating risk assessments into EIAs, the benefits can be substantial. In addition to alerting decision makers of possible dangers, a risk assessment can focus attention on risk reduction activities such as minimising the amount of waste generated in production processes, and it can also lead to the delineation of emergency response procedures in the event of accidents.

NEW CHALLENGES: EIA AND INTERNATIONAL ENVIRONMENTAL PROBLEMS

During the past decade, efforts have been made to expand EIA beyond the confines of country-specific decision making and into the arena of international environmental problem solving. These efforts are considered in three categories: (1) problems involving the global commons; (2) the use of EIA by development assistance organisations; and (3) the potential for EIA to inform decisions on international trade agreements.

Problems Involving the Global Commons

There is increased use of EIA to help identify how proposed projects will influence global climate change, depletion of the ozone layer, loss of biological diversity, and other international environmental concerns. Cumulative impacts are of central importance in this context, since even moderately scaled domestic projects can, collectively, have dramatic effects on the global commons.

Difficulties in using EIA to identify effects of domestic projects on global environmental problems are illustrated by considering, for example, requiring an EIS to examine how a proposed project affects biodiversity (Henderson et al. 1993). This raises many complex questions: How should biodiversity be defined? What indicators should be used to measure it? And what guidance can be offered on how to include impacts on biodiversity in EIAs? These questions notwithstanding, some agencies have made a solid start toward using EIA to examine a proposed project's effects on biodiversity (Hirsch 1993).

Attempts have also been made to use environmental impact statements to analyse how proposed actions influence production of greenhouse gases, such as carbon dioxide and methane (Cushman et al. 1993). In 1989, the US Council on Environmental Quality issued draft guidance to federal agencies on how to consider global climate change in the context of EIAs. Even those who advocate increasing the scope of EIAs to include global climate change are troubled by the "technical difficulties in making accurate predictions" (Cushman et al. 1993:460). There is also concern that the NEPA process has a minimal influence in reducing emissions of greenhouse gases, since NEPA applies to a limited set of actions (typically those involving federal agency projects, permits, grants or loans), and it has no influence on past actions that have contributed to the climate change problem.

EIA and Development Assistance

Some of the first applications of EIA in foreign aid were made by the US Agency for International Development (USAID). The agency did not embrace EIA voluntarily; it was forced to as a result of court actions brought by an environmental NGO in 1975. The reluctance of the USAID to use EIA for its foreign assistance projects is not surprising—foreign affairs personnel, like staff in other bureaucracies, typically resist infringements on their autonomy (Robinson 1992). But in response to the aforementioned court actions, USAID now routinely assesses environmental impacts of its overseas development activities.

During the late 1970s, many bilateral and multilateral aid agencies were pressured by NGOs to do EIAs for their projects. The need for EIAs for development assistance projects was compelling—it was increasingly evident that development aid organisations, such as the World Bank, were supporting projects with disastrous impacts on the environment (see, e.g., Payer 1982). As detailed by Kennedy (1988) and Mikesell and Williams (1992), multilateral and bilateral aid agencies (e.g., the Asian Development Bank and Japan's Overseas Economic Cooperation Fund) now require EIAs for many of the projects they fund.

Now that environmental assessment is required for much development assistance, the challenge is to implement the new EIA requirements in a way that is both productive and sensitive to the local context. There are numerous examples in which EIAs for development projects have turned into meaningless efforts only to satisfy procedural requirements (Hirji 1990). These EIAs did not affect decisions and only squandered time, resources, and hopes that EIAs could be applied productively to development aid projects. For EIAs to work well in development assistance, project lending officers of organisations like the World Bank need to give more attention to environmental factors in making their decisions. The World Bank's past failure to implement effectively its own EIA requirements suggests that getting project lending officers to consider environmental impacts will not occur quickly.

EIA and International Trade

Relationships between environmental protection and trade policy have recently increased in importance. As of 1991, there were 19 separate international agreements that concerned the environment and included measures affecting trade (US International Trade Commission 1991).

Debates in the United States on the environmental impacts of the North American Free Trade Agreement (NAFTA)—an agreement between Canada, the USA, and Mexico—demonstrate the need for EIA in reaching international trade agreements. The debates often lacked defensible predictions of environmental impacts, since no EIS was prepared. Confusion dominated. Anti-NAFTA environmental NGOs argued that Mexico should raise environmental standards and improve enforcement before any agreement was finalised. At the same time, pro-NAFTA environmental NGOs posited that the agreement would make Mexico wealthy enough to be able to afford investments in waste treatment facilities and other environmental protection measures. Citizens and Congress were served a steady diet of rhetoric, and analyses of environmental impacts were conspicuously absent.

The US experience with NAFTA represented a missed opportunity to set an example demonstrating the applicability and effectiveness of EIA in informing public debate on the environmental effects of international trade agreements. On a related point, some have argued that in the context of NAFTA, the USA was seen as a poor role model. In discussions of this point at the 1994 annual meeting of the International Association of Impact Assessment, some observers characterised the USA as being hypocritical in urging other nations to account for the environmental consequences of their actions while, at the same time, failing to implement the NEPA process in the context of NAFTA.

The failure to use EIA to inform the NAFTA debates resulted initially from a deliberate decision by the Bush administration. Environmental NGOs used the federal courts to try to force the Bush and later the Clinton administrations to apply the NEPA process, but they were ultimately unsuccessful. Examination of court decisions is instructive. In June 1993, the District Court for the District of Columbia found in favor of the environmental NGOs and concluded that an environmental impact statement "is essential for providing the Congress and the public the information needed to assess the present and future environmental consequences of, as well as alternatives to, the NAFTA when it is submitted to Congress for approval" [*Public Citizen* v. *Office of the United States Trade Representative*, 822 F Supp 21 (DDC 1993)]. The court ordered the Office of the US Trade Representative to prepare an EIS 'forthwith'. A few months later, the US Court of Appeals for the District of Columbia reversed this decision [*Public Citizen* v. *United States Trade Representative*, 5 F3d 549 (DC Cir 1993)]. In reversing, the appellate court held that there was no final action by a federal agency that could be challenged under either NEPA or the Administrative Procedure Act. The court reasoned that "the President has not even transmitted NAFTA to Congress for its approval; and if and when the President does submit the agreement and its implementing legislation, this would not qualify [as a final action of an agency] because. . . the President is not an 'agency'."

In contrast with the position of recent presidential administrations in the USA, the government of Australia has decided that its trade policies will be subject to environmental assessment. This policy decision was discussed at the 1993 Fenner Conference on the Environment in Canberra. At that meeting, it was announced that environmental assessment for trade policies would be carried out jointly by the Department of Foreign Affairs and Trade and the Department of Environment, Sport and Territories. Although it would take time to work out details on how to conduct environmental assessments for trade policies, the 1993 Fenner Conference report (Anon 1994) reflected strong government support for use of EIA in this context.

The momentum for free trade is increasing and the number of international agreements that involve both trade and the environment will probably increase in the future. Unfortunately, since trade policy is so politically charged, it is not clear whether countries will take advantage of the opportunity to use EIA to ensure consideration of environmental consequences in trade agreements.

CONCLUSIONS

Environmental impact assessment programs have changed the way project proponents and government agencies charged with approving projects do business. These changes have occurred in both projects and organisations. The most evident change is the inclusion of measures in project proposals to mitigate adverse environmental effects. A less common but significant project-level change is where EIAs have affected project type, size, and location.

What is arguably more significant but less widely studied is the influence of EIA on project proponents. While many project proponents have been marginally affected, others have changed fundamentally. These changes result from proponents hiring environmental specialists in order to meet requirements for EIA and in response to pressure for environmentally sound projects. Although many agencies have an initial tendency to meet EIA requirements with *pro forma* compliance, this sometimes changes to a situation where EIA is embraced as a standard operating procedure.

The shortcomings of EIA in practice are of two different types. One set of shortcomings stems from a systemic problem—EIA is typically conducted as a one-time exercise, whereas the process of project design is cyclical and iterative. Moreover, the EIA exercise is often conducted late in planning, often long after project proponents have become attached to a particular design concept. Under these circumstances, it is difficult to expect an EIA to affect fundamental decisions regarding the types of alternative projects given serious consideration or project scale or location. More typically, outcomes are either suggestions for mitigation measures or, far less frequently, outright rejection of projects.

A second set of shortcomings is less fundamental and thus more amenable to solution. These concern beneficial analyses and activities that could be more frequently conducted. Among these are strategic (or programmatic) EIAs, cumulative impact analyses, risk assessments, social impact studies, public involvement that is timely and meaningful, post-project monitoring, and follow-ups to ensure that proposed mitigations are implemented. Although these have been problematic areas in practice, there are numerous case examples in the literature demonstrating that progress can be (and is being) made.

In recent years, the potential applications of EIA have increased, particularly in the context of international environmental problems. Efforts are being made to apply EIA to a new range of problems, including the loss of biological diversity and global warming. In addition, many development assistance organisations view EIA as a linchpin in their efforts to facilitate development that is environmentally sound and sustainable. Finally, there are significant opportunities for EIA to ensure consideration of environmental consequences in trade agreements.

ACKNOWLEDGMENTS

The authors thank the following individuals for comments on an earlier draft of this chapter: Richard Gelting of Stanford University, Stanford, California; Wen-Shyan Leu, King's College, London, UK; and Bronwyn Ridgway, Monash University, Clayton, Australia.

REFERENCES

Abracosa, R.P. 1987. The Philippine Environmental Impact Statement System. Stanford University Ph.D. dissertation, Stanford CA.

Anon. 1994. "Report on the 1993 Fenner Conference: International trade, investment and environment." *ESCAP Environment News* 11(3):4–5.

Arquiaga, M.C., L.W. Canter, and D.I. Nelson. 1994. "Integration of health impact considerations in environmental impact studies." *Impact Assessment* 12(2):175–197.

Bass, R.E. and A.I. Herson. 1994. *CEQA Compliance: A step-by-step approach*. Point Area CA: Solano Press Books.

Beckwith, J. 1994. "Social impact assessment in Western Australia at a crossroads." *Impact Assessment* 12(2):199–213.

Blaug, E.A. 1993. "Use of the environmental assessment by federal agencies in NEPA implementation." *The Environmental Professional* 15(1):57–65.

Brown, A.L., R.A. Hindmarsh, and G.T. McDonald. 1991. "Environmental assessment procedures and issues in the Pacific Basin–Southeast Asia region." *Environmental Impact Assessment Review* 11:143–156.

Caldwell, L. 1982. *Science and the National Environment Policy Act*. University of Alabama Press.

Canter, L. 1993. "The role of environmental monitoring in responsible project management." *The Environmental Professional* 15(1):76–87.

Contant, C.K. 1984. Cumulative Impact Assessment: Design and Evaluation of an Approach for the Corps of Engineers Permit Program at the San Francisco District. Stanford University Ph.D. dissertation, Stanford CA.

Contant, C.K. and L.L. Wiggins. 1993. "Toward defining and assessing cumulative impacts: Practical and theoretical considerations," pp. 336–356 in *Environmental Analysis: The NEPA Experience*. S.G. Hildebrand and J.B. Cannon, eds. Boca Raton FL: Lewis Publishers.

Culhane, P.J. 1987. "The precision and accuracy of U.S. environmental impact statements." *Environmental Monitoring and Assessment* 8(3):217–238.

Culhane, P.J. 1993. "Post-EIS environmental auditing: A first step to making rational environmental assessment a reality." *The Environmental Professional* 15(1):66–75.

Cushman, R.M., et al. 1993. "Global climate change and NEPA analyses," pp. 442–462 in *Environmental Analysis: The NEPA Experience*. S.G. Hildebrand and J.B. Cannon, eds. Boca Raton FL: Lewis Publishers.

Dee, N. et al. 1972. Environmental Evaluation System for Water Resources Planning. Battelle Laboratory, Columbus OH.

Ensminger, J.T. and R.B. McLean, 1993. "Reasons and strategies for more effective NEPA implementation." *The Environmental Professional* 15(1):46-56.

Federal Environmental Assessment Review Office. 1988. *Manual on Public Involvement in Environmental Assessment*. Hull, Québec, Canada.

Fedra, K. 1991. Expert Systems for Environmental Screening. International Institute for Applied Systems Analysis, Laxenburg, Austria.

Ferester, P.M. 1992. "Revitalizing the National Environmental Policy Act: Substantive law adaptations from NEPA's progeny." *Harvard Environmental Law Review* 16: 207–269.

Fogleman, V. 1993. "Toward a stronger national policy on environment." *Forum for Applied Research and Public Policy* 8(2):79–84.

Formby, J. 1987. "The Australian government's experience with environmental impact assessment." *Environmental Impact Assessment Review* 7(3):207–226.

Formby, J. 1990. "The politics of environmental impact assessment." *Impact Assessment Bulletin* 8(1,2):191–196.

Gariépy, M. and F. Hénault. 1994. "Environmental Assessment and Organizational Culture, the case of Two Major Developers: Hydro-Québec and the Ministry of Transport," presented at the annual meeting of the International Association of Impact Assessment, 14–18 June 1994, Québec City, Canada.

Gibson, R.B. 1993. "Environmental assessment design: Lessons from the Canadian experience." *The Environmental Professional* 15(1):12–24.

Henderson, S., R.F. Noss, and P. Ross. 1993. "Can NEPA protect biodiversity?" pp. 463–472 in *Environmental Analysis: The NEPA Experience*. S.G. Hildebrand and J.B. Cannon, eds. Boca Raton FL: Lewis Publishers.

Hill, W.W. and L. Ortolano. 1978. "NEPA's effect on the considerations of alternatives: A crucial test." *Natural Resources Journal* 18(2):285–311.

Hirji, R.F. 1990. Institutionalizing Environmental Impact Assessment in Kenya. Stanford University Ph.D. dissertation, Stanford CA.

Hirsch, A. 1993. "Improving consideration of biodiversity in NEPA assessments." *The Environmental Professional* 15(1):103–115.

Holling, C.S., ed. 1978. *Adaptive Environmental Assessment and Management*. Chichester UK: John Wiley and Sons.

Irving, J.S. and M.B. Bain. 1993 "Assessing cumulative impact on fish and wildlife in the Salmon River Basin, Idaho," pp. 357–372 in *Environmental Analysis: The NEPA Experience*. S.G. Hildebrand and J.B. Cannon, eds. Boca Raton FL: Lewis Publishers.

Jain, R., et al. 1993. *Environmental Assessment*. New York: McGraw-Hill.

Kennedy, W.V. 1988. "Environmental impact assessment and bilateral development aid: An overview," pp. 272–285 in *Environmental Impact Assessment: Theory and Practice*. P. Wathern, ed. London: Unwin Hyman.

Lawrence, D.P. 1993. "Quantitative versus qualitative evaluation: A false dichotomy." *Environmental Impact Assessment Review* 13(1):3–11.

Lélé, S.M. 1991. "Sustainable development: A critical review." *World Development* 19(6):607–621.

Leon, B.F. 1993. "Survey of analyses in environmental impact statements," pp. 653–659 in *Environmental Analysis: The NEPA Experience*. S.G. Hildebrand and J.B. Cannon, eds. Boca Raton FL: Lewis Publishers.

MacDonald, M. 1994. "What's the difference: A comparison of EA in industrial and developing countries," pp. 29–24 in *Environmental Assessment and Development*. R. Goodland and V. Edmundson, eds. Washington DC: The World Bank.

Mandelker, D.R. 1993. *NEPA Law and Litigation*, 2nd ed. Deerfield IL: Clark Boardman Callaghan.

Mazmanian, D. and J. Nienaber. 1979. *Can Organizations Change? Environmental Protection, Citizen Participation, and the Corps of Engineers*. Washington DC: Brookings Institution.

Mikesell, R.F. and L.F. Williams. 1992. *International Banks and the Environment*. San Francisco CA: Sierra Club Books.

Nelson, R.E. 1993. "A call for a return to rational comprehensive planning and design," pp. 66–70 in *Environmental Analysis: The NEPA Experience*. S.G. Hildebrand and J.B. Cannon, eds. Boca Raton FL: Lewis Publishers.

Ortolano, L., B. Jenkins, and R. Abracosa. 1987. "Speculations on when and why EIA is effective." *Environmental Impact Assessment Review* 7(4):285–292.

Ortolano, L. 1993. "Controls on project proponents and EIA effectiveness." *The Environmental Professional* 15(4):352–363.

Payer, C. 1982. *The World Bank: A Critical Analysis*. New York: Monthly Review Press.

Rickson, R.E., et al. 1990. "Institutional constraints to adoption of social impact assessment as a decision-making and planning tool." *Environmental Impact Assessment Review* 10(1,2):233–243.

Ridgway, B.M. and G.P. Codner. 1994. "Design, Management and the Environment," presented at the International Association of Impact Assessment annual meeting, 14–18 June 1994, Québec City, Canada.

Robinson, N. 1992. "International trends in environmental impact assessment." *Boston College Environmental Affairs Law Review* 19(3):591–621.

Rosario Partidario, M. 1993. "Application in environmental assessment: Recent trends at the policy and planning levels." *Impact Assessment* 11(1):27–44.

Sadler, B. 1994a. "Progress Report prepared for the International Summit on Environmental Assessment and the IAIA 1994 Annual Conference." International Association of Impact Assessment annual meeting, 14–18 June 1994, Québec City, Canada.

Sadler, B. 1994b. "Environmental assessment and sustainability at the project and program level," pp. 3–19 in *Environmental Assessment and Development*. R. Goodland and V. Edmundson, eds. Washington DC: The World Bank.

Sánchez, L. 1993. "Environmental impact assessment in France." *Environmental Impact Assessment Review* 13:255–265.

Sinkule, B.J. 1993. Implementation of Industrial Water Pollution Control Policies in the Pearl River Delta Region of China. Stanford University Ph.D. dissertation, Stanford CA.

Smit, B. and H. Spaling. 1995. "Methods for cumulative effects assessment." *Environmental Impact Assessment Review* 15(1): 69–80.

Smith, E.D. 1993. "Future challenges of NEPA: A panel discussion," pp 81–98 in *Environmental Analysis: The NEPA Experience*. S.G. Hildebrand and J.B. Cannon, eds. Boca Raton FL: Lewis Publishers.

Taylor, S. 1984. *Making Bureaucracies Think*. Stanford CA: Stanford University Press.

Tu, S.L. 1993. Environmental Impact Assessment Implementation in Taiwan and Thailand— A comparative organizational examination of state-owned power companies. Stanford University Ph.D. dissertation, Stanford CA.

United States Council on Environmental Quality (CEQ). 1986. Regulations for Implementing the Procedural Provisions of the National Environmental Policy Act. 40 CFR 1500-1508.

United States International Trade Commission. 1991. Operations of Trade Agreement Program, 42nd report. USITC 2403.

van de Gronden, E.D. 1994. "Use and Effectiveness of Environmental Impact Assessments in Decision Making." Report of a pilot study by BCR Consultants, Rotterdam, the Netherlands (25 May 1994).

Yost, N.C. 1990. "NEPA's promise—Partially fulfilled." *Environmental Law* 20(3):681–702.

Chapter 2
Social Impact Assessment[1]

RABEL J. BURDGE
University of Illinois at Urbana/Champaign, USA
FRANK VANCLAY
Charles Sturt University, Australia

INTRODUCTION

About ten years ago, three state-of-the-art of social impact assessment (SIA) papers were written: Kurt Finsterbusch (1985), Bill Freudenburg (1986), and Steve Murdock, Larry Leistritz, and Rita Hamm (1986a, 1986b). Ten years before that, Charlie Wolf (1975, 1976, 1977) wrote a number of state-of-the-art papers. All of these focused on what SIA could offer in terms of predicting the consequences of projects. In the intervening years, much has happened, including the publication of many bibliographies[2] and the development of many national and international networks, both formal and informal, many of which have their own newsletters. However, the most significant event in recent SIA history, at least in the USA, was the agreement upon and subsequent publication of *Guidelines and Principles for Social Impact Assessment* by the [US] Interorganizational Committee on Guidelines and Principles for Social Impact Assessment (1994, 1995). The committee represented the breadth of the social science community in the USA, as well as the different US scholarly and professional organisations which focus on components of impact assessment. This publication is a milestone because the *Guidelines and Principles* represents a core body of knowledge and procedures about SIA that can be used as a starting point for modification and adaptation to the procedures of various land management and regulatory agencies as well as international aid organisations. Although it is clearly based on the regulatory framework operating in the USA, the document provides guidelines and principles for quality SIA that would be applicable under any jurisdiction.

---

[1]*Environmental and Social Impact Assessment* - Edited by F. Vanclay and D.A. Bronstein. Copyright © 1995 by the International Association of Impact Assessment. Published in 1995 by John Wiley & Sons Ltd. A version of this chapter will appear in *Impact Assessment*,the quarterly journal of IAIA.

[2] For example: Canter, Atkinson, and Leistritz 1985; Carley and Bustelo 1984; Carley and Derow 1980; Craig 1987; D'Amore and associates 1979; Freudenburg 1986; Leistritz and Ekstrom 1986; Scott 1978, 1981; Summers and Branch 1984; Vanclay 1989–95; White 1987.

SOCIAL IMPACT ASSESSMENT DEFINED AND DESCRIBED

Social impact assessment can be defined as the process of assessing or estimating, in advance, the social consequences that are likely to follow from specific policy actions or project development, particularly in the context of appropriate national, state, or provincial environmental policy legislation. Social impacts include all social and cultural consequences to human populations of any public or private actions that alter the ways in which people live, work, play, relate to one another, organise to meet their needs, and generally cope as members of society. Cultural impacts involve changes to the norms, values, and beliefs of individuals that guide and rationalise their cognition of themselves and their society.[3]

While SIA is normally undertaken within the relevant national environmental policy framework, it is not restricted to this, and SIA as a process and methodology has the potential to contribute greatly to the planning process. As an example, New Zealand health professionals have recently been planning the introduction of new health care systems in the indigenous Maori communities. They were looking at SIA to assist in the process of evaluation of alternatives and to help in their understanding and management of the process of social change (Association for Social Assessment 1994). These professionals realised that social change would occur as the result of the introduction of new health care delivery programs. They realised that they needed a way to involve and integrate Maoris in planning for the proposed programs. Although none of these health care professionals had formal training in managing social change or in undertaking SIA, they at least recognised the need to understand in advance what changes would likely occur depending upon the type of health care system that was implemented—precisely the type of task provided by a well-done SIA. This New Zealand example highlights opportunities for SIA. From the standpoint of a practitioner implementing social policy decisions, SIA research provides a direction for understanding the process, and guidance in the management of social change in advance of the implementation of the proposed changes. It thus facilitates a decision-making process to choose between alternative possibilities.

In general, the SIA process provides direction in (1) understanding, managing, and controlling change; (2) the prediction of likely impacts from change strategies or development projects that are to be implemented; (3) the identification, development, and implementation of mitigation strategies in order to minimise potential social impacts (i.e., identified social impacts that would occur if no mitigation strategies were to be implemented); (4) the development and implementation of monitoring programs to identify unanticipated social impacts that may develop as a result of the social change; (5) the development and implementation of mitigation mechanisms to deal with unexpected impacts as they develop; and (6) the evaluation of social impacts caused by earlier development projects, technological change, specific technology, and government policy.

---

[3] Largely based on the definition provided by the Interorganizational Committee (1994).

BENEFITS GAINED FROM CONDUCTING SOCIAL IMPACT ASSESSMENT

Often, the greatest social impact of many projects or policies, particularly those planned for community benefit (as in the New Zealand health care delivery program), is the stress that results from the uncertainty associated with it; for example, living near a major development, and being uncertain about the impacts that the project may have. Sometimes just experiencing a situation of rapid change is the cause of stress. By maximising community involvement in the SIA process— not just by consultation, but by directly involving locals in planning teams— uncertainty is reduced, the legitimacy of the SIA and the development project is enhanced, the accuracy of the SIA is increased, and the capacity for the SIA to mitigate impacts is maximised. Previous research has shown that local people from the affected communities have made substantial contributions to SIAs even though they may not be experienced in administrative procedures.

While the requirement to undertake SIAs may seem to be an unnecessary luxury adding to the costs of projects, there are substantial benefits to be gained from undertaking them, for governments, communities, and developers. SIAs that involve the community minimise local resistance to projects, and therefore reduce disruption; they increase project success; and they prevent major planning disasters and associated costs. They may well save money in the long run. It is particularly important that governments and communities insist on SIAs being undertaken because in the majority of cases, the costs of rectifying social and environmental impacts of development are borne by the public sector, not by the corporations that created them.

Even where there are mechanisms (e.g., regulatory or legal action) for extracting compensation from companies for the damage or impact they may create, the compensation is likely to only cover direct impacts, and not the vast amounts of indirect impacts. In local community settings, the compensation itself may have a considerable social impact. It is possible that some groups would be less affected by the development than by the compensation. Nevertheless, there are examples where compensation and other payments (e.g., mining royalties) to local peoples have been used in very positive ways for community development. The establishment of an Aboriginal-owned and -managed airline service, and an Aboriginal radio and television station in the Northern Territory, Australia, are examples.

In any case, the onus of proof to establish that a community, or certain groups within a community, did experience significant social impacts would rest with the community. For social impacts especially, it would be difficult to establish proof to the satisfaction of the courts. Furthermore, there are many impacts that cannot be mitigated or rectified so compensation is not necessarily a desirable strategy. Once local cultural life is affected, it is affected for good, and therefore it is important to prevent the majority of impacts before they actually happen. SIAs, therefore, should be required of all public and private activities (projects, programs, policies) that are likely to affect social life.

The costs of undertaking an SIA should be included as part of the costs of the project and should not be borne by the government or by the local community. However, care must be taken that the standard of the SIA undertaken is to the satisfaction of government and the community. Some review process is required to ensure that all SIAs, and EIAs, are up to a required standard.

A BRIEF HISTORY

Social impact assessment became formalised with the passage of the US National Environmental Policy Act (NEPA) legislation of 1969. It became evident that altering the environment of the natural ecosystem also altered the culture and social organisation of human populations. In 1973, after the decision had been made to build the Alaskan pipeline from Prudhoe Bay on the Arctic Sea to Valdez on Prince William Sound, an Inuit tribal chief made the following comment, "Now that we have dealt with the problem of the permafrost and the caribou and what to do with hot oil, what about changes in the customs and ways of my people?" (cited by Dixon 1978: 4; see also Berry 1975; McGrath 1977). Would the traditional culture and way of life be changed by such a massive construction project? What about the influx of construction workers that spoke a different dialectic (of English) and brought a distinctive lifestyle with them? Further, the state of Alaska had a very small population; consequently, few of the estimated 42,000 persons needed to work on the pipeline during peak periods would come from Alaska. Because of these impacts on human populations, the term 'social impact assessment' probably was first used in 1973 to refer to the changes in the indigenous Inuit culture due to the pipeline.[4]

The new field of SIA grew out of a need to apply the knowledge of sociology and other social sciences in an attempt to predict the social effects of environmental alterations by development projects that were subject to NEPA legislation in the USA and the Canadian Environmental Assessment and Review Process (EARP), passed in 1973. Most of the early SIA procedures were developed by social scientists located within federal, state, and provincial agencies, or by consultants hired by the engineering and architectural firms that prepared the larger environmental impact statements (EISs). These early impact assessors used social science labels in their EISs, but few of the concepts had a connection to prior literature on community and cultural change. US assessors opted for models that required such data as the number and types of new workers as an input to predict quantitative social changes in the geopolitical area of impact (Leistritz and Murdock 1981). Canadian assessors focused more on a social action model, with emphasis on helping the impacted population adjust to the impending change (Bowles 1981, 1982).

---

[4] In Europe, a 1973 study into the social impacts of the then-proposed Channel tunnel (which was not completed until 1994) represents one of the first European predictive SIA studies undertaken (Economic Consultants 1973).

The Inquiry by Chief Justice Thomas Berger of the Province of British Columbia (Canada) into the proposed Mackenzie Valley pipeline, from the Beaufort Sea in the Yukon Territory to Edmonton, Alberta, was the first case where social impacts were considered in project decision making (Berger 1977, 1983; Gamble 1978; Gray and Gray 1977). The inquiry was important because social impacts on indigenous populations were considered in depth. Furthermore, native populations were provided with funding to present their views and hearings were conducted in the native villages in the local dialects.

Of course, social impacts have been considered in different contexts throughout history. In anthropological analysis, retrospective analysis of social impacts has been a major feature of the discipline. Examples that form part of the literature of SIA include Cottrell (1951) and Sharp (1952). The social impact of tourism has been a major field of study in SIA as the international tourism market has expanded, with early anthropological analyses dating back to Forster (1964). Eric Cohen (e.g., 1971, 1972, 1979, 1984) has been a leading researcher in this area of study. The social impacts of mining has also been a major field of study for SIA, with social scientists being consulted to improve the design of mining towns in order to minimise social problems. An early Scottish example is Francis (1973); in Australia a number of studies were undertaken by the Pilbara study group, part of the Commonwealth Scientific and Industrial Research Organisation (CSIRO) (e.g., Pilbara Study Group 1974; Brealey 1974; Burvill 1975); while in Canada, the Institute of Social and Economic Research at the Memorial University of Newfoundland has undertaken and published numerous studies into the impacts of oil exploration and mining.   Development studies is another area with an interest in social impacts. An early study into the social impacts of relocation due to the construction of a dam in Africa is that by Colson (1971). In addition to tourism, mining and dams, nuclear power and new road (highway) construction have also provided the impetus for much SIA research.

The first international conference on SIA was held in Vancouver, British Columbia, in 1982 and gave academic and political credibility to the new field. Since then, the activities of this first conference have been combined into the International Association for Impact Assessment (IAIA) which held its first meeting in 1981 in Toronto, Canada.

By 1983, most US federal agencies had formalised environmental and social assessment procedures in agency regulations. The European Economic Community began to recommend EISs for their members in 1985, and by 1989 the recommendations became a requirement. In 1986, the World Bank decided to include both environmental and social assessment in their project evaluation procedures because liabilities were increasing for projects evaluated strictly on economic and financial criteria. Since then, SIA has become an important requirement (although to varying degrees) around the world as nations adopted and modified the original NEPA model.

In the USA, SIA reached its highest legitimacy when at the conclusion of the April 1993 'Forest Summit' in Portland, Oregon, President Bill Clinton mandated that a social assessment of each timber-dependent community in the Pacific

Northwest would be a required component in deciding among alternative management futures for old-growth forests. This directive was significant because it
formally recognised SIA as a component of the policymaking process. Although the
social assessment team of the Federal Ecosystem Management Assessment Team
(FEMAT) did not conduct a formal SIA for each of 300 communities under study,
they did use much of the literature on community change and cultural history
(particularly for indigenous populations) as a basis for making assessments of
community response to forest management alternatives (see Clark and Stankey 1994;
Stone 1993).

SIA IN THE LARGER CONTEXT OF ENVIRONMENTAL
IMPACT ASSESSMENT

SIA has become a part of project planning and policy evaluation and part of
environmental impact assessment (EIA) as a result of the recognition that social
considerations must be included alongside and even in lieu of solely economic
criteria in the evaluation and decision process. The definition of the environment in
impact assessment has been expanded to include a 'social component'. SIA now
increasingly carries equal weight with economic and environmental impact assessments in decisions to change policy or approve ecosystem alteration (USCEQ 1986).

However, there are few documented cases where SIA has actually made a
difference in the decision process. Two decades later, the Berger inquiry continues
to be cited as *the* case where the findings from a SIA provided the justification not
to proceed with the proposed development.[5] In that case, the SIA provided justification to stop the project. Similarly, there a few cases that point to SIA as a way to
enhance benefits or make a better policy decision. The spirit of NEPA was that
knowing about and understanding project effects in advance could make the proposal
better through the implementation of mitigation and monitoring procedures.
Although not benefiting from the same level of legal support as EIA, SIA has
achieved wide acceptance as evidenced by the following:

• The document, *Guidelines and Principles for Social Impact Assessment*, is
   important, not only because social scientists could agree on the content, but also
   because it is written to fit within the NEPA process and regulations now used
   by US federal and state agencies. SIA is tied directly to EIA by including public
   involvement, identification of alternatives, baseline conditions, scoping,

---

[5] There have been other examples. For example, in Australia, a major uranium and gold
mine at Coronation Hill in the Northern Territory, and a high temperature incinerator for
intractable waste, which was to have been built in Corowa in rural NSW, provide examples
of situations where social impacts stopped the projects even though they were arguably
acceptable on environmental grounds. However, these do not come close to the international
fame of the Berger inquiry.

projection, evaluation of alternatives, mitigation, and monitoring in the SIA process.

- In addition to North America, Australia, New Zealand, and the Philippines provide examples of countries that formally use SIA both as a part of environmental assessments and as a stand-alone assessment process. Some of the jurisdictions within these countries have agencies specially addressing social impacts. Unfortunately, most are subject to changing political winds.

- In early 1993, the US Council on Environmental Quality began to explore ways to formally incorporate SIA into their revised EIA regulations. While the 1978 Guidelines for NEPA (amended in 1986) have served as a model for project evaluation, they do not specifically require SIA. Rather, the courts have mandated that selected social components must be included and some federal agencies have included SIA in their regulations and handbooks. In addition, American Indian concerns and rights have been incorporated into the NEPA process particularly with regard to historic lands and spiritual places.

- The American Sociological Association held a professional workshop on SIA in August 1993 and has plans for another on integrative SIA.

- The number of universities listing courses in SIA is increasing, particularly in urban and regional planning departments as part of an environmental planning program. In addition, universities are incorporating SIA in curricula on community development, health and educational needs assessment, and as a component in policy analysis.

- The US Agency for International Development (USAID 1993) has continued to respond to the NEPA directives and has incorporated SIA-like procedures (which they term Social Soundness Analysis) into their project proposal and project identification documents (see chapter by Jiggins).

- Recent rulings by US courts have upheld the need for SIA in project evaluation procedures; for example, a February 1994 US [presidential] executive order expanded SIA to include the issues of environmental racism and justice; and the FEMAT project, referred to earlier, has highlighted the need for a 'social component' in all ecosystem management activity.

Despite these advances, the fact remains that in the two decades since SIA became a recognised subfield of research and policy application, there are few examples where its use has made a difference in the project/policy decision process. On the other hand, EIA has been shown to be one of the most far-ranging and significant methodologies to improve projects and policies. SIA is recognised as important, but has yet to be integrated sufficiently in the EIA process. Integration into the institutionalised policy and decision-making process will depend upon a proven track record of making projects and policies better, as well as an understanding by policymakers as to what SIA actually is all about. However, the SIA process must always help communities understand the impacts of external change and defend communities' interests.

CONCEPTUAL AND METHODOLOGICAL APPROACHES IN SOCIAL IMPACT ASSESSMENT

A Framework for Social Impact Assessment

There are many different models of the SIA process (e.g., Branch et al. 1984; Burdge 1994a, 1994b; Finsterbusch 1980; Interorganizational Committee 1994; Taylor et al. 1990; Vanclay 1989–95). All have roughly the same elements in varying numbers of steps and stages. The model presented here is based on the model adopted by the Interorganizational Committee (1994).

The basic SIA model is comparative and based on studying the course of events in communities where planned environmental change has occurred, and extrapolating from that analysis to predict what is likely to happen in another community where a similar developmental event or policy change is planned. Put another way, given similar predevelopment conditions and similar resource development projects, the social impacts resulting from a completed development project in Community A can be generalised to and predictive of what will happen in Community B. By identifying the probable undesirable social effects of development before they occur, recommendation can be made for mitigation. The SIA model also allows for alternative plans to be evaluated through analysis of the different impacts of different alternatives. The model also permits an evaluation of the impacts actually experienced by a particular impacted community. Thus three different tasks of SIA can be identified:

1. *Assessment and prediction* refers to the determination of the potential impacts of a specific action affecting a community before the commencement of any change. This information usually forms part of a conventional EIS, and is used in appraising the costs and benefits of projects for their social worth. The whole EIS, including a social impact statement, can be used to determine whether government approval should be given to the project.

2. *Mitigation and monitoring.* Mitigation involves both an initial statement about potential impacts and how that may be averted, and an ongoing role in the development process by all parties including the affected community, the developer, and the agency, in order to minimise any impact that does occur. Monitoring extends beyond the role of mitigation, checking that any change that occurs has been anticipated, and that appropriate mitigation strategies can be developed to deal with the consequences of any unexpected impacts.

3. *Audit and Analysis.* The process of SIA is heavily dependent on the use of prediction techniques during the assessment phase of the project. These prediction techniques may be highly culture bound, and are of varying reliability and accuracy. Therefore, each SIA that is undertaken, especially those for major projects, needs to contain an audit of methods and predictions. Furthermore, in order to advance the understanding of this area of study, analysis of the social impacts that have occurred as a result of past actions is necessary.

Assessment of social impacts should take place as part of any project requiring an EIS and/or any public or private activity (project, policy, proposal or process) that is likely to have significant social impacts. Mitigation and monitoring should be undertaken for any activities that proceed beyond the planning stage. Furthermore, since project planning itself can have major social impacts, for example in the form of increased concern in the community, and effect on the property (real estate) market, sometimes mitigation measures and public involvement programs may be required before a project commences operation, even if a project does not ultimately proceed. While it is obvious that mitigation and monitoring should be undertaken for any activity that has major social impacts as identified in the assessment phase, detailed monitoring and audit should be undertaken on at least some activities identified as having low or no social impact, to ensure the accuracy of methods. It may be that some of these activities have hidden impacts not considered by the assessment phase. All major projects should encompass all three tasks of SIA.

Sometimes the tasks of SIA may be undertaken individually. Obviously, often only assessment will be undertaken for projects that are not granted final approval. However, projects that are underway without government requirement for detailed EIA including SIA, can still be subject to monitoring, and possibly mitigation roles of SIA. Anthropologists and sociologists may undertake monitoring of social impacts independent of any formal government mandate to conduct an SIA. SIA can also be undertaken as a historical exercise with anthropologists, sociologists, and social historians examining the impact of certain activities on community life in the past.

There are some extra difficulties in conducting SIAs when they are not required or intended at the beginning of the planning and construction stage of the development. Research methodologies may lack pre-impact measures, thus preventing any longitudinal analysis of change over time. The research is likely to lack the large scale funding required for detailed assessment, and funds able to be provided by community groups are likely to be inadequate to cover the costs of a detailed study. The study may lack legitimacy in the eyes of the community, and responses to surveys, etc., may be low. The lack of requirement of compliance on behalf of the developer may mean that the developer is uncooperative in providing information about further plans, and in accepting any recommendation that the study may make in terms of alleviating impacts. Therefore, communities and governments should insist that SIAs are part of the whole development process and funded by funds lodged by the developer (see Buckley, 1991). The full potential of SIA is only achieved when SIA is incorporated in the development process from its inception to well past its completion. Despite the tasks of SIA sometimes being undertaken independently, the full concept of SIA is that it is a process, an integral part of project development, not an isolated step or hurdle to be overcome.

It is important that SIA be considered at all stages in project or planning development. Typically, development projects and policy development can be considered to have four stages: (1) planning or policy development; (2) construction/ implementation; (3) operation and maintenance; and (4) decommissioning or abandonment (Interorganizational Committee 1994). Social impacts will be different at the different stages. From the earliest announcement or rumour of a pending project or policy change, expectations and effects are experienced. The real estate market is affected through speculation or through reduced demand due to perceived undesirability of a project. If the project is likely to have major environmental, social or health impacts, people begin to worry, and may form interest groups, or use existing groups as a platform for discussing the project and mobilising action. Local governments and politicians begin to plan strategies and develop allegiances. Thus considerable impacts can occur simply in response to the information released about a project or policy, whether or not such information is correct, and long before the first shovelful of soil is turned.

Initial construction stages, particularly of large construction projects such as dams, power stations, and new mines, often create the most social impacts of any stage because they usually involve larger numbers of workers than the operation maintenance stage, and because appropriate infrastructure and procedures are not in place. It is also typically the time the largest number of changes take place. Construction workers tend to be separated from families, work long hours at hard work, and consequently develop a subculture that manifests itself in behaviours that are often disapproved of by the long-term residents of small communities, especially in association with the large quantities of alcohol that these workers tend to consume. Long-term residents may experience increases in price for housing and local services, community infrastructure may become over-stretched, there could be increased uncertainty about the future and a change in residents' feelings about their community. This may lead to resentment and friction between established residents and newcomers.

The operation and maintenance stage is the stage when, with appropriate planning and the implementation of mitigation and monitoring procedures, negative social impacts can be minimised and benefits maximised. It is a stage when communities return to a stage of 'normalisation'.

There are two forms of abandonment, decommissioning, or plant closure that can be conceived of: one is when a project that was scheduled to go ahead is cancelled, the other is when some long-term activity that was part of the everyday life of a community is shut down. With the former, impacts are likely to be less severe, but still important because of the rise in expectations that occurs when projects are announced. With the latter, impacts can be very severe, often with considerable job loss and flow-on economic decline for the community. For some operations, closure was not predictable and may be the result of economic downturn or uncompetitiveness (perhaps due to the age of the plant). In these situations, especially if a private sector operation, there is little that can be done. For public sector operations, mitigation measures to minimise the impacts during closure can be implemented. For projects being started that have a known project life of short-to-moderate term, contingency plans need to be developed for the closure of the project and need to be part of SIA and EIA activities at the beginning of the project.

The Steps in the Social Impact Assessment Model

As indicated before, there are many SIA models, all with different numbers of steps, but all having roughly the same content. The 10 steps presented here are adapted from the *Guidelines and Principles* (Interorganizational Committee 1994) which, in turn, emulate the EIA steps in the US CEQ guidelines. The Interorganizational Committee considers the steps to be logically sequential, but to overlap in practice.

1. *Public involvement*. Develop and implement an effective public involvement plan to involve all potentially affected publics.

2. *Identification of alternatives*. Describe the proposed action or policy change and reasonable alternatives. At a minimum, detail will be required about locations; land requirements; need for additional ancillary facilities such as roads, transmission lines, sewer and water plants, etc.; construction schedule; size of the workforce during construction and operation; facility size and shape; capacity to utilise locals; and institutional resources.

3. *Profile baseline conditions*. Document for the relevant human environment or area of influence the social baseline conditions prior to effects due to the current project. This will include developing an understanding of the relationships between the social and biophysical environment; historical background of the area; contemporary issues; political and social structures; culture, attitudes and social-psychological conditions; as well as basic population characteristics.

4. *Scoping*. Identify the full range of possible social impacts through a variety of means including discussion or interviews with numbers of all potentially affected.

5. *Projection of estimated effects*. Evaluate all possible impacts to determine the probable impacts.

6. *Prediction of responses to impacts*. Determine the importance of the identified social impacts to the affected publics.

7. *Estimate indirect and cumulative impacts*. Consider the flow-on ramifications of projects, including the second-, third- (and so on-) order impacts. Also consider how the impacts of one project may affect and be affected by other projects.

8. *Changes in alternatives.* Recommend new or changed alternatives and estimate or project their consequences.

9. *Mitigation.* Develop and implement a mitigation plan. Mitigation plans should firstly seek to avoid impacts; secondly seek to minimise unavoidable impacts; and thirdly utilise compensation mechanisms.

10. *Monitoring.* Develop and implement a monitoring program that is capable of identifying deviations from the proposed action and any important unanticipated impacts.

An eleventh step could be *Evaluation* (or *Audit*), that is, conduct a check of actual and mitigated impacts with those identified in the initial SIA. The Interorganizational Committee includes this activity in their tenth step. It is probably desirable to separate it from the tenth step because the task of undertaking monitoring tends to be assigned to a community or agency, while the evaluation or audit should be completed by the SIA practitioner.

Scoping identifies the type of the social impacts that are likely to be expected and clarifies the issues relevant to the project including: the frames of reference; the major issues; the key variables to be considered; the geographical area of most importance and other areas of likely impact; the units of analysis and methods of measuring or determining impact; interested parties or stakeholders (including those who have vested interests in the project or the affected community, and other groups who will suffer any impacts from the development); and identification of community leaders and spokespersons from the stakeholder groups. Undertaking a literature review to identify previous studies of a similar nature to the proposed action is an important step in the scoping process which should be commenced as early as possible (see Vanclay 1989–95). Scoping is a largely conceptual process undertaken by the SIA practitioner with assistance from discussions with interested parties.

Profiling, sometimes considered to be part of the scoping process, involves gathering information about the community in the pre-impact state to provide initial estimates for input into prediction models, and to provide baseline information with which to compare changes when they occur. Some of this data may already exist in the form of secondary data such as census and other government and community records, local histories, maps, newspapers, and telephone directories. Otherwise, a social survey may be required to collect this essentially quantitative data. Interviewing long-term residents is also a valuable source of information. Attempts should be made to consider the nature of changes that are inherent in the cultural setting and likely to effect change on the community independent of the current development project. The likely impact of other development projects in the proximity of the current development and any cumulative impacts should be considered.

The potential for mitigation of impacts is the greatest when the SIA practitioner is included in project formulation from the earliest stages possible. Issues such as the siting of the development, and the type of technology used, as well as implementation decisions such as the availability of workforce, and the location of construction workers, have profound social impacts. While these issues are often regarded as economic concerns, the costing of social impacts could easily alter economic decisions made that ignored social impacts, especially if the social impacts differed markedly between alternatives. Extra expense during planning stages and the selection of a more costly alternative, may reduce social and environmental impacts and be economically advantageous in the long run.

Audit refers not just to a review of predictions or methods, but a review of the whole procedure of SIA as it was applied to the specific project. Although it is important to review predictions, in many cases, predictions will be different to outcome because SIA has been successful in its role of mitigation and monitoring. This does not make the predictions wrong, nor does it make original assessments invalid. It does mean that care needs to be placed on the interpretation of predictions at the assessment stage, and during any audit.

Data Considered and Methods Used in Social Impact Assessment

The practice of and research in SIA develop concepts and methodologies that may be used to understand likely social change before the event occurs. Like other impact assessment procedures, SIA is anticipatory in nature, and uses data and methodologies on an *ex-ante* basis. In addition, the SIA assessor must consider how different concepts relate to each other and needs to formulate predictions based on complex and speculative interactions of many social and environmental variables. The concepts used in SIA are drawn from a wide research literature on rural community change, natural resource development and alteration (mostly reservoir and other public works projects), rural industrialisation, displacement and forced relocation, combined with literature on community infrastructure needs, among other areas (Elkind-Savatsky 1986; Finsterbusch 1980; Freudenburg and Gramling 1992; Burdge 1994a, 1994b).

It is almost impossible to catalogue the true dimensions of social impacts, change itself creates other changes, impacts multiply and proliferate, and impacts can occur at a considerable distance from, and time after, a project has commenced or has been completed. Furthermore, societies and culture are themselves dynamic, thus it is difficult or even impossible to determine what changes are due to a project, and what changes would have occurred in a community anyway. Furthermore, development does not occur in a vacuum, projects often occur in conjunction with other changes in a society or a particular geographical location. Thus it may be difficult to predict or identify the changes that are due to each project or process. However, social scientists have identified the basic dimensions that can be measured which reflect the fundamental and important characteristics of a community. Studied over time, dimensions such as the unequal distribution of benefits and consequences of a project, changing power structures, family disruption, racial and cultural diversity, disintegration of community cohesion, give us insight as to how social structures and communities will be altered when development occurs. Therefore the basic conceptual strategy of SIA becomes one of identifying likely future impacts based on reconstructing the social impacts of past events.

The major categories of social impacts identified in the literature include: population change, community and institutional structures, political and social resources, individual and family changes, and community resources (Interorganizational Committee 1994). Similar collections of labels include lifestyles; attitudes, beliefs, and values; and social organisation as used by Taylor et al. (1990); and community resources, community social organisations, and indicators of individual and community well-being (Branch et al. 1984). Within each of these clusters of social impacts are a series of SIA concepts and variables which point to measurable change in human populations, communities, and social relationships, which result from development projects or policy change. Examples of identified SIA variables include: relocated populations, size and structure of local government, historical experience with change, distribution of power and authority, identification of stakeholders, perceptions of risk, health and safety, attitudes toward the policy/project, and change in community infrastructure, to name but a few. The comprehensive computer database SIA bibliography compiled and maintained by Vanclay (1989–95) uses over 400 keywords to describe different aspects and interests of SIA.

PROBLEMS CONFRONTING SOCIAL IMPACT ASSESSMENT

Despite the advances of SIA, some conceptual, procedural, and methodological difficulties remain. These can be grouped into four major categories:

1. Difficulties in applying the social sciences to SIA:
   - Units of analysis, theoretical models, and the language of the various social science disciplines are sometimes contradictory or inconsistent, making interdisciplinary communication difficult.
   - Social science traditions, especially sociology, tend to be critical and discursive, rather than predictive and explanatory. Thus the core theoretical disciplines which comprise SIA fail to provide background in the processes of developing conceptual frameworks or valid measures for testing the inter-relationships among variables.

2. Difficulties with the SIA process itself:
   - Data are often poorly collected, and therefore projections are based on inadequate information which is often isolated, not systematically collected and therefore lacks validity checks. Estimates about the consequences to human communities of likely future events should be based on conceptual relationships developed from theory and previous research supported by data collected utilising the appropriate methods and subject to empirical verification.
   - The methodologies for assessing social impacts are numerous and complex, and exist as a process as much as a discrete entity. Consequently, they are difficult to document and to evaluate.

3. Problems with the procedures applying to SIA:

- SIAs are often done by consultants who do not know relevant social and economic theory, and who may not be trained in either SIA or social science methodology. There is no registration of suitably qualified and experienced SIA practitioners, and some over-zealous consultants have claimed expertise that they did not have.
- Regulatory agencies and corporations have not checked the credentials of consultants who undertake SIAs or insist that SIA consultants have appropriate social science training.
- There is little evaluation or audit of SIA reports, and agencies and corporations receiving SIA statements seldom take the time to determine the validity and reliability of the contents of these reports.
- Relevant literature on SIA is hard to find, and often not accessible. Many valuable resources are not published, but exist only as consultancy reports. Because of both litigation and commercial secrecy concerns, consultants, proponents, and government agencies often prefer not to publish or make widely available many reports. Where reports are published, they often do not provide the detail necessary to fully evaluate the methodologies used and the validity of the claims.
- SIA is seen as a single event, as a discrete statement of impacts, not as a process which develops its full potential in the mitigation of impacts, and as a process which governs the planning and development process. The regulatory frameworks under which EIA–SIA are undertaken (including NEPA and NEPA-like structures) impose this discrete event mentality.
- Because of its project-based conceptualisation, SIA, when undertaken according to the regulatory guidelines, although not to its full potential, cannot address cumulative impacts resulting from multiple projects.
- While mitigation is part of the project-based conceptualisation, the potential for the development and implementation of effective and ongoing mitigation strategies is limited by the failure to see SIA as a process.
- Impact statements tend to be used to determine whether a project should go ahead or not; and if approval is given under what conditions, such as what mitigation strategies and/or what compensation should be paid. The failure to utilise SIA as a process with effective monitoring, mitigation, and management means a reliance on the use of SIA as an approval mechanism and to determine the level or form of compensation. Thus, approval may be denied to projects that potentially could be acceptable provided that certain mitigation strategies were in place. And other projects are approved, with compensation paid, even though the project and the compensation (or royalties) itself may create considerable social impact that appropriate mitigation and planning may have avoided (Altman 1983; Connell and Howitt 1991; O'Hare 1977; Swartzman, Croke, and Swibel 1985; Tatz 1984; Turnbull 1979).
- In some countries, there are statutory requirements to undertake SIAs, but even in these countries there is seldom a requirement for the results of SIAs to be seriously considered. SIAs often go unread, at least unheeded, and mitigation measures are seldom taken seriously.

- As a component of the policymaking process, SIAs will come under increased scrutiny in the adversarial setting of the public hearing and judicial review process, therefore, the assessment must be based on rigour and at least a minimal level of quantification. If an assessment is questioned in a legal setting, it will be by another social scientist hired to critique the methods and conclusions. Because of the nature of public settings, data and ideas will be evaluated in the context of special interests. In the USA and Canada, the various reviews of SIAs are done in the setting of a formal hearing. As such, the SIA practitioner needs some peer-supported guidelines and principles for justification of the general methodological approach and sociological content (i.e., social variables) of the study. These are provided by, in the USA at least, the *Guidelines and Principles for Social Impact Assessment*.

4. There is what can be described as a prevailing 'asocietal mentality'—an attitude that humans do not count—amongst the management of regulatory agencies and corporations (proponents undertaking the proposed development) which commission SIAs. This mentality also extends to politicians at all levels of government, public officials, physical scientists, engineers, and even economists and some planners. Persons with this mindset do not understand—and are often antithetical to—the social processes and social scientific theories and methodologies which are very different in form from those in the physical sciences in which these people are often trained. Because of the power of this mentality within the regulatory and administrative subcultures, even when sympathetic individuals join, they are often socialised into this mentality. The implications of this asocietal mentality for SIA are:
   - A failure to accept the need for SIA in the first place. The mentality naively assumes that development is good and that there are no social (and sometimes no environmental) consequences of development.
   - There is no recognition of the need for special skills or expertise to assess social impacts. Since no credence is given to society as a special entity, it is assumed that anyone can determine the social consequences of development. The legitimacy and unique knowledge of the applied field called SIA is not yet fully recognised, understood, and accepted.
   - Since there was no understanding of the nature of potential impacts, or of the concerns that community members might have, there was no expectation that SIA statements should provide anything other than a statement about the change in the number of jobs, and the number of children going to school. SIA was little more than primitive demographic impact assessment (see chapter by Becker) and fiscal impact assessment (see chapter by Leistritz). With this expectation, it was not in the interests of consultants to provide more, even if they were capable of doing so.
   - Persons not familiar with SIA have difficulty in understanding the use and integration of public involvement in the SIA process. In some organisations and agencies, public involvement has been equated with SIA. The problem comes when administrators or decision makers believe that because they have

done public involvement, they have done SIA. Public involvement is a component of the SIA process and may be used to collect data on key SIA variables. Public involvement is also part of the initial scoping (and is required under the NEPA regulations) and must be incorporated throughout the entire process, but is not social impact assessment!

- Consultants who intended to undertake a thorough SIA were thwarted because of the lack of understanding about how long it would take and how much it would cost to do the job adequately. Reputable consultants were underbid in the tender process by charlatan (at least with respect to SIA) consultants intending to do a superficial analysis.
- There is a lack of understanding, and often disagreement with the results of SIA studies. Because individuals possessing this asocietal mentality do not understand social processes, they often rejected the results of bona fide consultants whose results often contradicted their notions of common sense.
- Another problem is articulating the complex stakeholder network (both corporate and community-based) in which SIA and EIA is conducted. Special interest groups will define problems and see results of studies from their point of view, and attempt to use SIAs to their particular advantage, possibly distorting the intent of the study or the specific result in the process. In a litigious and/or confrontationist situation, altruism and concern for such global (and even regional) goals as a quality environment and the future welfare of an impacted community are seldom part of the debate.
- Because in the physical sciences generally there tends to be clearly defined problems for which singular solutions can be identified with the appropriate analysis, there is a belief that social issues are similar, and an expectation that SIA statements will deliver clear statements of social impacts and that singular mitigation strategies can be identified.
- There is a complete lack of recognition of the complexity and heterogeneity of society, and how the impacts of developments benefit and disadvantage different components of society in different ways.

FUNDAMENTAL ISSUES IN SOCIAL IMPACT ASSESSMENT

In addition to the difficulties confronting SIA, there are a number of more complex, fundamental issues affecting SIA, which are problematic in most situations where SIA is to be applied. These issues are best expressed as questions to which definitive answers cannot be easily given.

Who have legitimate interests in the community? How is the 'affected community' to be defined and identified?

It is fundamental to SIA that in all development projects, the distribution of costs and benefits is not equal across the community (Elkind-Savatsky 1986; Freudenburg 1984). One of the tasks of SIA is to identify the stakeholders, the winners, and the losers in any development. Usually, those examining social impacts are concerned about the social distribution of costs and benefits, usually in terms of social class and ethnic minority groups. A further concern of SIA is to predict how the nature of the community will change as a result of a specific project. However, 'community' is a reified concept in sociology. In a stable community (one in which the rate of change of members is low) faced with a single project development, it is relatively easy to identify bona fide members of the community. Most projects bring newcomers to a community, and development itself promotes additional growth in service industries (the regional multiplier). In a community experiencing rapid growth, newcomers to the community may be very different in values, attitudes, and behaviours to the existing community members. Their concerns vis-a-vis any development project may be very different to the established local community. If a community is experiencing rapid growth, should newcomers be regarded as part of the community and their concerns be included in any impact assessment—or are they part of the problem?

Rural rezoning and rural–urban fringe development provide many examples of situations where newcomers, predominantly middle-aged professionals, have very different concerns with respect to further development than the pre-existing (often farming) communities. Attractive locations, such as coastal zones, which are subject to tourist development and inundation of new settlers provide additional examples of such situations. The environmental assets of many coastal locations attract city leavers. In the case of Port Douglas in northern Queensland (Australia) and in other areas such as northern NSW (Australia), successive waves of newcomers have each arrived, each causing their own impact. At the time when each wave of new settlers arrives, these new arrivals typically either want no further development and no more settlers (thus pulling up the ladder), or want the opportunity to develop income-producing activities that may have significant social (and environmental) impacts.

Because of the rate of growth experienced by the community, and the successive waves of immigration to the region, at any point in time when an SIA is to be undertaken, how is the SIA consultant to establish what the views of the community are? Whose views are entitled to be considered? The problem may be further compounded by the seasonal nature of the residence of many new arrivals. In many cases, the original inhabitants are forced out of the community by rate and rent increases; by local councils that become dominated by socially and politically astute new arrivals who set certain building standards that exclude, in the extreme case, the hermit-type existence of some original inhabitants; and simply by their own desire to escape from the development in order to find another place where they can regain some of their lost solitude (where doubtless the same process will occur again in a few years' time).

There are often conflicts, especially in locations of tourist potential or ecological value, between the local community (defined as residents living and working in the area for the majority of the year) and other sections of the community such as holiday makers and the broader community that experiences vicarious satisfaction from knowing about an area. Thus, when an ecologically significant old-growth forest that was intended to be logged is protected from logging, the local community may experience social impacts in the form of job loss, forced migration from their home in search of work, or long-term unemployment; loss of identity and self-worth; loss of their sense of their community; etc. But had the logging proceeded, the wider public may have experienced social impacts in the form of lost opportunities for holiday-making or ecotourism to that location, as well having experienced grief at the knowledge of further environmental destruction.

Although not usually considered in SIA, future communities comprising generations not-yet-born perhaps ought to be considered as part of the public whose interests ought to be protected. Future communities will suffer social and environmental impacts as the result of present human activities (see chapter by Goodland and Daly).

SIA can not deal with these questions, nor should it. These questions are political. The role of SIA is to identify how different sections of the broader community are effected by development projects (and what can be done to minimise these impacts). SIA tends to, and probably correctly so, pay more attention to local concerns over the concerns of the broader distant public; but in so doing, SIA practitioners, and local communities must accept that broader concerns may outweigh purely local concerns in the ultimate decision about whether a project or policy ought to proceed.

What should be the role of community participation in the SIA?

This question raises many issues about the extent and validity of the knowledge and opinion of local communities, and about the right of local communities to determine their own destinies independent of outside interference. While one might take the view that community involvement is an intrinsic good or right and that community involvement will always lead to an increased knowledge of the project by the community and therefore reduce potential impact caused by uncertainty (see chapter by Roberts), there are two situations likely to be of concern in the SIA process: one where the public is opposed to the project, yet by independent assessment the project is likely to be beneficial; the other, where the public is in favour of the project, but independent assessment of the project suggests that the social (and/or environmental) problems are likely to outweigh the benefits.

The general community does not necessarily know what the likely effects of development will be. The public may be manipulated by advertising, and may be deceived by promises of economic prosperity. Public support for, or opposition to, a project may simply be a matter of timing, the role of the media and public relations exercises by the developer.

Strong public support for a project does not mean that there will not be any major social impacts, or that a project should necessarily proceed. Independent, non-partisan, expert assessment of likely impacts needs to be undertaken. On the other hand, where the public is opposed to the project, it is possible the public perception of risks associated with a project may be over-inflated, and actual impacts may be slight—except that fear itself and associated psychosocial stress are major social impacts, although they can be mitigated by careful management, usually through a public involvement process. Research into residents' concerns about nuclear power stations (Travis and Etnier 1983; Brown 1989; Gwin 1990; Health and Safety

Executive 1992)—and probably other major and different projects, such as high temperature toxic waste incinerators for intractable waste—indicates that the fear of danger associated with these plants far exceeds the actual probability of the risk involved. However, for everyday risks, such as those associated with the effects of smoking, excessive alcohol consumption, and probability of road accident, the risks are under-perceived. Perception of risk is an emotive issue and does not correlate with the actual risk (expressed as a probability of occurrence) (see chapter by Carpenter).

The other major concern affecting public participation is that the nature of the public participation methodologies used may mean that the view gained from the so-called participation is not representative of the community. Public meetings, sadly often the only format of public participation that is used, constitute neither participation nor representation. They are not participative because they usually consist of one way information transfer, and they are not representative because only certain groups go to such meetings, and only some individuals representing some of these groups say anything, and almost none of them have any attention paid to them. Other forms of participation do not necessarily guarantee representation either. Invariably, elite or power groups, the very same groups that tend to benefit most from developments, are also the most likely to gain representation through the various avenues that are used to involve the public and certain groups of people, especially social underclasses, tend to be excluded from public participation exercises. The chapter by Richard Roberts discusses these issues further.

Public participation, not matter how carefully undertaken with respect to community concerns, is not a substitute for a thorough SIA using appropriately qualified SIA professionals, although it remains an essential component of SIA.

What impacts are to be considered?

SIAs are usually undertaken at the behest of a community group, local or regional government, or the developer. Each one of these bodies has vested interests that they are keen to promote or protect. Consequently, the regulatory framework under which SIAs and EIAs are undertaken affects the integrity of the SIA or EIA (Buckley 1991). Where EIA–SIA consultants are engaged directly by developers, with no review procedure other than public comments, the consultants tend only to give a prodevelopment line, with any negative or critical comments couched in very masked terms. Consequently the impacts that are considered (both perceived as impacts and/or measured) are those that are politically or socially determined at the time the study is done. Many potential impacts are excluded from consideration because they may not be regarded as important at the time. Therefore, unless SIA is an ongoing process undertaken by truly independent and professional individuals, all SIA statements will be inadequate. There needs to be a procedure for ensuring that all potential impacts are considered. Adoption of, and promotion of the understanding contained in, the *Guidelines and Principles* is a step toward ensuring appropriate and professional SIA practice.

How should impacts be weighted?

Certain impacts, such as changes to the nature or character of a community may be perceived as negative by some members of the community, and as positive by other members. Thus impacts are not simply positive or negative in themselves (such as job growth is positive; job loss is negative), but are subject to the value judgements of individuals. For example, one of the consequences of the siting of a new prison in a rural community might be the movement of previously city-based families of prisoners (i.e., wives and children) to that community. Some existing members of the community may view this as a negative experience and may be concerned about the loss of community integration, the changing nature of the community, and may have concerns about their personal safety and the security of their belongings. Other members of the community might believe that the community was too narrow-minded to begin with, and that the intrusion of new and a largely different type of people might be good for the community because it will lead to a broadening of the mental horizons of the more conservative members of the community. Thus the same consequence of development is both a positive impact and a negative impact depending on the perspective of individuals.

This situation of whether consequences are positive or negative is even more problematic because individuals may change their mind over time. A consequence may be a negative impact for a period, and a positive impact thereafter, or vice-versa. In this situation what should be the position of the SIA practitioner?

Other concerns about the weighting of impacts include difficulties for SIA about different strengths of feelings different individuals attribute to impacts. Some individuals may regard an impact as a mild inconvenience or unpleasantness, while for other individuals, the same impact may create a major change in life. Many individuals over time will adapt to a new environment, even if the (romanticised) past is reflected on as having been preferable. However, there are certain vulnerable groups, particularly the aged and the socially disadvantaged, who are unlikely to be able to adapt and who bear most of the social impacts. These groups need special attention by SIA practitioners.

SIA cannot judge. It can merely report the how different segments of a community are likely to respond to development projects or policies, and advise on appropriate mitigation mechanisms.

Who judges?

A nation's regulatory framework usually specifies the role of the SIA or EIA in decision making and planning. In many countries, while EIAs and SIAs may be compulsory, in most cases, they are perfunctory, and regulatory bodies tend not to be bound by the outcome. In any case, because of the issues outlined above, and cost benefit analysis and other economic decision-making techniques notwith-standing, SIA cannot, except in very obvious cases, make definitive decisions about whether a given project ought to go ahead or not. In any case, SIA is not, of course, accorded regulatory power sufficient for this to happen even if it could decide. Decisions are about whether a project should proceed, or what compensation a developer should pay, are ultimately and inherently political. Even in the examples cited as being successes of SIA in stopping a project (the Berger Inquiry, Coronation Hill, and the Intractable Waste Incinerator), it could be argued that it was not the SIA study that stopped the project, but other political pressures, with the SIA providing the convenient excuse (see Toyne 1994). Consequently, it could be argued that SIA is no better than the normal political process, complete with social and power inequalities that are vested in the political system. However, to the extent that SIA provides information for informed decision making, media commentary, and public discussion, SIAs can do no harm, and have the potential to contribute. It is unlikely, however, that SIAs can change the inherently political process of decision making and planning.

Even in a benevolent political system where SIA was genuinely desired as a decision-making tool, SIA could not deliver a decision-making mechanism. The inherent social inequalities of costs and benefits of projects will mean that on simple cost–benefit analysis, where benefits are greater than costs, projects will proceed even if the same social groups will always be adversely affected. Even if mitigatory action is taken to minimise these impacts or if affected groups are compensated, it is unlikely that such compensation will cover the full extent of costs actually experienced. It is also unlikely that any government would have sufficient courage or conviction to enforce full compensation, and no developer would engage in such practices voluntarily.

Decisions about projects are ultimately and inevitably political.

WHAT SIA SHOULD DO TO ADDRESS THE PROBLEMS IT FACES

With acceptance comes expectations—a general cliche that is particularly appropriate for the field of SIA at the present time. With general agreement within the SIA field on procedures and content, as well as the general conceptual orientation of SIA becoming more widely accepted, attention must now shift to the development of conceptual and methodological issues which will strengthen the field. Two over-riding issues are (1) the application of SIA in the larger policy context, and (2) whether SIA can be successfully integrated into the planning and development process to improve projects and planning generally, rather than being seen solely as methodology to provide only a statement of potential impacts to determine whether a project ought, or ought not proceed. Such a narrow view wastes so much of the potential of SIA because many possible impacts can be easily avoided by simple and cost effective mitigation strategies that can turn development projects with negative social impacts into projects with positive impacts, at least for many members of the community. Such an enlightened approach to SIA has been recognised in certain industries, particularly the mining industry, where through the use of SIA and community development consultants, practical social strategies and social design concepts can have a profound influence on community wellbeing which flows through to reduced costs and enhanced productivity for the company. Conversely, early mining developments which had no social planning often were social disasters with severe social problems which had, not only social impacts on the workforce and surrounding (often indigenous) populations, but also on productivity through lost work due to sick leave, alcoholism, strikes, and 'slow days'.[6]

SIA needs to address the issue of scale, i.e. how the concepts and procedures of SIA can be applied to larger geographical regions than the immediate vicinity of a specific project, such as a large river basin, an ecosystem, or even regional and national political units. Some assessors are even calling for the analysis of global processes, such as the effects of world trade and GATT negotiations on agricultural and rural restructuring in peripheral and semi-peripheral nations (e.g., Lawrence and Vanclay 1994). The social effects of some developments can be extremely dispersed from the original site of the development. This is perhaps best illustrated by the impacts of satellite television on locations which had no previous exposure to outside cultures (see O'Rourke 1980). Social and biophysical EIAs rely on localised, project-level measures to predict impacts. As the scale is expanded, it becomes more difficult to establish significance because larger geopolitical units tend to wash out project-level impacts. Burdge's (1994b:2) axiom is pertinent:

> The social benefits and consequences of project development, consolidation, and closure (abandonment) always occur, can be measured, and are usually borne at the community and local level—but the rationale for projects and the decisions are justified and sold on the basis of regional and national economic goals.

There are two general approaches to SIA, a generic one and the project-level approach. The strengths of each need to be examined. The generic approach to SIA sensitises people to general social change. It assumes the presence of major impacts and a rather wide policy application. Project and policy impacts are seen as leading to radical shifts in the distribution of the population and in turn producing recognisable changes in how human groups relate to each other. Implicit in the generic approach to SIA is the notion of understanding social change through experience. Being sensitive to the existence of social impacts is seen as more important than actually being able to identify them. Often the objective of the generic type of SIA is to get the social science point of view across to the non-social scientist.

The project-level approach to SIA assumes that social change is ubiquitous, but that a new project or policy change alters the normal flow of social change. Furthermore, this approach stresses that impact events will vary in specificity, intensity, duration, and a variety of other characteristics. It then becomes important to understand what will be the actual social impacts of a particular development rather than only being aware that social change will take place. The researcher or

---

[6] There are many examples. For those highlighting the negative side of mining towns, see Birrell, Hill, and Stanley 1982; Williams 1981. For references stressing benefits of planning, see Australian National Commission 1976; Bowles 1981, 1982; Brealey and Newton 1978; Brealey et al. 1988; Bulmer 1976; Burvill 1975; Burvill and Kidd 1975; Gribbin and Brealey 1980; Lea and Zehner 1986; Parker 1987; Pribble 1984; Sturmey 1990.

practitioner uses past social science research to better understand what is likely to happen to human populations given different development events.

SIA will be most successful when fully integrated with planning at the appropriate level of jurisdiction where project development or a proposed policy development will occur. When this integration is accomplished, both social and environmental factors become central to planning decisions, rather than being treated as external or peripheral to the planning process. Achieving such integration requires a sound understanding of the nature of planning on the one hand, and how advances in knowledge about impact assessment and its many methodologies can fit into modern planning models on the other. Additionally, functional integration of the key components of the planning process, from project inception to postdevelopment monitoring, is an important goal of modern comprehensive planning. This kind of integration is essential because, as a dynamic process, planning requires data collection across time and ongoing revision of plans to ensure that planning goals are being met (Armour 1990). Similarly, SIA requires successful integration of all phases from scoping to monitoring, mediation to mitigation, as well as continual and cumulative assessment of results.

The methodology for measuring, and the substance of, cumulative effects in SIA need to be researched. Certain cumulative effects are obvious—such as basic infrastructural needs generally provided by local government and utility companies. Infrastructure payments and other financial arrangements can generally satisfactorily compensate existing communities for these impacts (see chapter by Leistritz). Other, still basic, questions are less obvious, for example, does increased population require an increased size of local government? At what stage should the assessment consider the community infrastructure needs? Finally, and most importantly, there are a whole range of questions relating to cumulative impacts that the SIA process raises that it cannot answer. Communities have a basic resilience and can accept a certain amount of change or impact. Impacts become important when the number or extent of changes exceeds a certain threshold. This threshold is likely to be unknown and unknowable for any community. Very large projects with major social impacts may exceed this threshold, but so may many different small projects. When the SIA procedure is undertaken on a project by project basis, it is very difficult to determine if this threshold will be exceeded, and of course, no specific project was individually responsible for exceeding that threshold. Overall advance planning is required to ensure that the threshold is not exceeded for any community, and to maximise the benefits and minimise the negative impacts for each project.

SIA achieves its greatest benefit to society through its ability to advise on mitigation of impacts. However, not only must mitigation procedures be developed and improved, but appropriate political procedures must be established to determine who is responsible for mitigation and monitoring at each step of a project, not only during construction and operation of projects, but also for the decades and centuries after the project has been abandoned or the policy implemented, especially in the case of projects with long-term impacts (such as nuclear waste repositories where the wastes have radioactive half-lives of thousands of years!).

These issues imply both policy and administrative procedures as well as mechanisms to pin point responsibility. A well thought out public involvement program is a must for the mitigation and monitoring steps of the SIA process.

As pointed out earlier, early SIA assessors faced ideological resistance (the asocietal mentality) as well as political and legal obstacles, not only in including SIA within EISs, but also in deciding which variables to use in the analysis. Social variables were often considered suspect. To ensure the presence of a social component in an EIA statement, SIA was changed to socioeconomic indicators, and as a result many US federal agencies adopted the term 'socioeconomic' impacts. In practice, the 'social' part of the word was not done and socioeconomic became an economic impact assessment, with a concentration on demographic changes and regional multipliers. The linkage persists today and a goal for SIA must be to separate further the 'social' from socioeconomic impact assessment, and to enhance the legitimacy of purely social concerns.

The SIA research community needs to publish more widely, making available the really good case studies which point out where SIA actually made a difference in the decision process, not only cases in which the SIA stopped the development, but the cases for which SIA substantially improved it. Now that there are agreed upon methods, long range research on case studies using good SIA practice needs to be undertaken. Well conceptualised, long-range research projects will provide the legitimacy for the verification of existing—and the identification of new—SIA variables.

SIA needs to be considered in the broader policy context. In the FEMAT project in the Pacific Northwest of the USA (see Clark and Stankey 1994), the researchers could not differentiate between impacts to an individual community and social impacts that occurred over the entire region as a result of proposed alternatives in levels of timber harvesting. The social assessment team was asked to do an SIA for a region without the benefit of data on any one of the impacted communities. Thus a regional data base to study social impacts which is ongoing and cumulative is needed. The FEMAT social scientists had no longitudinal data base (other than census information) as a starting point to measure social impacts. They needed an agreed upon list of SIA variables and the funds to maintain this information over time. As part of the data base problem, the FEMAT exercise highlighted the problems of integrating qualitative and quantitative data. Qualitative SIA indicators are just as valid and in many cases are more insightful and provide a more holistic perspective than quantitative indicators. However qualitative data are more difficult to store on a cumulative basis and are difficult to sell to non-social scientists. Furthermore, both types of social science indicators when used in SIA, are usually in some form of small area analysis, and consequently face the problem of 'statistical significance'. The FEMAT team of social assessors were repeatedly asked to defend the significance of the social impacts identified in the assessment process and probability levels seemed the only acceptable answer, even though these were impossible to provide (Clark and Stankey 1994).

All types of assessment face the problem of integration. How do EIA and SIA fit together in providing a comprehensive picture of likely project impacts? At present, most SIA statements are stapled to an EIA, and the total recommendations are the sum of the parts. No attempt is made to integrate and interpret the collective findings. This needs to be improved. Higher order impacts are possible. Thus environmental impacts can have social impacts, and social impacts can turn into environmental impacts. Mitigation strategies (both biophysical and social) can also have their own environmental and social impacts which may not be considered in impact statements because impact statements tend not to go past first order impacts.

The relationship between SIA and public involvement needs clarification, especially to managers of agencies and corporations who often confuse them. Public involvement is an important process which goes on throughout the EIA, SIA and planning process, but it does not tell us what social impacts will occur as a result of a policy or future project (Burdge and Robertson 1990; see chapter by Roberts).

The broadening of, and increased acceptance of, a cultural understanding of society and social processes may be the most important contribution of SIA to the assessment process. The science and research of EIA assumes a logical-positivist model in the analysis and implementation of the planning process. The model is generally reductionist and therefore lacks a holistic approach. It also assumes that all interested and affected parties have the same perspective and goals. In order to be acceptable to the EIA masters, some SIA practitioners have adopted a similar perspective in their approach to social impacts. Such an approach is inadequate in dealing with many of the real social impacts that may arise, that are purely cultural, and that will be identified if such a perspective is adopted.

For example, to many individuals and groups, particularly but not exclusively indigenous peoples, there are many spiritual and religious issues that need to be considered in SIA (Greider and Garkovich 1994). Concepts such as attachment to the land and identification with place (Burdge 1994b) are very difficult to quantify and easily discounted in the formal decision process, yet are the most important factors in determining project success and probable acceptance by local populations. The Berger inquiry cited earlier is certainly a classic example of the importance of culture in assessments. Other examples, particularly relating to the Australian Aborigines abound.[7]

One dimension of culture is spirituality, which is profoundly evident in the many indigenous cultures of the world—although anthropologists would argue it applies equally to all cultures, even if it may take on various forms which many people may not recognise as spiritual. The reference to and the use of the 'spiritual' in understanding human interaction with ecosystems is a jolting revelation for decision makers and runs counter to traditional western positivist thought. The alteration of sacred ground is a good example. A new airport built in Denver, Colorado (USA), was completed and scheduled to be opened in 1993. However, an automated baggage system could never be perfected and a series of other mechanical failures

---

[7] Berger 1977, 1983, 1985; Berry 1975; Chase 1990; Connell and Howitt 1991; Coombs et al. 1989; Macintyre and Janson 1990; Peterson and Langton 1983; Reynolds and Nile 1992; Wilmsen 1989.

have delayed the opening indefinitely. When the airport was first proposed, the American Indian tribe who lived in the area said the site was on sacred ground and that the 'spirits' would never allow the project to be completed. The natives were right, at least up until the time this paper was published!

The SIA for a uranium mining project in Australia (Coronation Hill) revealed a similar story, where if the sacred ground was disturbed, the spirit of Bula would rise up and destroy the world (Lane et al. 1990). In this case, the SIA resulted in a federal government decision not to allow mining. Unfortunately, most EIA–SIA statements allude to the importance of culture, but never really admit that cultural history is a key component in the decision process. Again much more legitimacy must be placed on the social and cultural factors of society.

PRINCIPLES FOR SOCIAL IMPACT ASSESSMENT

Because of the various issues discussed in the preceding sections, the Interorganizational Committee (1994) found it desirable to articulate a number of guiding principles that should govern the use of SIA. SIA practitioners, those commissioning SIAs and EIAs, and those evaluating impact statements ought to be mindful of these principles. The following list of the nine principles includes abridged summaries, substantially quoted, of the descriptions in the original document.[8]

1. *Involve the diverse public. Identify and involve all potentially affected groups and individuals.*
   Public involvement and conflict management should be implemented to complement and fit within the SIA process by identifying potentially affected groups, and by interpreting the meaning of impacts for each group. Public involvement should be truly interactive, with communication flowing both ways between the agency or corporation, and affected groups.

2. *Analyse impact equity. Clearly identify who will win and who will lose, and emphasise vulnerability of under-represented groups.*
   Impacts should be specified differentially for all potentially affected groups and not just measured in aggregate. For all projects, there are winners and losers. However, no single group, particularly those that might be considered more vulnerable or at risk as a result of age, gender, ethnicity, race, social class, or other factors, should have to bear the brunt of adverse social impacts. SIA has a special duty to identify those whose adverse impacts might get lost in the aggregate of benefits. Practitioners must be attentive to the groups that lack political efficacy—such as groups low in political or economic power that often are not heard or do not have their interests strongly represented.

---

[8] Abridged from *Impact Assessment* 12(2), 1994, by permission of the authors and publisher.

3. *Focus the assessment. Deal with issues and public concerns that 'really count', not those that are just 'easy to count'.*
   Because of time and resource constraints, it is important for SIA practitioners to focus on the most significant impacts in order of priority, and for all significant impacts for all impacted groups to be identified early using a variety of rapid appraisal or investigative techniques. Clearly, impacts identified as important by the public must be given high priority, but because of limitations of public participation in terms of representativeness, additional methods of scoping and impact assessment must also be used to ensure that the most significant impacts are addressed—whether or not they are identified by the public.

4. *Identify methods and assumptions and define significance in advance. Describe how the SIA was conducted, what assumptions were used, and how significance was determined.*
   The methods and assumptions used in the SIA should be made available and published prior to a decision in order to allow decision makers as well as the public to evaluate the assessment of impacts. This should be done at least to the standards required under the relevant jurisdictional regulations and agency commitments (e.g., in the USA to the satisfaction of the US Council on Environmental Quality 1986 *Regulations for Implementing the Procedural Provisions of NEPA*). In general terms, they should be specified sufficiently to allow other SIA practitioners to evaluate the work and to determine whether the procedures followed would be what would likely have been undertaken by another practitioner.

5. *Provide feedback on social impacts to project planners. Identify problems that could be solved with changes to the proposed action or alternatives.*
   Findings from the SIA should feed back into project design to mitigate adverse impacts and enhance positive ones. The impact assessment, therefore, should be designed as a dynamic process involving cycles of project design, assessment, redesign, and reassessment.

6. *Use SIA practitioners. Trained social scientists employing social science methods will provide the best results.*
   The need for professionally qualified, competent people with social science training and experience cannot be overemphasised. Experienced SIA practitioners know the data, are familiar and conversant with existing social science evidence pertaining to impacts that have occurred elsewhere which may be relevant to the impact area in question. This breadth of knowledge and experience can prove invaluable in identifying important impacts that may not surface as public concerns or as mandatory considerations in the applicable jurisdictional regulations or compliance procedures. Social scientists will be able to identify the full range of important impacts and to select appropriate methodologies and measurement procedures.

7. ***Establish monitoring and mitigation programs.*** *Manage uncertainty by monitoring and mitigating adverse impacts.*

Crucial to the SIA process is the development and implementation of mitigation programs, and the over-time monitoring of significant social impact variables and the mitigation mechanisms for their efficacy. Monitoring and mitigation should be a joint agency and community responsibility. These activities should occur on an iterative basis throughout the project life cycle. While responsibility for long term monitoring and mitigation is not easily defined, and may depend on the nature of the project and time horizon, local communities should be provided resources to assume a portion of the monitoring and mitigation responsibilities.

8. ***Identify data sources.*** *Use published social scientific literature, secondary data and primary data from the affected area.*

SIAs should draw on the published social scientific literature pertinent to SIA [see the references for this chapter; the mini-bibliography appended to the Interorganizational Committee (1994) paper, or larger SIA bibliographies such as Vanclay (1989–95)]. Existing documentation is useful in identifying which social impacts are likely to accompany a proposed action. Since the best guidance for future expectations is past experience, caution is needed when an SIA study proposes to present a conclusion that contradicts published literature. In such cases, the reasons for the difference should be explicitly addressed.

9. ***Plan for gaps in data.***

SIA practitioners often have to produce assessments in the absence of all relevant and sometimes necessary data. In general, gaps in the data should only be permitted when relevant information cannot be obtained because the overall costs of obtaining it are exorbitant or the means to obtain it are not known. In such cases, a statement of the relevance of the incomplete or unavailable data, a summary of the existing literature on the issue, and an evaluation of the likely and possible impacts based upon theoretical approaches or research methods generally accepted in the SIA community, should be provided. Three points are provided as acceptable to the SIA community when there are shortages of resources to do the desired data collection:

(a) It is more important to identify likely social impacts than to precisely quantify the more obvious social impacts—it is better to be roughly correct on important issues than to be precisely correct on unimportant issues.

(b) It is better to be inclusive rather than exclusive in reporting likely social impacts. If the evidence for a potential type of impact is not definitive in either direction, the conclusion reported should be that the impact cannot be ruled out, not that there is no evidence to support the existence of the impact.

(c) The less reliable data there are on social impacts, the more important it is for SIA work to be performed by competent professional social scientists. There are only two situations in which it may be appropriate to proceed without professional social scientist involvement: (a) In cases where proposed actions are considered by persons within the agency with social science training, and by those in the potentially affected community, to likely cause only negligible or ephemeral social impacts; and (b) In cases where a significant body of empirical findings is available from the social science literature, which can be applied fairly directly to the proposed action in question, and is referenced, summarised and cited by the person(s) preparing the SIA section of the EIS. Not to undertake an SIA using professional social scientists when either (a) or (b) did not apply, would be imprudent for both the agency and affected groups and communities.

CONCLUSION

The progress of the field of SIA has been remarkable. There have been some major agreements: a shared definition and understanding of the SIA process; a basic framework; and an outline of what ought to go into an SIA. However, more longitudinal research of SIA (and EIA) case studies is needed, particularly to evaluate (audit) past studies and predictions. There is widespread consensus that human or social impacts should be considered as part of the environment. In particular, the SIA process has raised awareness of how projects and policies and political change alter the cultures of indigenous populations. Experience has provided a realistic appraisal of what is likely to happen in the future as a result of particular policies or actions. SIA is beginning to be fully integrated into the EIA process, and EIA (and SIA) into the planning process.

Despite the success of the last ten years, the big issue ahead is how project specific knowledge about social, economic, and environmental impacts can be used for larger policy assessments. SIA has yet to bridge the schism between project level research findings and the larger scale assessments which are needed for regional and national policy decisions.

SIA should also be considered an integral part of the development process, not a step or hurdle to be overcome. Done poorly, SIA may be nothing more than a public relations exercise for illegitimate development by unscrupulous corporations. SIA is not designed to hamper development, but is designed to maximise the potential benefit for all parties associated with the development. For the community this means minimising social impacts on the community and maximising community benefits. For the developer it means minimising social impacts and therefore the costs of rectification of these impacts in the future. Effective SIA increases the legitimacy of the development, and may well facilitate the development process. SIA removes uncertainty from the process, for both the community and the developer. To a small extent, SIA reduces impacts on the workforce and has the potential to increase productivity and reduce disruption.

The effectiveness of SIA rests on the integrity of the SIA practitioner. Community participation is essential, as is community evaluation of any report or recommendations. However, public participation exercises are not of themselves social impact assessments. Governments should consider appropriate measures to ensure that SIA and EIA that is undertaken are to a satisfactory standard. Furthermore, it must be accepted that SIA cannot be an ultimate guide in decision making. Decisions were always and will always, of necessity, be political. Nevertheless, SIA can be a useful tool in providing information that will assist in that process.

Predicting the future based on the past is tricky, but that is what impact assessment is all about. Charlie Wolf's analogy of the binoculars with a lens pointed to the front and another to the back, with a weather vane at the top to give an indication of which way the (political) wind is blowing, is exemplary. Perhaps the analogy should be expanded by attaching a wheel to remind SIA practitioners not to reinvent it! The challenge ahead is to make sure that the field of SIA can deliver what it promises, and at the same time present a realistic picture of what the field of SIA can provide in the planning/decision making process.

REFERENCES

Altman, J. 1983. *Aborigines and Mining Royalties in the Northern Territory*. Canberra: Australian Institute of Aboriginal Studies.

Armour, A. 1990. "Integrating Impact Assessment into the Planning Process." *Impact Assessment Bulletin* 8(1/2): 3–14.

Association for Social Assessment (New Zealand). 1994. "Social assessment and Maori policy development." *Social Impact Assessment Newsletter* 35: 10–11.

Australian National Commission for UNESCO Seminar on Man and the Environment. 1976. *New Towns in Isolated Settings*. Canberra: Australian Government Publishing Service.

Berger, T.R. 1977. *Northern Frontier, Northern Homeland: The Report of the Mackenzie Valley Pipeline Inquiry* (2 vols). Ottawa: Ministry of Supply and Services Canada.

Berger, T.R. 1983. "Resources Development and Human Values." *Impact Assessment Bulletin* 2(2): 129–147.

Berger, T.R. 1985. *Village Journey: The Report of the Alaskan Native Review Commission*. New York: Hill and Wang.

Berry, M.C. 1975. *The Alaska Pipeline: The Politics of Oil and Native Land Claims*. Bloomington: Indiana University Press.

Birrell, R., D. Hill, and J. Stanley, eds. 1982. *Quarry Australia?: Social and Environmental Perspectives on Managing the Nation's Resources*. Melbourne: Oxford University Press.

Bowles, R.T. 1981. *Social Impact Assessment in Small Communities*. Toronto: Butterworths.

Bowles, R.T., ed. 1982. *Little Communities and Big Industries*. Toronto: Butterworths.

Branch, K. et al. 1984. *Guide to Social Assessment*. Boulder CO: Westview Press.

Brealey, T.B. 1974. "Mining towns are for people." *Search* 5(1/2): 54–59.

Brealey, T.B., C.C. Neil, and P.W. Newton, eds. 1988. *Resource Communities: Settlement and Workforce Issues*. Melbourne: Commonwealth Scientific and Industrial Research Organisation.

Brealey, T.B. and P.W. Newton. 1978. *Living in Remote Communities in Tropical Australia*. Melbourne: Commonwealth Scientific and Industrial Research Organisation, Division of Building Research.

Brown, J. ed. 1989. *Environmental Threats: Perception, Analysis and Management*. London: Belhaven.

Buckley, R. 1991. "Should environmental impact assessment be carried out by independent consultants?" *Environment Institute of Australia Newsletter* 16, June 1991: 12–13.

Bulmer, M.I.A. 1976. "Sociological models of the mining community." *Sociological Review* 23: 61–92.

Burdge, R.J. 1994a. *A Conceptual Approach to Social Impact Assessment: Collection of Writings by Rabel J. Burdge and Colleagues*. Middleton WI: Social Ecology Press.

Burdge, R.J. 1994b. *A Community Guide to Social Impact Assessment*. Middleton WI: Social Ecology Press.

Burdge, R.J. and R.A. Robertson. 1990. "Social impact assessment and the public involvement process." *Environmental Impact Assessment Review* 10(1/2): 81–90.

Burvill, P.W. 1975. "Mental health in isolated new mining towns in Australia." *Australian and New Zealand Journal of Psychiatry* 9(2): 77–83.

Burvill, P.W. and C.B. Kidd. 1975. "The two town study: A comparison of psychiatric illness in two contrasting Western Australian mining towns." *Australian and New Zealand Journal of Psychiatry* 9(2): 85–92.

Canter, L., B. Atkinson, and F.L. Leistritz, eds. 1985. *Impact of Growth: A Guide for Socio-Economic Impact Assessment and Planning*. Chelsea MI: Lewis.

Carley, M.J. and E.S. Bustelo. 1984. *Social Impact Assessment and Monitoring*. Boulder CO: Westview.

Carley, M.J. and E.O. Derow. 1980. *Social Impact Assessment: A Cross-Disciplinary Guide to the Literature*. London: Policy Studies Institute.

Chase, A. 1990. "Anthropology and impact assessment: development pressures and indigenous interests in Australia." *Environmental Impact Assessment Review* 10(1/2): 11–23.

Clark, R.N. and G.H. Stankey. 1994. "FEMAT's Social Assessment: Framework, Key Concepts and Lessons Learned." *Journal of Forestry* 92(4): 32–35.

Cohen, E. 1971. "Arab boys and tourist girls in a mixed Jewish–Arab community." *International Journal of Comparative Sociology* 12(4): 217–233.

Cohen, E. 1972. "Towards a sociology of international tourism." *Social Research* 39(1): 164–182.

Cohen, E. 1979. "Rethinking the sociology of tourism," *Annals of Tourism Research* 6(1): 18–35.

Cohen, E. 1984. "The sociology of tourism." *Annual Review of Sociology* 10: 373–392.

Colson, E. 1971. *The Social Consequences of Resettlement: The Impact of the Kariba Resettlement upon the Gwembe Tongo*. Manchester: Manchester University Press.

Connell, J. and R. Howitt, eds. 1991. *Mining and Indigenous People in Australasia*, Melbourne *(sic)*: Sydney University Press (in conjunction with Oxford University Press).

Coombs, H.C. et al. 1989. *Land of Promises: Aborigines and Development in the East Kimberley*. Canberra: Centre for Resource and Environmental Studies, Australian National University.

Cottrell, W.F. 1951. "Death by dieselization." *American Sociological Review* 16(3): 358–365.

Craig, D. 1987. *Social Impact Assessment Bibliography*. East Kimberley Working Paper Number 17. Canberra: Centre for Resource and Environmental Studies, Australian National University.

D'Amore, L.J. et al. 1979. *Social Dimensions of Environmental Planning: An Annotated Bibliography*. Ottawa: Environment Canada.

Dixon, Mim. 1978. *What Happened to Fairbanks: The Effects of the Trans-Alaska Oil Pipeline on the Community of Fairbanks, Alaska*. Boulder CO: Westview Press.

Economic Consultants Ltd. 1973. *The Channel Tunnel: Its Economic and Social Impacts on Kent*. Report presented to the Secretary of State for the Environment, London: Her Majesty's Stationery Office.

Elkind-Savatsky, P., ed. 1986. *Differential Social Impacts of Rural Resource Development*. Boulder CO: Westview Press.

Finsterbusch, K. 1980. *Understanding Social Impacts: Assessing the Effects of Public Projects*. Beverly Hills CA: Sage.

Finsterbusch, K. 1985. "State of the art in social impact assessment." *Environment and Behavior* 17(2): 193–221.

Forster, J. 1964. "The sociological consequences of tourism." *International Journal of Comparative Sociology* 5(2): 217–227.

Francis, J. 1973. *Scotland in Turmoil: A Social and Environmental Assessment of the Impact of North Sea Oil and Gas*. Edinburgh: St Andrews Press.

Freudenburg, W.R. 1984. "Differential impacts of rapid community growth." *American Sociological Review* 49(5): 697–705.

Freudenburg, W.R. 1986. "Social impact assessment." *Annual Review of Sociology* 12: 451–478.

Freudenburg, W.R. and R. Gramling. 1992. "Community impacts of technological change: Toward a longitudinal perspective." *Social Forces* 70(4): 937–955.

Gamble, D.J. 1978. "The Berger inquiry: An impact assessment process." *Science* 199 (March): 946–952.

Gray, J.A. and P.J. Gray. 1977. "The Berger Report: Its impact on northern pipelines and decisionmaking in northern development." *Canadian Public Policy* 3(4): 509–514.

Greider, T. and L. Garkovich. 1994. "Landscapes: The social construction of nature and the environment." *Rural Sociology* 59(1): 1–24.

Gribbin, C.C. and T.B. Brealey. 1980. "Social and psychological wellbeing and physical planning in new towns." *Man-Environment Systems* 10: 139–145.

Gwin, L. 1990. *Speak No Evil: The Promotional Heritage of Nuclear Risk Communication*. New York: Praeger.

Health and Safety Executive. 1992. *The Tolerability of Risk from Nuclear Power Stations*. London: Her Majesty's Stationery Office.

Interorganizational Committee on Guidelines and Principles for Social Impact Assessment. 1994. "Guidelines and principles for social impact sssessment." *Impact Assessment* 12(2): 107–152; reprinted in *Environmental Impact Assessment Review* 15(1): 11–43.

Lane, M. et al. 1990. *Social Impact of Development: An Analysis of the Social Impacts of Development on Aboriginal Communities of the Region, Kakadu Conservation Zone*. Inquiry, Resource Assessment Commission Consultancy Series. Canberra: Australian Government Publishing Service.

Lawrence, G. and F. Vanclay. 1994. "Agricultural change in the semiperiphery," pp 76–103, in *The Global Restructuring of Agro-Food Systems*. P. McMichael, ed. Ithaca: Cornell University Press.

Lea, J.P. and R.B. Zehner. 1986. *Yellowcake and Crocodiles: Town Planning, Government and Society in Northern Australia*. Sydney: Allen and Unwin.

Leistritz, F.L. and B.L. Ekstrom. 1986. *Social Impact Assessment and Management: An Annotated Bibliography*. New York: Garland.

Leistritz, F.L. and S.H. Murdock. 1981. *The Socioeconomic Impact of Resource Development: Methods of Assessment*. Boulder CO: Westview Press.

Macintyre, S. and S. Janson, eds. 1990. *Through White Eyes*. Sydney: Allen and Unwin.

McGrath, E. 1977. *Inside the Alaska Pipeline*. Milbrae: Celestial Arts.

Murdock, S.H., F.L. Leistritz, and R.R. Hamm. 1986a. "The state of socioeconomic impact analysis in the United States: Limitations and opportunities for alternative futures." *Journal of Environmental Management* 23(2): 99–117.

Murdock, S.H., F.L. Leistritz, and R.R. Hamm. 1986b. "The state of socioeconomic impact analysis in the United States." *Impact Assessment Bulletin* 4(3/4): 101–132.

O'Hare, M. 1977. "Not on my block you don't: facility siting and the strategic importance of compensation." *Public Policy* 25(4): 407–458.

O'Rourke, D. 1980. Yap: How did you know we'd like TV? Canberra: Ronin Films. (video)

Parker, P. 1987. *Resource Development and Local Government*. Canberra: Australian Government Publishing Service.

Peterson, N. and M. Langton, eds. 1983. *Aborigines, Land and Land Rights*. Canberra: Australian Institute of Aboriginal Studies.

Pilbara Study Group. 1974. *The Pilbara Study*. Canberra: Australian Government Publishing Service.

Pribble, L.W. 1984. *Planning and Construction of Remote Communities*. New York: Wiley.

Reynolds, H. and R. Nile, eds. 1992. *Indigenous Rights in the Pacific and North America*. London: Sir Robert Menzies Centre for Australian Studies.

Scott, A. 1978. *The Social Impact of Large Scale Industrial Developments: A Literature Review*. Glasgow: Planning Exchange.

Scott, A. 1981. *The Social Impact of Large Scale Industrial Developments*. London: Social Science Research Council.

Sharp, L. 1952. "Steel axes for stone age Australians," pp. 69–90 in *Human Problems in Technological Change*. E. Spicer, ed. New York: Russell Sage Foundation.

Stone, R. 1993. "Spotted owl plan kindles debate on salvage logging." *Science* 261: 287.

Sturmey, R.I. 1990. *Women and Services in Remote Company-Dominated Mining Towns*, Armidale: Rural Development Centre, University of New England.

Summers, G.F. and K. Branch. 1984. "Economic development and community social change." *Annual Review of Sociology* 10: 141–166.

Swartzman, D., K. Croke, and S. Swibel. 1985. "Reducing aversion to living near hazardous waste facilities through compensation and risk reduction." *Journal of Environmental Management* 20(1): 43–50.

Tatz, C., ed. 1984. *Aborigines and Uranium*. Consolidated Report to the Minister for Aboriginal Affairs on the Social Impact of Uranium Mining on Aborigines of the Northern Territory. Canberra: Australian Government Publishing Service.

Taylor, C.N., C.H. Bryan, and C.G. Goodrich. 1990. "Social assessment: Theory, process and techniques." *Studies in Resource Management* No. 7, Centre for Resource Management, Lincoln University, New Zealand.

Toyne, P. 1994. *The Reluctant Nation: Environment, Law and Politics in Australia*. Sydney: Australian Broadcasting Commission.

Travis, C.C. and E.L. Etnier, eds. 1983. *Health Risks of Energy Technologies*. Boulder CO: Westview.

Turnbull, S. 1979. *Impact of Mining Royalties on Aboriginal Communities in the Northern Territory*. Canberra: Australian Government Publishing Service.

United States Agency for International Development (USAID). (September) 1993. Handbook No. 3, "Project Assistance," Appendix 3F: Social Soundness Analysis, Washington DC.

United States, Council on Environmental Quality (USCEQ). 1978. National Environmental Policy Act—Regulations. *Federal Register* 43(230): 55979-5600-7. Washington DC: Government Printing Office.

United States, Council on Environmental Quality (USCEQ). 1986. Regulations for Implementing the Procedural Provisions of the National Environmental Policy Act (40 CFR 1500-1508). Washington DC: Government Printing Office.

United States, The National Environmental Policy Act of 1969, 42 USC 4321 (1994).

Vanclay, F. 1989–95 (updated regularly). *Social Impact Assessment Bibliography: A Database and Interrogation Program for DOS-Based Personal Computers*. Wagga Wagga: Centre for Rural Social Research, Charles Sturt University.

White, I. 1987. Annotated Bibliography on Tourism and Aborigines. East Kimberley working paper # 13. Canberra: Centre for Resource and Environmental Studies, Australian National University.

Williams, C. 1981. *Open Cut: The Working Class in an Australian Mining Town*. Sydney: George Allen and Unwin.

Wilmsen, E.N., ed. 1989. *We Are Here: The Politics of Aboriginal Land Tenure*. Berkeley: University of California Press.

Wolf, C.P. 1975. "Social impact assessment: The state of the art," pp 1–44 in *Man–Environment Interactions* Part 1. D.H. Carson, ed. Stroudburg: Dowden Hutchinson and Ross.

Wolf, C.P. 1976. "Social impact assessment: The state of the art restated." *Sociological Practice* 1: 56–70.

Wolf, C.P. 1977. "Social impact assessment: The state of the art updated." *Social Impact Assessment* 20: 3–22.

Chapter 3
Technology Assessment[1]

ALAN L. PORTER
Georgia Tech, USA

The meaning of technology assessment (TA) is changing in important ways. TA is an essential tool of the 1990s to enable sustainable development and facilitate management of technology. Yet, it is a tool in need of serious repolishing. This paper tracks the evolution of TA; juxtaposes TA with other forms of impact assessment; briefly considers doing and using TA; and poses four serious issues for the future of TA.

TA: AN EVOLVING DEFINITION

It should not shock us that two general, widely used, and ambiguous terms— 'technology' and 'assessment'—when combined, do not yield a singular meaning. Nonetheless, we can track and even, perhaps, make sense of the usage of 'technology assessment' (TA). The initiation of TA in the late 1960s in the USA engendered lively discussion along two distinct streams (reviewed in Porter et al. 1980). The more direct sought to devise an effective policy analysis mechanism to help the US Congress better cope with Executive Branch proposals. The other, philosophical in bent, concerned the broad roles of technology in society, seeking to help society better manage technology. Both streams struck fear in those committed to technology-based free enterprise, as expressed in charges that TA meant 'technology arrestment'.

The basic TA model in the scholarly community traces back to the National Science Foundation (NSF) program led by Joe Coates in the 1970s that supported development of a body of knowledge about doing TA. An enduring TA definition reflects this perspective (Coates 1976):

> . . . the *systematic* study of the *effects on society*, that may occur when a technology is introduced, extended, or modified, with emphasis on the impacts that are *unintended, indirect, or delayed* [italics added].

[1]*Environmental and Social Impact Assessment* - Edited by F. Vanclay and D.A. Bronstein. Copyright © 1995 by the International Association of Impact Assessment. Published in 1995 by John Wiley & Sons Ltd. A version of this chapter will appear in *Impact Assessment,* the quarterly journal of IAIA.

Note the italicised terms. Systematic study entails an approach that is as orderly and comprehensive as possible. The effects on society are the impacts of the changed technology across all sectors. The unintended, indirect, or delayed impacts are those not traditionally considered in economic analyses. Figure 1 illustrates the notion of indirect impacts. Coates' companion retrospectives on the automobile and the refrigerator also end in sixth-order divorce. These semi-serious analyses show both the notion of 'effects of effects' and the complex nature of technology–society interactions. The latter imply that TA cannot describe the future with certainty; it can only aspire to reduce future uncertainties.

Figure 1.   **The effects of technology**

At times, technologies have unintended consequences that combine to have serious impacts undreamed of by the creators of the technology. The following example (adapted from Coates 1971) demonstrates how television may have helped to break down community life.

CONSEQUENCES OF TELEVISION

1st order: People have a new source of entertainment and enlightenment in their homes
2nd order: People stay home more, rather than going out to local clubs and bars where they would meet their fellows
3rd order: Residents of a community do not meet so often and therefore do not know each other so well (also, people become less dependent on other people for entertainment)
4th order: Strangers to each other, community members find it difficult to unite to deal with common problems; individuals find themselves increasingly isolated and alienated from their neighbors
5th order: Isolated from their neighbors, members of a family depend more on each other for satisfaction of most of their psychological needs
6th order: When spouses are unable to meet heavy psychological demands that each makes on the other, frustration occurs; this may lead to divorce

---

This model—call it the comprehensive TA model, the NSF model, or the western TA model—has evolved into a variety of forms. For one, the scope of TAs varies greatly according to user needs and study resources, ranging from comprehensive, full-scale studies involving interdisciplinary teams for several person-years to quick-and-dirty (say, one person-month) micro assessments (Rossini et al. 1976).

Erik Baark (1991) distinguishes four schools of TA thought that aid in tracking TA evolution. These draw from different perspectives on the technological innovation and diffusion process:

Regulatory. Perceiving technological development as deterministic, expects the state to set limits on the use of a particular technology (e.g., food additives) or legislate requirements (e.g., safety precautions); TA is to help the state exert control, in a reactive sense, over actual or projected impacts.

Promotional. Again, the course of technological development is pretty much given, guided largely by market forces; but TA should help formulate suitable policies to promote technological innovation in the interest of national competitiveness or development.

Constructive. Does not accept that the course of technological development is deterministic; rather, seeks to tune that development in response to social and political priorities; this suggests proactive state intervention (e.g., devising incentives to encourage alternative energy sources); seeking to 'constructively' redirect the process of technical change, uses TA to clarify interests among technology developers and users.[2]

Experimental/Participative. This extension of the constructive approach requires active intervention; a wide spectrum of parties at interest participate in testing technological alternatives and/or performing social experiments to improve the design of the innovation.

This framework helps plot the evolution of TA. Studies by the US Office of Technology Assessment (OTA) represent a largely regulatory stance (cf., figure 2 topics), with occasional forays into the promotional (e.g., 'technology opportunities for economic conversion'). 'Development TA' (Baark 1991) favours promotional over regulatory issues. Recent European efforts explore the constructive orientation.

Baark's framework emphasises public sector TA. Private sector TA is done by the firm, or industry association, to identify threats and opportunities for the sake of promoting development. Impacts are typically considered from a more limited, parochial point of view. TA may also entail public–private collaboration. For instance, the British Centre for the Exploitation of Science and Technology (CEST) assesses technological opportunities.

Continuing to stretch the promotional school model, one can fruitfully think in terms of a combination of technology foresight activities—*technology monitoring, forecasting, and assessment.* According to the UN/OTA workshop report (1991: 4):

> Technology assessment ultimately comprises a *systems approach to the management of technology* reaching beyond technology and industrial aspects into society and environmental domains. Initially, it deals with assessment of effects, consequences, and risks of a technology, but also is a forecasting function looking into the projection of opportunities and skill development as an input to strategic planning. In this respect, it also has a component both for monitoring and for

---

[2] Arie Rip and colleagues initiated this social construction of technology perspective on TA (cf., Smits and Leyton 1991; den Hond 1993).

scrutinizing information gathering. Ultimately, TA is a policy and consensus building process as well [italics added].

The 1993 UN expert group meeting further embeds TA with technology monitoring and forecasting (Porter and Weisbecker 1993; Weisbecker and Porter 1993). Monitoring consolidates available knowledge on a particular technology and its context (technological and social). Technology forecasting anticipates future developments. TA is integrated with monitoring and forecasting, rounding out a systems approach. 'TA' emphases vary enormously depending on the needs of the study users. For example, in many developing countries, such studies may emphasise monitoring (keying on what is available from developed countries and the impacts experienced there) more than forecasting or future-oriented impact assessment. In sum, TA has evolved into a set of foresight analyses adapted to particular users.

I must add that the systems view is stretched unnaturally far to accommodate all usages of 'technology assessment'. Many, if not most, users of the term refer to *technical evaluations* that bear little resemblance to the TA types just catalogued (see figure 2). Of course, with stretching, the systems view could encompass these forms. Evaluation of alternative technologies may entail forecasting, as in the design of a future combat aircraft. Or in the case of an important TA form, medical (or health) technology assessment seeks to help decision makers deal with the development, acquisition, and utilisation of healthcare technologies (Menon 1993). While somewhat narrowly cast toward technical equipment evaluation, medical TA can entail predictive methods and weigh clinical and economic impacts (e.g., as in the assessment of computed tomography's appropriate applications) (Wells et al. 1991).

TA AND EIA: DEVELOPING ON SEPARATE TRACKS

Both TA and Environmental Impact Assessment (EIA) were legislated into existence in the US to improve the social management of technology. We need not belabour the history of the US National Environmental Policy Act of 1969 (cf., Caldwell 1970), nor the development of viable mechanisms to implement its required environmental impact statements (EISs). Neither need we reiterate the sequence of studies and deliberations, led by Congressman Daddario, resulting in creation of OTA in 1972 (cf., Porter et al. 1980), nor its trying evolution into a well-respected staff arm of the US Congress.[3]

---

[3] Vig (1992) considers OTA's institutional development and the evolution of parliamentary TA institutions in Europe.

Figure 2. 'Technology assessment' according to the databases

I searched four electronic databases for 'technology assessment' in December 1993. ACAD (Expanded Academic Index) primarily tracks social science journals, containing some 1.3 million items beginning in 1988. BUSI (Business Index) emphasises management and industry sources, containing some 1.6 million items beginning in 1988. ENGI (Engineering Index) contains some 1.5 million items commencing in 1985. NTIS (National Technical Information Service) keys on US government research reports, with 1.7 million items from 1964.

'Technology (adjacent to) assessment' occurred 369 times in ENGI. In a small sample of those, almost half referred to narrowly defined technical evaluation, with another third denoting medical TA, and a very small percentage pertaining to either technology forecasting or impact assessment. Amazingly, the term appeared 11,500 times in NTIS. Disappointingly, in a small sample, these overwhelmingly concerned technical evaluation, largely in the military and nuclear technology arenas. Perhaps 15 percent evidence impact assessment.

'Technology assessment' appeared 286 times in ACAD. I inspected a small sample of these abstracts, finding that over 90 percent referred to an Office of Technology Assessment report or activity.

'Technology assessment' appeared 426 times in BUSI, showing the most diversity in usage. To explore usage further and to see if there are notable changes over time, I consolidated the 164 articles dated 1988 or 1989 as one 'early' bunch, and the 81 articles dated 1992 or 1993 as a 'recent' bunch (incomplete 1993 coverage contributes to the reduced number). This yields the following tabulations of index terms in those articles:

1988–89 TA: 392 keywords, dominated by 'Office of Technology Assessment' (131), with frequent mentions (4 or more, most likely reflecting foci of OTA studies) of hazardous waste, industrial research, environmental policy, biotechnology, medical technology, magnetic recorders, hospitals, balance of trade, superconductivity, genetic engineering, refuse, digital audio tape, and manufacturing industry

1992–93: 151 keywords, led by OTA (37), with frequent (3 or more) mentions of pharmaceuticals, defence, environmental policy, automobiles, developing countries, and air quality.

In sum, 'our' meaning of TA differs from 'their' meanings. In some literatures, TA means technical evaluation. Our meaning is primarily represented in the literature by OTA references.

Beginning in the early 1970s, the NSF TA program funded some two dozen comprehensive TAs and related studies (Rossini et al. 1978). These nurtured an academic and professional TA community. That community formed a professional association, the International Society for Technology Assessment (ISTA), which flourished in the mid-1970s, expiring abruptly due to financial difficulties. In the late 1970s, Charlie Wolf, Fred Rossini, and I moved to establish a successor professional association. I phoned Walter Hahn, founding president of ISTA, for his advice—the key element of which was not to use the term 'technology assessment'. Walter felt that TA failed to draw in those professionals engaged in EIA, etc. Hence was born the International Association for Impact Assessment (IAIA).

IAIA has consistently sought to bring together those practicing TA, EIA, social impact assessment (SIA), and (less successfully) those doing risk assessment. These forms of impact assessment (IA) share many essential features, thus warranting close scholarly and professional interchange among their practitioners (Porter et al. 1980). Most importantly, all purport to *address the future, in terms of the potential effects, of technological changes, for the purpose of informing policy making*. Distinctions among the forms of IA often blur. TA keys on an input consideration— 'technology'. Yet on the one hand, most IA concerns the potential effects of technological developments, and, on the other hand, some 'TA' focuses on problems (e.g., alternative energy choices) or policies (e.g., implications of telecommunications standards).[4] Risk assessment keys on a particular form of concerns, implying performance of TA to get at those risks. Risk assessment in practice tends to be narrower in focus than does TA and more quantitative (e.g., nuclear plant fault tree analyses, epidemiological cancer profiles).

TA generally contrasts with EIA in *not being driven by explicit governmental EIS requirements, and not being localised to a particular site development*. Also, EIA policy considerations are typically more circumscribed; participatory procedures are more developed and routinised; the impact horizon is shorter; and data sources are more likely to be primary (direct data collection as opposed to relying upon literature). Furthermore, those who perform TA (e.g., OTA staff) are rather distinct from the larger contingent who perform EIA (e.g., professional consultants). The distance is even greater between EIA practitioners and those who perform more promotional forms of TA, involving monitoring and forecasting, private sector foresight studies, etc.

TA and other forms of IA thus overlap considerably. We might all have been spared considerable confusion had technology assessment been named instead 'Technology Impact Assessment'![5]

Returning to institutionalisation matters, IAIA has advanced IA by providing meeting grounds, written media, and networking (Porter 1989). During the 1980s,

---

[4]   'Integrated impact assessment' addresses a class of technologies, sometimes for a geographic area (e.g., US Environmental Protection Agency sponsored TA of western US energy, focusing on the policy issues involved).

[5]   This title was chosen by David Minns in organising a session at the Canadian Chemical Engineering Conference, Ottawa, October 1993.

IAIA became thoroughly international through a series of successful meetings outside North America—in Australia, the Caribbean, China, and Europe.

Through meetings and resultant publications (cf., UN 1991 and 1993), the UN Branch for Science and Technology for Development has significantly advanced TA. The branch earlier had established the Advance Technology Alert System (ATAS) and now bears responsibility for TA in the UN system. They focus on development of TA capabilities in developing countries. The branch has also spurred a new 'association of associations' to promote TA—the International Association of Technology Assessment and Forecasting Institutions.[6]

DOING TA

Figure 3 gives our 10 steps for TA (Porter et al. 1980). Briefly, the assessors first bound the study to focus on the main concerns of their intended users. The next four steps describe the technology and its context, then anticipate how these are likely to evolve over the time horizon of interest. Technology description and forecasting emphasise the functional capabilities of the changing technology and its direct applications. Societal context description and forecasting seek to ascertain the key influences upon the technology's development. Note how readily these steps fit the 'systems approach' and the inclusion of monitoring and forecasting with TA.

The next three steps [impact identification, analysis, and evaluation] are the soul of TA. These progress through asking 'what might result?' to 'how much is likely?' to 'so what?'. In principle, these steps consider all potential impacts[7] and how they affect all parties at interest. Extending the notion of *all* impacts through a cascade of indirect impacts (figure 1) conveys the challenge of IA.

Impact analyses are followed by policy analyses to elucidate current decision options and their prospective consequences. Lastly, communication to all interested parties is essential.[8]

---

[6] For information on IATAFI, contact Gary Williams, Environmental Assessment and Information Sciences Division, Argonne National Lab, 370 L'Enfant Promenade SW, Suite 702, Washington DC 20024.

[7] I like the generic checklist of the acronym, 'THISPECIES'—Technology, Health, Institutional, Social, Political, Economic, Cultural, Individual, Environmental, and Security (Porter et al. 1991).

[8] And should begin at the *problem definition* step. The steps should be iterated fully or partially, as warranted, not treated linearly.

Figure 3. **Problem definition/bounding**

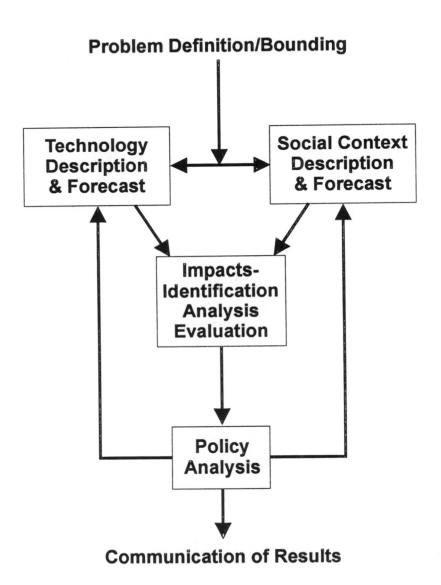

Problem Definition/Bounding

Technology Description & Forecast

Social Context Description & Forecast

Impacts- Identification Analysis Evaluation

Policy Analysis

Communication of Results

Each of the 10 steps receives more or less emphasis depending on the orientation of the TA (technology development versus impact appraisal), the interests of the target audience,[9] data availability, and the study resources available. To accomplish the steps, assessors draw upon a variety of qualitative and quantitative methods. A survey of practitioners (Lemons and Porter 1992) shows considerable overlap in the methods used in TA, SIA, and EIA in both developed and developing settings. The table in figure 4 tabulates the more prominent TA methods. Expert opinion and monitoring are most widely used. Note that most TA relies on multiple methods.

Figure 4. **TA methods use (%)**

	Developed countries	**Developing countries**
Expert opinion	72	87
Nonexpert opinion	25	36
Monitoring	64	70
Trend extrapolation	48	84
Scenarios	41	65
Modeling (qualitative)	34	15
Modeling (quantitative)	18	10
Checklists	16	26
Matrices	13	17

NOTE: 'Use' is the mean percentage of studies reported to use each method, averaged for three time periods: 1980–84, 1985–96, and planned for 1990–94. For developed countries, 14 respondents described 157 studies averaging 3.3 methods/study; for developing countries, 5 respondents described 28 studies, averaging 4.2 methods/study.

This paper frames TA as a single study of a single technological opportunity. Instead, one may establish an ongoing TA process. That implies capabilities to (1) monitor emerging technologies to determine when TA is warranted, (2) arrange for such TAs to be done, and (3) update information and revise projections as new

---

[9] For instance, OTA's Congressional users prefer to receive comparisons of policy options, not explicit recommendations.

information becomes available and concerns arise. In that TA is often needed 'immediately', an institutionalised TA process, with effective memory, can better respond to the occasional windows of opportunity to resolve policy issues.

USING TA

TA should inform decision making. Such information, however, can extend over a wide spectrum. At one end lies the early warning, *intelligence* function, alerting parties to be 'at interest' concerning an emerging technology. At the other, lies the *decision aiding* function. This function can be served by framing policy options, helping to weigh the merits of such options, and/or providing ammunition for stakeholders to support their existing positions. 'Constructive TA' goes further in striving to improve the options under consideration.

The most intriguing application arena for TA is the developing country—the thrust of the UN TA initiative. Developing countries, after several years of hesitation that TA was being foisted upon them as a device to stall their development, have warmed to the challenge of adapting western TA approaches to their contexts. This adaptation favours a promotional 'systems approach', inclusive of monitoring and forecasting as well as IA. In particular, an indigenous TA capability should aid in technology acquisition decisions. Issues include whether to import a particular advanced technology, versus a possibly more 'appropriate' technology; whether to buy components or entire systems; as well as whether to accept the environmental and social impacts attendant on introduction of the technology and siting of facilities. Clark (1990) suggests that development TA must deal with three broad concerns: (1) policies to promote the relevant science and technology base, (2) those concerned with economic applicability—that is, assessing where resources should be targeted to optimise development; and (3) those concerned with the institutional changes needed for effective adoption (e.g., regulations, venture capital availability).

Discussions at the UN expert group meeting (1993) also affirm that TA in developing countries is inextricably wedded to development concerns. Figure 5 sketches three TA dimensions. *Level* stretches from the specific (technology, site) to the global; *Tech Foresight* stretches from TA (as regulatory IA) to a systems approach (monitoring, forecasting, and assessment). The third dimension ranges from TA as analysis (the focus throughout this article) to TA as development activity (institutionalisation, promotion, etc.).

Figure 5. **Three dimensions of technology assessment**

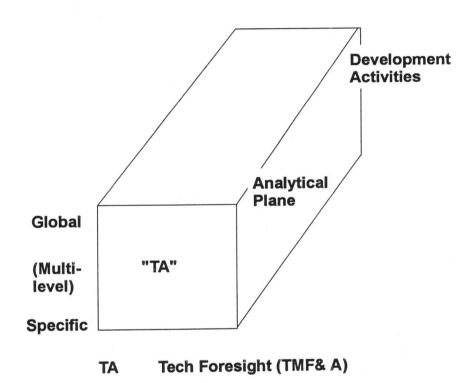

EMERGING ISSUES IN TA

I spotlight four issues deserving serious consideration:

1. Stretching TA to incorporate life cycle analysis prompted by concerns for sustainable development
2. Devising TA mechanisms for transnational issues
3. Adapting TA to new thrusts in the management of technology and technology policy
4. Developing frameworks and support for knowledge generation and dissemination about the TA process

TA enhances sustainability considerations by directing attention to technological options (through monitoring and forecasting) and broadening consideration of potential impacts beyond the natural environment. We need to inform the sustainable development community about TA and IA so that these are included in their frame of reference (e.g., 'Design for Environment' guidelines being crafted with US Environmental Protection Agency support).

Conversely, sustainability criteria enrich TA. Sustainable development objectives crystallise concern about the long-term viability of given technologies. They provide a readily grasped benchmark (Can the environment support this activity indefinitely?) to help assess proposed technologies. Most important, sustainability criteria challenge TA to address the full technological life cycle. Upstream, this means that we need to consider (ideally, categorise and measure) the resources required for the technological development (e.g., raw materials, habitat affected). Downstream, it means we need to address (ideally, categorise and measure) the residuals. We need to reach beyond the technology (as product, process, or service) itself to assess attendant waste disposal or utilisation, and *re-* opportunities (*re*use, *re*clamation, *re*manufacturing, *re*cycling).

Transnational issues arise as certain impacts (such as global warming) transcend national boundaries. The severity of such potential effects cries out for TA, but the difficulties are daunting (Porter 1987). Who holds responsibility to do transnational assessments? Who will provide the resources? Where are the policymakers to act on the findings? The UN is tackling some transnational concerns. In particular, the United Nations Conference on Environment and Development (UNCED) held in Rio de Janeiro in 1992 focused attention on global change and sustainability issues. The Montreal Accords on limiting chloroflourocarbons to protect the ozone layer offer encouragement that nations can mobilise rapidly to address a difficult transnational impact.

TAs or integrated IAs are needed to confront a growing list of transnational issues attendant to: human biology (population control, AIDS, increased life spans), information age (network management, global village, changing work patterns), and the globalisation of the economy (while relying upon national controls).

'Management of Technology' (MOT) is growing explosively in academia. MOT is prompted by concerns about enhancing competitiveness, seen as highly dependent on technology. Usually based in business schools, MOT engages engineering in providing the skills needed to manage technologically based enterprise. The list of such skills includes analysis of technological innovation processes; technology monitoring, forecasting, and assessment; and risk assessment. These foresight analyses are vital to succeed in "managing the present from the future" (Porter et al. 1991). That is, we need to develop sound (not certain) projections of alternative futures and use these to inform the decisions we are making today. TA needs to be operationalised as an MOT tool to address specific technologies. But, in addition, we might do well to revisit the 'big picture' philosophical TA discourses of the 1960s; TA could also help society rethink what its larger objectives are, with an eye toward the tradeoffs among impacts (Harman 1992).

Technology policy opens up another critical venue for a systems approach to TA. The legitimation of proactive governmental technology management suggests a strong role for promotional, constructive, or experimental/participative TA. How is this to be institutionalised?

Lastly, we need to develop the TA (and IA) knowledge base. The nurturing role played by NSF in the 1970s has not been taken up by another institution.[10] As a result, there is no research base on how best to perform TA/IA. How valid and useful are TA/IAs? What makes some better than others? What methods work best—where, when, and why? Given the vital national interests in the policies that TA can inform, there would be tremendous payoff from modest, but well-led, support for such studies.

As a consequence of the lack of a research base, there is no teaching base for TA. To the best of my knowledge, the last TA textbook was published in 1980 (Porter et al.). There are no clearcut intellectual homes for this interdisciplinary activity in the university. Yet someone needs to ascertain ways to model technology life cycles with regard to impact assessment, to advance creative participatory mechanisms for TA, and so on. Initiatives to promote the collation and dissemination of knowledge on TA/IA are desperately needed. There is an outrageous mismatch between the importance of the activity and the inattention to training people to do it well.

In conclusion, TA needs attention. Through the 1980s, the practice of TA was visible mainly in the fine work of OTA. Thanks to formation of TA offices by a number of countries (Vig 1992) and efforts by international bodies, especially the UN, TA is coming back into scholarly and professional prominence.[11] That is rightful in that TA deals with the really big issues—the implications of emerging technologies on entire societies and environments. We need to devote resources to improving the TA process and effectively integrating it into MOT and sustainable development efforts.

---

[10] Only the risk assessment area continues to receive any NSF support.

[11] Unfortunately, the UN Branch for Science and Technology for Development was disbanded in 1994. It is not clear which, if any, UN agencies will assume responsibility for TA.

REFERENCES

Baark, E. 1991. "Development Technology Assessment—Some theoretical and method-ological issues." Paper presented at the United Nations/Office of Technology Assessment Workshop on Technology Assessment for Developing Countries, Washington DC, November [UN Branch for Science and Technology for Development].

Caldwell, L.C. 1970. *Environment: A Challenge to Modern Society*. Garden City NJ: Anchor Books, Doubleday & Co.

Clark, N. 1990. "Development Policy, Technology Assessment and the New Technologies." *Futures* (22): 913–931.

Coates, J.F. 1971. "Technology assessment, the benefits...the costs...the consequences." *The Futurist* 5(December): 225–231.

Coates, J.F. 1976. "Technology assessment—A tool kit." *Chemtech* (June): 372–383.

den Hond, F. 1993. "The Assessment of Environmental Impact in Technology Forecasting." Paper presented at the *R&D Management Conference on Technology Assessment and Forecasting*, Zurich, Switzerland, July.

Harman, W. 1992. "Memo to Earth Dwellers." Sausalito CA: Institute of Noetic Sciences.

Lemons, K.E. and A.L. Porter. 1992. "A comparative study of impact assessment methods in developed and developing countries." *Impact Assessment Bulletin* 10(3): 57–66.

Menon, D. 1993. "Technology assessment and biomedical engineering education." *Medical and Biological Engineering and Computing* 31 (1): HTA33–HTA36.

Porter, A.L., ed. 1987. *Impact Assessment Bulletin* 5(3), Special Issue on International Impacts of Technology.

Porter, A.L. 1989. "IAIA: The first decade." *Impact Assessment Bulletin* 7(4): 5–15.

Porter, A.L. et al. 1991. *Forecasting and Management of Technology*. New York: John Wiley.

Porter, A.L. et al. 1980. *A Guidebook for Technology Assessment and Impact Analysis*. New York: North Holland.

Porter, A.L. and L.W. Weisbecker. 1993. "Issues in the Use of Technology Assessment for Development." UN Expert Group Meeting on Technology Assessment, Monitoring and Forecasting, Paris.

Rossini, F.A. et al. 1978. "Frameworks and Factors Affecting Integration Within Technology Assessments." Report to the National Science Foundation, Washington DC (NTIS-PB 294 607 & 294 608).

Rossini, F.A., A.L. Porter, and E. Zucker. 1976. "Multiple technology assessments." *Journal of the International Society for Technology Assessment* 2: 21–28.

Smits, R. and J. Leyton. 1991. "Technology Assessment, waakhond of speurhond?" Zeist: Kerkebosch: 307–318.

UN Branch for Science and Technology for Development. 1991. United Nations Workshop on Technology Assessment for Developing Countries. Hosted by the Office of Technology Assessment, Washington DC.

UN Branch for Science and Technology for Development. 1993. Report of the Expert Group Meeting on Technology Assessment, Monitoring and Forecasting [draft]. Hosted by UNESCO, Paris.

Vig, N.J. 1992. "Parliamentary technology assessment in Europe: Comparative evolution." *Impact Assessment Bulletin* 10(4): 3–24.

Weisbecker, L.W. and A.L. Porter. 1993. "Issues in Performing Technology Assessment for Development." UN Expert Group Meeting on Technology Assessment, Monitoring and Forecasting, Paris.

Wells, P.N.T., J.A. Garrett, and P.C. Jackson. 1991. "Assessment criteria for diagnostic imaging technologies." *Medical Progress through Technology* 17(2): 93–101.

Chapter 4
Policy Assessment[1]

PETER BOOTHROYD
University of British Columbia at Vancouver, Canada

INTRODUCTION

Public sector policymaking draws on a number of assessment and evaluation traditions. We can group these traditions into two general categories: policy analysis and evaluation (PE), and impact assessment (IA). PE is intrinsic to policy design: PE continuously analyses and evaluates policies, as they are planned and implemented, in terms of the objectives (or goals) they are designed to meet.[2] IA tends to be additive to design. Through IA, plans are assessed in terms of their undesigned-for impacts (or, in other words, their unintended consequences, externalities, spin-offs, or side-effects). IA asks: regardless of whether the motivating objectives are met, what else occurs?[3]

This chapter proposes that policy assessment (PA) is best seen as an emerging synthesis of PE and IA, i.e., as a process concerned with assessing all the intended and unintended outcomes of policies being planned, proposed, implemented or reviewed.[4] In order to show the need and potential for PA, this chapter (1) traces the respective evolutions of PE and IA, analysing their inherent tensions; (2) discusses the need for assessment of policy impacts; (3) argues that simply doing IA at the policy level (PIA), either as inhouse policy vetting or as more public strategic environmental assessment, is inadequate; (4) identifies the forms that a fuller PA could take; and (5) concludes that the formal, heuristic form of PA is probably most practicable and effective.

---

[1] *Environmental and Social Impact Assessment* - Edited by F. Vanclay and D.A. Bronstein. Copyright © 1995 by the International Association of Impact Assessment. Published in 1995 by John Wiley & Sons Ltd. A version of this chapter will appear in *Impact Assessment,* the quarterly journal of IAIA.

[2] 'Goals' and 'objectives' are used synonymously in this paper.

[3] Some theorists see IA as properly assessing impacts defined as both objectives and side effects. The mainstream of practice, however, is concerned only with the latter.

[4] For stylistic reasons, the text will often connote simply the predictive forms of PE, IA, and PA. The arguments and conclusions, however, also apply to retrospective PA which aids decision making by generating empirical knowledge of consequences.

THE EVOLUTION OF POLICY ANALYSIS AND EVALUATION

The evolution of PE is complex, but essentially it has involved a progressive broadening of the social goals that frame it. Through the course of the twentieth century, policy goals have been explicated in roughly this chronological order:

1. *Particularistic goals.* Policy is analysed and evaluated in terms of goals declared and established by the powerful according to their interests, values, or whims.
2. *Efficiency goals.* The more democratic goal of aggregate welfare, particularly material welfare, is posited; policy is analysed and evaluated according to the principles of utilitarianism, welfare economics and Pareto optimality. Cost–benefit analysis (CBA) becomes a PE tool (Prest and Turvey 1965).
3. *Equity goals.* Distributional goals, especially poverty alleviation, are introduced; welfare states develop 'social' policy analysis and evaluation tools (e.g., Gini coefficient).
4. *Quality-of-life goals.* Qualitative considerations are added to numerical calculations of efficiency and equity; attention focuses on social indicators and urban planning techniques (e.g., Hill 1968; Lichfield et al. 1975) for determining and responding to varying aspects and conceptions of quality of life.
5. *Sustainability goals.* Science shows the need to live within global carrying capacity; heuristic tools are developed for coping with system complexity (e.g., Holling 1978; Wackernagel et al. 1993); policies and their consequences are seen holistically.

This outline of PE's substantive evolution, which was essentially complete by the early 1970s,[5] refers to the trajectory of PE's theoretical treatment in the literature and its rhetorical treatment by planning agencies and interest groups. Day-to-day, inhouse PE is still mostly particularistic, with the universals of welfare economists enjoying primary respect.

Even the theoretical and rhetorical evolution has not been completely linear; for example, attention to equity has risen and fallen with the fortunes of the welfare states, and conservation concerns at the turn of the century, though soon overshadowed by efficiency-oriented CBA, presaged today's sustainability movement.

TENSIONS INHERENT IN POLICY ANALYSIS AND EVALUATION

Throughout its evolution, PE has been suspended by a basic methodological tension between positivistic and heuristic conceptions of how policy-relevant knowledge should be acquired and applied. The tension is not over the worth of science or the

---

[5] By the early 1970s, United States water resource planning was to be governed by four accounts. Beneficial and adverse effects were to be listed (not necessarily quantified) in terms of 'national economic development' and 'environmental quality' (identified as the two major objectives) plus 'social well-being' and 'regional development' (United States 1973).

importance of rigour, but over the nature and role of science and the meaning of rigour.

Positivists seek ever-more precision and replicability in their explanations and predictions, being prepared to sacrifice as necessary holistic understanding. At worst, they seek to know more and more about less and less. Trusting only well-validated information, they attempt to restrict judgements to those scientifically supportable. The result is that technical issues come to dominate their PE. As in the case of CBA, complex issues are avoided or improperly simplified. Method determines question, tool becomes master.

Those favouring heuristic analysis seek PE methods that give insight into the most important systemic processes influenceable by policy. Science is employed but does not limit inquiry. In addition to deductive logic and empirical research, PE is informed by metaphor, reflection, and moral judgement. Arbitrariness is checked not only by the numbers of objective science but also by collaborative inter-subjectivity. Good questions are rigorously developed and answered the best they can be.

Positivism is perhaps most clearly revealed in welfare economics; the potential and necessity of heuristics is most clearly revealed in sustainable development perspectives. But regardless of the goal PE is oriented to, either methodological approach can be taken.

Very much secondary in PE, is procedural tension over how much the timing and responsibility for analysis and evaluation activities (setting questions, answering, and reporting) should be formalised. It is generally assumed by PE practitioners and those that direct them, that PE should remain a fluid informal art internal to government, and to the extent the public is involved at all, it should be through consultation processes (trial-balloons, white papers, surveys, hearings, etc.) developed on an *ad hoc* basis.

THE EVOLUTION OF IMPACT ASSESSMENT

This section traces the evolution of IA in terms of its expansion in three dimensions: (1) substantive breadth, (2) systemic depth, and (3) hierarchical range.

Broadening the Substance of Impacts Assessed

In terms of the goal-related substance of impacts assessed, IA's evolution has roughly parallelled the evolution of PE. The difference between PE and IA is that at each common stage in their parallel evolutions, PE has involved analysing and evaluating the effectiveness of political choices directed to the then-emerging goal, while IA has involved assessing the impacts on that goal of initiatives directed to other goals. For example, when PE was focusing on whether policies directed to equity would (or did) help achieve that goal, IA was interested in how growth-orientated policies were affecting equity. The full history of IA as a practice—not only its recent history as a named institutionalised practice—can be summarised as having evolved through the following stages.

Particularistic IA
The most basic and continuously influential form of IA (though not usually recognised as such) has from the beginning of modern political systems been the assessment of decision implications for the power-holders' own interests. *Particularistic IA* is essentially the assessment of impacts on the maintenance of power and wealth distributions within the society. In its current forms, *particularistic IA* relies on lawyers to assess legal subtleties and precedents, media experts to assess images, and pollsters to assess political 'fall-out'.

Efficiency IA
The modern epoch in *de facto* IA has been characterised by systematic assessment of impacts relative to assumed universal criteria. It was initiated by liberals calling for *all* decisions to be accountable to the utilitarian principle of welfare efficiency (i.e., the greatest aggregate good for the greatest numbers), not just those decisions initially designed to have general benefit. Thus, what might be called *efficiency IA* questioned the presumed benefits of market-distorting 'pork barrel' policies that serve special interests. *Efficiency IA*, although unnamed and unrecognised, is still very influential.

With the rise of the welfare state and equity-oriented policies, and the strengthening of community-based groups fighting to protect quality of life or the environment, *efficiency IA* became a reactionary force. It is now concerned with drawing attention to the impacts on productivity—and putatively on gross consumption—of decisions related to other goals; for example, it is concerned with the negative impacts of minimum wage laws, community stabilising local-hire preferences, or environmental regulations. Through *efficiency IA*, weaker interests are kept on the defensive—being required to prove that their policy choices for equity, quality of life or sustainability will not weaken productive efficiency of ostensible universal benefit. The continuous influence that *efficiency IA* has had derives from its simultaneous pandering to the powerful as well as its mystification to the public through liberal cant, esoteric utilitarian concepts, and quantitative techniques.

Equity IA, Socioeconomic IA
IA oriented to equity complements proactive social policy PE and became respectable in liberal circles with the recognition by welfare economists in the 1950s that efficiency-oriented decisions can have 'externalities'—i.e., public costs and costs to other individuals—or what most people would now call 'impacts'. Early *equity IA* focused on societywide impacts of efficiency-oriented decisions. The influence of IA with this focus peaked about 1970; its steady decline since is one manifestation of the dying welfare state. Since the 1970s, *equity IA* has thrived as 'socioeconomic' IA (SEIA), the component of project social impact assessment (SIA) concerned with the distribution of economic benefits (e.g., new jobs) and costs (e.g., price inflation, lost jobs). *Equity IA* is also increasingly concerned with

the negative impacts on some people of policies designed to enhance overall quality of life (e.g., neighbourhood preservation) or sustainability (e.g., forest protection).

Quality-of-life IA, Social IA

Assessment of impacts on quality of life stemming from land use and infrastructure decisions has been a central feature of urban planning for decades (in some senses, since the beginning of cities). Patrick Geddes (1915) brought to public discussion the health, social, and aesthetic impacts of industrial cities and conurbations which, if planned at all, considered only production efficiency. His legacy was to make quality of life a proactive planning goal. Jane Jacobs (1961) then showed that planning's modernistic conception of quality of life as cleanliness, predictability, and good services had its own negative impacts—fast streets, monumental housing projects, office complexes, and shopping malls were killing the zest and civility of self-created organic neighbourhoods. Elites have always protected their own quality of life through *de facto* IA. In modern times, groups of middle-class homeowners (ratepayers), have also increasingly done so—often to the distress of elites and planners concerned about the barriers to development posed by NIMBY (not-in-my-backyard) parochialism. However, *quality-of-life IA* by community-based interest groups is often more than NIMBY expressions. For example, their assessments of urban freeway impacts have often considered the systemic impacts on whole cities, and not just the immediately impacted houses and individuals.

With the institutionalisation of environmental impact assessment (EIA) and concern that the burgeoning EIA industry was overlooking quality of life impacts, social impact assessment (SIA) was developed as an identifiable separate field of practice (see Wolf 1974 and chapter by Burdge and Vanclay). Although SIA has broad application, for the most part it has been practised as an adjunct to biophysical project EIA. In this role, SIA has extended quality of life concerns from urban design to resource management. At the same time, IA's focus on individual sensory life (e.g., safety, health, and aesthetics) has been broadened to illuminate changes in cultural life (i.e., experiences that provide identity, meaning, joy, and security such as sharing knowledge, mutual caring, and organising). At its best, SIA has shown the links among well-being, economy, and environment.

As SIA deals with people who, unlike the beings or things that concern *bio-physical IA*, are potentially capable of speaking for themselves, it faces distinct methodological and procedural challenges (see chapter by Burdge and Vanclay). The studied have become part of the studying processes, sometimes directing them (see chapter by Roberts). They influence the procedures that determine when IA is done, with what scope, on what initiatives, and with what products. Furthermore, SIA must address fundamental conflicts in lifestyle values, whereas EIA as distinct from SIA primarily addresses tradeoffs between sustainability and efficiency values, both of which almost everyone claims to hold. SIA has become a more transparently political process (see chapter by Burdge and Vanclay).

Environmental IA, Sustainable Development IA

Recognition of environmental hazards and ecological stresses in the 1960s (e.g., Carson 1962) provided the impetus and rationale for institutionalising EIA. For example, the United States National Environmental Policy Act (NEPA) of 1969 (soon followed by equivalents in other jurisdictions) not only marked a resurgence of American interest in conservation and a growing awareness that efficiency-oriented CBA did not address all social goals and needs, but also was a major conceptual breakthrough in explicating the complex, nonlinear relationship of ends and means, and was a major procedural innovation in planning and decision making. The conceptual breakthrough was the realisation that planning to serve one goal at a time, can lead to net negative consequences taking all goals into account. To put it another way, it became apparent that the abstracted metaphorical decision-tree, through which linear hierarchies of ends and means are unambiguously deduced (as in a bureaucratic organisational chart in which a subordinate reports only to one person), does not reflect the reality of complex systems. While previously it had been realised that working to one goal can impinge on efforts to reach other goals, this realisation had not been conceptually systematised in planning and decision making. Prior to NEPA, pre-decision IA (though not PE) was only intuitive and casual.

NEPA's procedural innovation was its action-forcing mechanism: the requirement that, prior to a federal government decision on a potentially environmentally significant proposal, an environmental impact statement (EIS) be prepared and publicly vetted. Formalised procedures were established to implement this mechanism, e.g., on EIS timing (after design), screening criteria (potential impact significance), and determination of scope. The tough-looking and sometimes tough-acting EIS mechanism was accepted by a society still fixated on economic growth (as were similar mechanisms by similar societies) because it was seen as concerned with production, particularly site-specific development projects and their local impacts, rather than with impacts on natural capital of aggregate consumption/ production. EIA did not represent a fundamental threat to a consumption-oriented way of life. It was accepted as a mitigation or design tool that need not threaten increasing throughput (harvesting, mining, burning) of natural resources. The global ecological impacts of policies were scarcely considered, and if they were, then judged or claimed too difficult to assess.

EISs and related court cases somewhat democratised IA. It was no longer left to politicians, bureaucrats, and occasionally lawyers fighting over property rights, to assess impacts and determine their weight in decision making. Now, all manner of interest groups, including groups caring for the public interest such as 'environmentalists', could become involved. Project planners were forced to pay attention to the fact that an effective project in terms of certain objectives may be dysfunctional from the perspective of others.

In each jurisdiction, the way EIA has been institutionalised is the outcome of pressure and deliberation by many actors including government, industry, the courts, and citizens' organisations—all who seek to shape the EIA process to fit their various needs and interests. The outcome of this process in many jurisdictions is a highly systematic step-by-step procedure, intricate in its internal checks and balances, which provides for much technical and consultative activity, but which has little direct influence on forms of development.

At its best, EIA's immediate output is mitigation of identified impacts, with longer term benefits including social learning for preventative work at future design stages. At worst, EIA procedures obfuscate the fundamental sustainable development issues (as identified by WCED 1987; Rees 1992) lulling people into a complacency that costly technical EIA activity is guarding against deleterious development, while giving the impression of genuine public involvement and sensible environmental management (see chapter by Ortolano and Shepherd).

Deepening the Systemic Awareness of Impact Chains

The substantive broadening of IA has not just meant that impact assessors have to consider longer checklists. Increasing substantive breadth through and within the stages identified above has also led to a deeper awareness of the relationships within and among the various natural, social and constructed systems and sub-systems that mediate decisions and their impacts. The result of this awareness is that the horizon of impacts assessed has been progressively extended from the systemically immediate to the systemically distant—or, in the usual language of IA: from the direct, isolated, local, and certain, to the indirect, cumulative, synergistic, global, and uncertain. *Efficiency IA*, for example, has deepened systemic thought by adding consideration of indirect costs and benefits. *Equity IA* began questioning whether in real socioeconomic systems the poor actually do benefit from development programs as was presumed in the simplistically efficient, socially naive, 'trickle-down' development paradigm.

Further broadening of IA to include a consideration of ecological relationships and quality of social life during in the 1970s, roughly coincided with an intense period in the development of systems theory and its application to social (Georgescu-Rogen 1971; Laszlo 1974; Beer 1975; Bennett 1976; Churchman 1979) and ecological (Odum 1971) issues. General systems theory and cybernetics in their softer, heuristic forms (as opposed, say, to mathematical operations research) have provided IA with potentially powerful conceptual tools for analysing complex and uncertain impact chains—though the potential seems hardly recognised. The concepts include: positive and negative feedback, homoeostasis and entropy, open and closed boundaries, hierarchy and recursion, control and variety (Beer 1975; Heylighen et al. 1990, Wolf 1990; Boothroyd 1994). Systems theory not only enables more accurate IA by lengthening and therefore broadening the identification of chains of impacts, it also enables IA to consider cultural and ecological systems holistically— i.e., to consider their vitality, stability, and fragility—and therefore to determine which decisions are likely to make them healthy and which are likely to make them sick, or dead. For example, some IAs have referred to the impact of northern Canadian industrial resource exploitation—e.g., hydro-electric power in Quebec and petroleum in the Mackenzie Valley—on the viability of indigenous peoples' 'way of life'—a holistic concept connoting the organic nature of ecology, economy, culture, and their linkages (Berger 1977; Berkes 1988). Arms-length IAs of World Bank mega-projects have taken similar perspectives (Morse and Berger 1992).

System-sensitive IA, whether or not it references general systems theory, is still the exception, however. IA is still mostly *systematic* (mechanical and reductionist) rather than *systemic*. As IA begins to pay more attention to systems thinking, it will deal better with the important, literally vital, impacts of decisions. Increasing the systems-sensitivity of IA will affect not only what is studied and how—but also by whom. The more we realise that IA faces complexity and uncertainty, the more we realise that many kinds of knowledge must be considered if we are to understand and respond appropriately to the full extent of all possible impacts. This is the systems theory rationale (the 'law of requisite variety') for IA to be broadly participatory and—what amounts to the same thing—politicised.

Stretching the Application of SIA and EIA to Higher Decisions

The deepening of IA to systems-sensitive, politicised SIA and EIA has been accompanied by a stretching of its application to higher, and therefore more general, decisions. Roughly speaking, SIA and EIA have become progressively interested in resolving the intellectual and political difficulties inherent in assessing—
- Projects;
- Explicit programs of projects;
- Disguised programs (as revealed by the logic of implied project completion and complementarity);
- Explicit sectoral policies (e.g., technology assessment, see chapter by Porter);
- Implicit sectoral policies (as revealed by analysis of options presented for assessment, and cumulative effects assessment); and now,
- Overall development policies.

However, while EIA is theoretically, and legally in many jurisdictions, as applicable to policies as to physical projects, EIA has been applied primarily to the latter for several reasons:
- Projects are tangible, dramatic, highly organised, discrete geographically and temporally, and for all these reasons amenable to systematic IA.
- Localised negative impacts of individual projects are perceived to be able to be mitigated, or are regarded as insignificant in comparison to project benefits, thus EIA can be applied rigorously on a case-by-case basis without threatening the current unsustainable development path and those most benefiting from it.
- Policymaking is secretive, or at least guarded. Power-holders feel threatened by increasing explicit systematisation and the public accountability it produces or seems to imply. Rejection of a well-defined policy proposal because of an EIS would be regarded as causing loss of public confidence in the proponent—the higher the proponent, the more serious this loss.
- The most important policies are unwritten, often unspoken, certainly not reviewed, and are therefore not assessable.

It is in only in the last five years or so that there has been much talk of the need for EIA of policy and questions raised about how to do it. In the years after the initial flurry of project EIAs in the 1970s, but before the recent interest in policy EIA, there were major efforts directed to program EIA, sometimes under the rubric of developing tools for assessing cumulative effects. Some of the problems encountered in assessing programs were the same as those now seen as facing policy assessment—e.g., the difficulties involved in tracing, let alone predicting, complex cause–effect relationships.

Increasing concern in the mid-1980s with the development of methods for assessing cumulative impacts (CEARC, USNRC 1986) is now giving way to recognition that rather than attempting to assess the aggregative, synergistic, threshold-crossing, and sometimes surprising effect on a natural or social system of a *collection of activities* as Sonntag et al. (1987) proposes, it could be more effective to assess the *policy sources* of these activities. Rees (1995:15), for example, argues, "Cumulative environmental assessment should. . . become a more proactive planning tool to ensure *no net loss of natural capital* as a routine development objective" [original emphasis].

Summary of the Evolution of IA

The progressively comprehensive nature of IA in terms of decision levels and the impacted systems considered, reflects and reinforces IA's broadening comprehensiveness in terms of substantive attention to equity, quality of life, and sustainabilty. Taken together, the dimensions of increasing IA comprehensiveness have meant that IA, although reactive in nature, has become increasingly overlapping with the also increasingly broadening proactive PE.

TENSIONS INHERENT IN IMPACT ASSESSMENT

The relative importance of the methodological and procedural tensions in IA is the reverse of that in PE. The fundamental tension in IA, as it has culminated in EIA, is procedural not methodological. The tension is created by the polar answers to the question: "To what degree of detail and with how much authority should IA procedures be explicitly prescribed so that IA fulfils its current mandate as a public, accountable and transparent scrutinising process interposed between design and decision making to ensure proponents and decision makers consider externalities as they plan to meet specific sectoral objectives?" In short, how formal should IA be?

Highly formal IA has three characteristics: (1) *explicitness* of procedures which are detailed in their prescriptions; (2) *accountability* requirements imposed on those responsible for conducting the procedures; and (3) *enforceability* of procedures, accountability mechanisms, and of negotiated outcomes such as mitigative works or compensation. Informal IA is vague and flexible, requiring little or no accountability, and offering no legal or specified administrative basis for enforcement. Formality–informality is a continuum: at one extreme IA procedures can be legally prescribed in detail; at the other, there may be only vague guidelines; in between there may be various combinations of legalities and exhortations to good practice.

Legislation formalises IA because it enhances accountability and enforceability. Procedures that prescribe consulting with or informing the public with regard to planning designs, assessments, research, or case-specific processes, are more formal than procedures which govern inhouse (internal government) processes because public involvement expands accountability and the likelihood of enforcement. For this reason, procedures that maximally empower the hitherto least powerful publics can be regarded the most formal. Of course, public involvement in IA can also occur without procedures that actually require it; i.e., public involvement can be informal.

Methodological tension is secondary in IA. In IA generally and EIA especially, positivism reigns. The need to rely on scientific laws and empirical research so as to give as precise quantitative predictions as possible is assumed in most IA (there are some attempts to overcome this in SIA: see the chapter by Burdge and Vanclay). Accuracy of prediction or explanation is sought rather than full systemic insight. Since accuracy is most attainable when only spatially and temporally immediate systems are considered, IA avoids the big questions, e.g., of global warming or development paradigms, and finds it difficult to deal with questions of cumulative effects.

EIA has favoured technical science over conceptual wisdom not only because of a predisposition to positivism but also as an almost inevitable outcome of common EIA procedures and practices:

- EISs, in being open to public scrutiny, encourage assessors to limit speculation and to stick to 'hard' verifiable data.
- Proponents, who are responsible for EISs, find it in their interest to focus on factual detail rather than on big questions.
- It is conceptually difficult for the EIS, being an add-on to the design process, to return to first principles—technical mitigation rather than systemic assessment is the *de facto* focus of the EIS.
- Consideration of objectives is not really seen as within the purview of IA (whether or not there is a mandate to consider the need for the project or the no development option).
- The project orientation of IA virtually precludes consideration of fundamental (i.e., policy) as opposed to technical issues.

The limitations to fundamental relevance posed by the technicality of IA are rarely, if ever, overcome by public involvement. Few publics raise fundamental issues in IA because of the way issues are framed by proponents, and/or by regulators, who tend to see the world the same way as proponents. Indeed, technicality inhibits public involvement: the more technical the studies and impact statements, the greater their credibility but the less opportunity they provide for average people to participate, and the less their public education value. However, to the extent public involvement is mandated and occurs, it does make the asking of timely big questions possible—even if officials declare these questions outside the mandated process. Thus formal procedures which require public involvement help promote fundamental debate and heuristic thought and ought to be part of all impact assessment protocols (see chapter by Roberts).

ASSESSMENT OF THE IMPACTS OF POLICIES

The Need for Assessment

Awareness of the need for systematic assessment of policy impacts arises from experience with project EIA/SIA together with a deepening understanding of and commitment to the sustainability imperative. IA's potential to enhance decision making has been established, but so has its inability at the project level to deal with the big issues. As Partidario (1993: 1) puts it (with reference to CEMP 1990):

> The need for an environmental assessment in support of policies, plans and programmes is becoming widely recognised. Existing literature emphasises the importance of introducing environmental considerations at earlier stages of the decision-making process as part of a strategy for strengthening the tools for environmental management.

Furthermore, at a workshop at the 1993 IAIA meeting in Shanghai, there seemed to be a consensus that sustainable development—if defined as quality of life improvement through development that stabilises global consumption and therefore energy throughput within the earth's carrying capacity—cannot be achieved without policy impact assessment, project impact assessment being incapable of dealing with the cumulative impacts of growth (Boothroyd 1993).

Wood and Dejeddour (1992) provide 12 reasons, reducible to four, for moving EIA 'upstream' from the project level to the policy level:

1. Involving more government and public actors in assessment of key decisions
2. Increasing assessment efficiency by applying it to one macro policy rather than many individual projects
3. Widening the range of options assessed
4. Improving assessment of cumulative, synergistic, and secondary impacts

Besides dealing with cumulative project impacts, assessment of policy could consider the impacts of regulated activities, such as logging practices, which usually are outside the scope of EIA. Most optimistically, assessment of policy impacts might help policymakers proactively consider the effects of their implicit policies, such as the *laissez-faire* approach to farmers' management of soils and wetlands (Bregha et al. 1990).

While the need for systematic assessment of policy impacts is increasingly recognised, such assessment is conceptually, practically, and terminologically inchoate and a number of issues remain unresolved.

- Conceptually, there is no agreement on what assessment of policy impacts should or could be in a systematised form, specifically:

    a. Whether to apply EIA-like procedures to policies, or something entirely different.

    b. Whether the process should be public or closed, centrally controlled or independent, environmentally focused or holistic, applied only to legislation or broadly to all executive decisions including budgets, programs, and regulations. Coherent and consistently named bodies of concepts are only just beginning to be developed and accepted.

- Institutionally, although the practice of assessing policy impacts is growing, it is diffuse, being undertaken even in a single given jurisdiction by a number of actors following a variety of fluid meta-policy procedures unique to that jurisdiction. This is unlike project-focused EIA which is often legislated under a specific act following one of a limited set of models (e.g., NEPA) and administered or overseen by a specific bureaucracy. There are not yet clear, coherent bodies of practice that utilise an agreed-upon framework.

- Politically, because power holding is threatened by explicit systematisation, there is not a strong demand from the top for, and often resistance to, creating policymaking approaches with well-articulated identities.

Policy Impact Assessment (PIA) versus Policy Assessment (PA)

Different meanings attached to the words *impact*, *policy*, and *assessment* underlie differing conceptions of policy impact assessment and policy assessment. In order to facilitate conceptual clarity and communication, it is proposed here that—

- *Policy* be seen as a general prescription for guiding decision making which can be very broad or narrow, explicit or implicit, with layers of policy forming a complex ends–means hierarchy.

- *Policy assessment (PA)* be seen as the assessment of *all* outcomes of policies being planned, being proposed, or already in existence, including assessment of (1) whether objectives will be or have been met—not just externalities and unintended consequences; and (2) whether objectives are 'right' given the ultimate goals policy is to serve, not only if objectives will be met or have been met.

- *Policy impact assessment (PIA)* be seen as the process that asks: "What outcomes additional to the meeting of objectives result from implementing the policy?"

PIA thus becomes a component of PA: PIA is the application of IA that bridges to PE; PA is the attempted integration of PIA with PE.

Commonalities between PIA and PA

There are a number of commonalities between PIA and PA that relate to the goals, methods, difficulties, and benefits of PIA and PA.[6]

---

[6] The characteristics of PIA/PA listed in this section are adapted from those identified during a workshop at the 1992 IAIA meeting in Washington DC. This workshop was part of an IAIA consultancy to the Indonesian Ministry of Population and Environment (see Boothroyd

Goals

The two goals of both PIA and PA are (1) to help decision making be comprehensive, integrative, systematic, rational, and reliably informed; and (2) to increase social learning about social and ecological systems and the consequences of decisions for these systems (Boothroyd 1992).

Methods

The method generally employed to assess impacts of policy is a comparison of 'with policy' and 'without policy' scenarios. In order to do this, PIA and PA must both deal with the basic methodological questions of—

- Who does the assessment (i.e., policymaking agency or external consultant)?
- How does one determine the significant outcomes to focus on (i.e., what should be the scope of the assessment)?
- How are outcomes to be compared (issues of common metrics, decision-rules, etc.)?
- What are the differential outcomes (who wins, who loses what) and how are they to be resolved?

Difficulties

Because assessment at the policy level encourages opening up to systematic scrutiny the most important (broadest, most fundamental) decisions taken by government, PA and PIA face more difficulties than project assessment (Boothroyd 1992). They—

- Arouse more political sensitivities.
- Take place in an osmotic rather than clearly defined decision-making procedural structure.
- Involve more difficulty (i.e., less certainty) in determining outcomes because
  a. Policy is not concrete.
  b. Policy is less site- and time-specific and is usually expressed qualitatively, often vaguely or in value-symbolic terms values such as 'fairness', rather than quantitatively and precisely, if policy is expressed at all.
  c. It is difficult for policy impact assessments to learn from previous exercises to the extent project IA can.
  d. Projects and project contexts tend to be more similar than policies and policy contexts.

Benefits

PIA and PA provide common benefits. They both can—

- Be proactive as well as reactive, i.e., they can put issues on the agenda as well as reacting to explicit proposals or implementations; they can start from identifying and analysing a problem, e.g., global warming, then proceed to

---

1992 in Marshall 1992). Participants were Kristi Branch, Rabel Burdge, Kurt Finsterbusch, Larry Leistritz, Roy Rickson, Richard Roberts, Barry Sadler, Michaela Smith, Charlie Wolf, and Peter Boothroyd.

assess implicit and explicit causal policies, or they can start from identifying implicit policies that produce certain kinds of programs and projects, or they can start by asking, "Given this situation what kind of policy change is needed?"
- Reduce uncertainty about what actual policy is by encouraging explication of policy assumptions and goals.
- Involve formulation of policy alternatives to be assessed (whereas project design and even program formulation usually requires technical expertise impact assessors may not possess).
- Culminate not only in recommendations for disposition or modification of a proposal but also in identifying new policies that may be needed to deal with identified problems or opportunities.
- Enhance coordination among agencies, improve program planning, and identify need for projects.
- Assess real and optional implementations (subpolicies, programs, projects) of policy being assessed.
- Focus on cumulative impacts.
- Assess assessment processes, e.g., by asking: are we thinking about the problem in the right way? are we asking the right questions?

POLICY IMPACT ASSESSMENT (PIA)

Under PIA, a policy proposal is first formulated to meet certain social goals and then assessed in terms of its impacts on other goals. To make sure assessments are as efficient, objective, and useful to decision makers as possible, procedures are established to guide or regulate the assessment process. Support for development of PIA comes from both the PE and IA traditions. While PIA is being theorised as an extension of IA from the project level, it is in fact mostly being practised as an adjunct to PE.

PIA as an Adjunct to PE: Policy Vetting (PV)

Within IA circles, PIA tends to be seen only as an extension of project EIA. However, a *de facto* form of PIA oriented to equity, quality of life, and sustainable development was actually introduced about two decades ago by governments refining their PE meta-policies on internal proposal-writing in order to respond to increasing awareness of unintended consequences, i.e., awareness of the complex and significant consequences (e.g., for gender equity) of what not-so-long-ago were regarded as fairly straightforward decisions in any given sector (e.g., transportation). Policy analysts, in both design and assessment roles, began to be required to vet high-level inhouse proposals for projects, programs, or policies—distinctions often unclear in practice—against a growing checklist of policy desiderata reflecting usually overlooked social goals. In this form, which may be called policy vetting (PV), PIA was initiated as an adjunct to PE oriented in the first place to particularisms or economic efficiency.

The Canadian government, for example, has required that its departments presenting proposals to Cabinet note the likely impacts of a proposal on, for instance, the environment, federal–provincial relations, and indigenous peoples. Departments in turn have passed responsibility for making such assessments to the organisations they fund, in some cases explicitly or implicitly elaborating the list of social concerns to be addressed. A number of central agencies—such as the advisory and financial management bodies serving Cabinet—have responsibility for assessing the adequacy of departmental PV products.

When conceived as an extension of PE, PIA automatically tends, like PE itself, to be informal. Little attention is given to devising procedural checks and balances to ensure public accountability because (a) it is assumed such procedures would pre-empt legitimate power; (b) such procedures would not map well with the complex reality of policymaking where sources of initiative and assessment are diffuse and continuous; and (c) such procedures would compromise the presumed need for confidentiality. The only requirement is for policy analysts to address, explicitly and tersely, the full checklist of specified impact categories—how they come to their conclusions is left to them. In contrast to the usual EIA approach, there are no PV rules governing whom to consult, when, in what form, and with what product.

Because of its low degree of procedural formality, PV in practice has been slow in living up to the spirit of its requirements. At its worst, the practice has been a cynical process in which departments ritualistically and perfunctorily predict no negative impacts. Even then, however, PV does represent a step in the inhouse formalisation of PE. As it has matured, moreover, PV has begun to encourage departments and agencies to broadly and heuristically consider impacts when formulating policy proposals. Increasingly, policies are vetted against a checklist of thought-provoking questions, and conceptual assistance is provided to help with the answers.

PV focuses attention on the tradeoffs between positivistic accuracy and heuristic insight in developing and selecting assessment methods. Assessment of impacts on the status of women illustrates the methodological tension. Practice has continuously been improved with the development, dissemination and acceptance of general empirical knowledge and gender-sensitive conceptual frameworks. Policy analysts have progressively reacted: (1) from initial incredulity at the requirement for gender IA; through (2) scorning its perceived triviality; (3) mechanistically responding to the requirements; to (4) appreciating the challenge; (5) discovering and relying on gender specialists; to (6) finally maturing by incorporating relevant perspectives into daily work. For any given proposal, the major challenge now facing the gender-sensitive analyst is to determine how detailed and accurate the assessment of impacts on gender relationships should be, and therefore how many hours and words to devote to it.

Notwithstanding PV's contribution to the progress of PIA, its informality allows analysts to remain unaccountable for their often lack of methodological rigour in either the positivistic or heuristic sense. As a result, PV is rarely as informative for assessors or readers as it could be. When making and stating required impact predictions, proponents rarely reference systems theory, state-of-the art substantive concepts, or empirical data. There is nowhere near the effort that is put into project cost–benefit analysis or EIA/SIA. Also, because PV is inhouse, it makes no contribution to broader social learning.

PIA as an Extension of EIA: Strategic Environmental Assessment (SEA)

Since the 1980s, impact assessors frustrated with the inefficiency and limited effectiveness of project-focused EIA have been calling for formal, publicly accountable EIA to be extended 'upstream' to the policy level where generic impacts can be considered.[7] Advocates of PIA formalisation see this as an advance over the inhouse often-confidential rather ritualistic checklist approach of PV.

Strategic Environmental Assessment (SEA) is the increasingly common term in IA circles for EIA at the program, plan, or policy levels. In introducing SEA to the general literature,[8] Wood and Dejeddour (1992: 11) address the worries of those concerned about the difficulty of extending EIA from projects to higher order decisions with the claim that:

> . . . the vast majority of the tasks involved in SEA are identical to those in project-level EIA. It follows that many of the methods employed are directly transferable, though many will differ in detail and level of specificity.

Similarly, a United Nations' Economic Commission for Europe report (ECE 1992) concluded that environmental assessment procedures for policies, plans, and programs should, as much as possible, reflect project EIA principles related to assessment initiation, scoping, external review, public participation, documentation, decision making, and monitoring.

A positivistic bias to SEA seems implicit, or at least not rejected, by its promoters who look to the EIA tradition with its emphasis on science-informed EISs. This bias is suggested by the methodological conclusion of the ECE report.

> Qualitative information and analysis have usually been used for environmental assessment of policies, plans and programmes. [This use] needs to be supplemented with quantitative information, since reliable quantitative data strengthen the evaluation of environmental factors (ECE 1992: v).

The central PIA issues from the SEA perspective relate to the level and nature of procedural formalisation, i.e., how detailed, strict, and enforceable procedures,

---

[7] O'Riordan and Sewell (1981) were among the first to identify an emerging interest in formal 'policy review' as a necessary supplement, and perhaps extension, to project EIA. They foresaw the democratic benefits of formalization, and the political obstacles.

[8] Lee and Walsh (1992) presented a concurrent introductory overview of SEA.

particularly those governing public involvement, might reasonably and effectively be. There are calls for SEA to offer much more formality to PIA than PV offers, though few suggest SEA should have as much formality as project EIA. Proponents of SEA are not opposed to PV, but rather champion allocations of more energy and power *to* the policy-vetting role. They point to the history and promise of better decisions resulting from intensive *public* vetting, such as evidenced by NEPA, its clones or lookalikes. They wish to ensure that the vetting actually vets, that the process is incorruptible, that checks and balances prevent PIA becoming captured by proponents. Sadler (1994) provides the rationale for public involvement: it is the best countervail against political, bureaucratic and technocratic distortion. Cognisant of "the reluctance to open policymaking to systematic assessment [which] extends to all contentious issues," Sadler (1994: 11) sees public involvement as "the litmus test of the utility and effectiveness of SEA."

Canada's inhouse environmental review of the North American Free Trade Agreement (NAFTA), prepared pursuant to Canada's new (1990) highly informal environmental assessment (EA) process for policy and program proposals, provides one example of how, in the absence of formal procedures to ensure public involvement, proponent-conducted PIA can be the captive of the proponents. The review document (Canada 1992b), which finds no environmental problems relating to NAFTA, was prepared by the interdepartmental committee formed to ensure environmental considerations would be taken into account throughout the NAFTA negotiations. The document is deficient in information and systems sensitivity. It was released after it and the draft agreement had gone to Cabinet. Even then, no process was established to encourage public comment. The review is regarded as a Canadian milestone in that EA of a crucial policy was supposedly integrated with the design of the policy and the results were publicly reported, but it also shows that in the absence of procedural safeguards, PA by proponents can be narrowly self-serving.

The putative upside of SEA is the objectivity it induces. The downside is that it may settle for politically effete technical analysis.

SEA discussion focuses on policies governing physical projects. For instance, while Wood and Dejeddour (1992: 4) note that various international bodies, such as the World Bank and the World Commission on Environment and Development (the Brundtland Commission), consider that 'environmental issues must be addressed as part of overall economic policy', the examples used by Wood and Dejeddour are only of physically oriented policies, plans, and programs—primarily roads, power plants, coastal developments, and other infrastructure, though also agriculture, forestry, fishing and mining, and land use. Perhaps as a consequence of their physical focus, their substantive focus is on the environment as an externality, rather than the complete human-ecology system.

The ECE (1992) report on the application of EIA Principles to Policies, Plans and Programs is also physically oriented. It uses 10 case studies of environmental assessment to show that upstream EIA can enhance efficiency. Seven of the cases involved substantive planning: (a) for energy supply—electricity in Ontario, power-plant and industrial fuels in the USA, coal technology demonstrations in the USA, offshore petroleum exploration and development in Norway; (b) for wetlands retention and restoration—Canadian prairies; (c) waste management—The Netherlands; and (d) an urban commercial development—Germany. The remaining three cases involved planning of procedures to enhance consideration of environmental impacts in substantive planning of water management (Czechoslovakia), municipal operations (Finland), and roads (Sweden).

Limitations of Strategic Environmental Assessment

The PIA potential of SEA, as proposed and practised, is limited by its: (1) positivism; (2) binding but unempowering formality; and (3) narrow scope. These limitations result from SEA being an extension of reductionist, linear, environmental IA. SEA's broader PIA potential is also constrained by three other limitations which reflect both SEA's unique character as a form of PIA and the nature of PIA in general. Those limitations relate to SEA's: (4) process; (5) policy level; and (6) impact horizon.

1. *Positivism*. The limitations of positivism in PE as described earlier apply equally to positivism in SEA.

2. *Formality*. The theoretical formality proposed for SEA, while less than that contained within EIA, is still unattractive to governments. As Wood and Dejeddour (1992: 13) put it:

> While SEA is methodologically feasible, there are bound to be institutional and political resistance to its wider use and complaints about its cost. Indeed, in the United States there has been a slight movement away from [SEA], perhaps as result of the level of generality of the EISs produced and their apparent lack of influence on decision making.

Wood and Dejeddour (1992: 13) nevertheless conclude that:

> There is every likelihood that SEA, if implemented judiciously and at the appropriate level in the various planning processes, will establish itself in the same way as EIA has as a cost-effective tool of environmental management.

As support for this conclusion, they point to what they perceive as a trend in several countries, such as Canada, towards acknowledgement of 'the irrefutable logic of SEA'. The Canadian example, however, is not so supportive. Canada introduced, in 1990, an environmental assessment process for policy and program initiatives, but it was not to be legislated and it was restricted to—
  • Proposals to Cabinet that are 'environmentally relevant' with a public statement on anticipated environmental impacts being released at the time of announcement of the policy or program;

- Enrichment of environmental considerations as part of the Regulatory Impact Analysis Statement process; and
- Assessments by ministers on their own authority where these are considered to warrant an environmental statement and, if appropriate, a public statement (Canada 1992a, 1993).

Exemptions were to be possible for: emergencies, national security matters, matters already assessed under a previous proposal, and proposals that had an inherent environmental benefit (Canada 1992a, 1993). This is far from formal EIA, even by Canada's loose standards of formality.

Formalisation, and therefore democratisation, of SEA will continue to be opposed by politicians and policy analysts as unrealistic. Even if opposition were overcome, the spirit of the procedures would be mostly honoured in the breach, and public involvement would be a largely ineffectual add-on to the design process, just as it is at the project level. This is not because meaningful public involvement cannot be fostered by formal procedures but because the formal EIA/SEA model still keeps proponents with narrow goals in the driver's seat with positivistic assessors in the navigator's seat.

3. *Narrowness.* SEA is limited to externalities related to the environment. Tradeoffs and synergies between sustainability and equity or nonenvironmentally defined quality-of-life are excluded from consideration. Not only do social impacts get short shrift but so do those impacts' environmental impacts—i.e., the second-order environmental impacts of first-order social impacts. Moreover, SEA, in practice, and usually as proposed, does not apply to economic, cultural or defence policies— which may have the greatest impacts on aggregate resource consumption, global pollution, greenhouse gas, and ozone-depleting substance emissions—e.g., taxation policies that determine motor vehicle usage or mining rates. SEA is limited to policies governing the building and operation of physical projects—policies that are in fact primarily induced by economic policies.

4. *Linear partialising process.* Reflective of the IA process in general, SEA involves addressing major social goals sequentially, partially and therefore invidiously. The sectoral goals of a line department or its client instigate policy, program or plan design. It is only then that the consequences for other goals—in this case, environmental—are assessed. This limitation of the IA process is illustrated by a recent US interdepartmental committee's *Guidelines and Principles for Social Impact Assessment* as part of EIA under NEPA (United States 1994). SIAs of proposed projects, programs and policies—which the guidelines treat generically— are to depend on "data requirements needed from the project proponents to frame the SIA . . . [such as] on locations, land requirements, needs for ancillary facilities. . . construction schedule, size of the work force. . . facility size and shape, need for a local work force, and institutional resources" (p.11). SIA is thus left in the well-known weak position of identifying desirabilities for redesign or mitigation (see chapter by Burdge and Vanclay).

While Wood and Dejeddour (1992: 15) call for integrating "assessment into the formulation of the action [i.e., policy, plan or program] from the earliest possible stage, rather than assessing the action when it is almost determined," the design-enhancement role they envisage for SEA, as indicated by their 8-step model, still involves sectoral proponents or 'lead authorities' initially shaping proposals to meet their mandated goals before inviting others to comment on environmental impacts. However, Wood and Dejeddour do introduce a step that could lead to more proactive comprehensive PA—reviewers may comment 'where feasible' on the objectives of the proposal as well as on its externalities.

SEA is linear and partialising, not only within a given design-assessment decision exercise, but also through the whole process of governance. From the SEA perspective, decision making itself is seen as linearly deductive. Wood and Dejeddour (1992: 8) believe that "generally. . . there exists a tiered forward planning process which starts with the formulation of a policy at the upper level, ... followed by a plan at the second stage, and by a programme at the end." This conception of governance is usually at odds with reality and probably with the ideal. If it were so straightforward, the central planning of the Soviet Union would be still functioning. In a complex, uncertain, changing world, projects are generated independently of explicit policy, explicit policy is not always implemented, and policies contradict each other.

Woods and Dejeddour's neatest example of tiering is a national transport policy that guides a long-term national roads plan, which guides a 5-year building program, which guides construction of a motorway section. Even here, the process may have lots of room for sideways interventions at downstream levels that induce change upstream. Parochial political pressure, national budget constraints, or indeed project EIAs, may change the policy, plan, or program for the existence, location, timing, or design of roads. Truncated freeways all over the world testify to this. EIA-modelled SEA could conceivably have changed freeway policy before neighbour-hoods discovered and reacted to project plans, but it is not likely that it would have done so. As legions of project EIAs have shown us, even if the technical difficulties of impact prediction, the political-power obstacles to open information, and the timing problems limiting meaningful public involvement are all overcome—and they rarely are—the momentum and disposition of powerful proponents usually prevail when they initiate design according to their mandated goals. Citizens, decision makers, and in many cases experts, find it difficult to evaluate all the social and environmental impacts of any design, but especially policy designs. Actual impacts of freeways have to be experienced before many citizens are motivated to fight. Until then, the perceived, claimed and real benefits of freeways (e.g., heightened mobility, reduced travel time, reduced road trauma, sales of cement, jobs, and political monuments) outweigh concerns about negative impacts (e.g., lack of public money for public transport, noise, urban sprawl, physically splitting and adding noxious traffic to neighbourhoods). Thus, transport policy assessment must be a continuous process, undertaken not only before, but also during and after project IA. As illustrated by the ECE (1992) study of the German urban development assessment process, downstream assessment can instigate and shape upstream policymaking. The three assessed options in that case all involved the same location. The EIS finding that all three options were inadequate led the town council to go to a higher decision level and reject the site (though not the project).

5. *Low policy level.* Perhaps as an inevitable consequence of its being applied only to explicit proposals, SEA in practice so far seems well downstream from the cultural assumptions that are the real policy source.[9] Broad development options, e.g. reducing and redistributing consumption vs trickle-down growth, were not assessed in the ECE cases. The four energy supply cases all involved assessment of option sets constrained by implicit, and in some cases explicit, policies that put little or no emphasis on energy demand reduction. While, the analysis of The Netherlands waste management process did include waste reduction as part of the environmentally preferable option, the extent of envisaged changes to lifestyle was not made clear.[10] The wetlands options seemed constrained by agricultural policy, the commercial development options by urban growth policy. The options considered in the ECE cases may be best thought of as program options constrained by implicit policy. Program SEA can lead to more efficient identification of aggregate regional resource destructions than that offered by individual project EIAs—but it will not lead to sustainability so long as it operates within the implicit development policy paradigm that promotes universal unending growth in material consumption. Paradigmatically constrained program SEA will help us get more outputs with smaller increases of resource consumption, but not a better, different quality of outputs with absolute declines in resource throughput.

6. *Short impact horizon.* Just as the most important policies and options tend to be ignored by SEA, so too the most important impacts. For example, in her study of EIA of waste management policy, electricity supply programs, and land use planning in The Netherlands, van Eck (1993: 8) found that "the EIA approach for projects appeared to work for plans and programmes with only a few alterations... mainly due to the existence of a scoping phase in the EIA procedure" that allow EISs to be 'made to measure'. Once again, fundamental ecological issues related to natural capital, let alone the equity and quality of life issues that drastic changes in policy would raise, were not considered in her cases. It appears the EIAs 'worked' because they were immediately focused and narrowly scoped. The short impact horizon seems partly explainable by SEA, and PIA in general, being reactively concerned with partial externalities rather than proactively with integrative visions.

---

[9] Indeed, after reviewing SEA practice and procedures in various countries, Therivel (1993: 162) found that 'SEAs prepared to date have been carried out exclusively for plans and programs, rather than policies'.

[10] Most historical forecasts about future lifestyles have proved to be erringly conservative. There is no reason to doubt that current assumptions about future changes will turn out to be equally so.

Inherent Shortcomings of Policy Impact Assessment

Like EIA, PIA as institutionalised IA is supposed to help level the political playing field on which powerful interests use PE to set policy agendas. PIA's function is to scrutinise explicit policy proposals for their externalities. In actuality, because PIA continues IA's generic shortcomings, weaker interests are considerably disadvantaged. There are three fundamental shortcomings of PIA in both its PV and SEA forms. These relate to PIA's (1) focus, (2) application, and (3) function. By nature, PIA cannot assess all important outcomes of any policy, nor any outcomes of all important policies, and it has little bearing on the most important decisions. These shortcomings cannot be overcome by changing how PIA is done procedurally or methodologically.

1. *Outcome focus.* PIA overlooks outcomes related to objectives. Therefore PIA cannot address relative effectiveness or efficiency of a proposed policy in relation to alternative policies. Nor can it address policy-effected tradeoffs among social goals, or the need for a policy, program or project. Because IA is restricted to externalities, proponents can successfully argue that issues of need, and for that matter feasibility, are outside the mandate and competence of impact assessors; thus IA provides no check on proponents' justifications.

2. *Policy application.* PIA ignores implicit (i.e., informal, *de facto*, or secret) and existing policy. The policy to expand airports to meet demand regardless of cost-benefit ratios is, for example, not assessed because it is never considered—nor are policies that support agricultural monoculture or commodity accumulation or automobile driving. Policies developed through the courts, e.g., judgements of moral standards in areas such as pornography, are also not assessed. The problem is that policies (as opposed to programs) are for the most part not often officially created and when they are, say in the form of legislation, the most important steps in the policy-making process are taken behind closed doors. Often the nature of policy has to be induced from the programs and projects it supports.

3. *Process role and timing.* PIA as PV or SEA is only seriously applied to proposals after they have been designed and put forth—rhetoric about integration of assessment with design notwithstanding. This means PIA is not likely to have much impact on decision making other than to identify mitigations that do not threaten existing implicit basic tradeoffs among social goals. As is widely known from EIA experience, IA of completed designs, whether public or inhouse, by formalised or *ad hoc* procedures, is often ineffectual, costly, or both. Completed designs entail commitment by powerful proponents who have put much energy into them. Procedures may call for alternatives to the proposed design to be comparatively assessed but this is a practical and theoretical impossibility in any meaningful sense.

Conclusion

The differences between what the PE and IA traditions have produced or proposed for PIA reflect their different approaches to scientific method and planning procedure. PV takes for granted PE's minimisation of procedural formality and focuses on methodological issues; SEA takes for granted EIA's maximisation of positivistic rigour and focuses on procedural issues. Each of the two forms of PIA has its strengths and weaknesses but neither ensures that the most important outcomes of the most important policies are effectively assessed. A more systemic approach to policy assessment is needed.

POLICY ASSESSMENT

Conceptual Need for Policy Assessment

Despite the policy planning advances made through PE, IA, and the two forms of PIA they have respectively spawned (PV and SEA), there are still three major lacunae in the conceptualisation of policymaking as a rational, democratic process. (Practice, of course, always falls short of the ideal, however it is conceptualised.)

1. There is a need for the conception of a process that comprehensively assesses outcomes, i.e., that simultaneously involves assessment in terms of objectives, higher goals and externalities, as distinct from PE and IA which are only partial in this regard.
2. There is a need for the conception of a process that comprehensively assesses fundamental policies, implicit as well as explicit, existing as well as proposed, and that thereby broadens the possibilities for initiating assessment, as distinct from PE and IA, which concern only explicitly proposed policies.
3. There is a need for the conception of an assessment process that is both integrated with policy design and scrutinises designs, as distinct from PE and IA which each play only one of these two roles.

Conceptions of these three kinds, it is proposed, will in due course lead to improvements in the conscious practice of policymaking. These improvements will be changes that create more rational and democratic policies, policies that produce more of the outcomes society needs and wants. It is proposed that the three conceived processes be collected under the name of 'Policy Assessment' (PA) so that PA is defined as the process by which fundamental policy options are continuously identified and assessed in terms of all highest level societal goals.[11] PA as comprehensive fundamental continuous assessment thus fuses and goes far beyond PE and IA (see Table 1).

---

[11] Under PA, one goal could initiate the design process, but immediately others would be brought to bear on the formulation of the problem. Higher goals could be stated by the proponent, but they would be assessed against more universal values as interpreted by the assessor.

Table 1.   **Policy assessment compared with policy analysis and evaluation (PE), and impact assessment (IA)**

	PE	IA	PA
Outcomes assessed	Intended	Externalities	All
Policies assessed	Explicit	Explicit	All
Relationship to design	Intrinsic	Scrutinising	Both

Through PA thus defined, equity, quality of life, and sustainable development which are highlighted but still marginalised in PIA, become defining goals equal to economic efficiency and growth. The three conceptual needs met by PA distinguish it from PIA (see Table 2).

Outcomes assessed. PIA takes as given the assessed policy's objectives and design for meeting these objectives and restricts its focus to assessing (predicting, evaluating, and monitoring) externalities or unintended consequences. PA (like technology assessment) considers all policy outcomes, not just externalities, but also effectiveness and efficiency in meeting objectives and consistency of objectives with higher goals, whether or not the goals are explicated. Thus PA stretches further up and down, and more widely across, the ends–means hierarchy than does PIA. It combines PE's and PIA's concerns. Like PIA, PA asks: what are the systemic implications (externalities) beyond those likely considered initially by the proponent focused on finding a solution to a particular problem? Unlike PIA, PA also asks does the proponent's solution in fact solve the real problem?

Policies assessed. PIA is restricted to assessment of official policy proposals and decisions. PA applies to policies that are informal as well as formal, existing or possible as well as proposed.

Process. PIA is a scrutinising adjunct to design, occurring only after and external to design. PA is a complete process involving both integration of assessment with policy design from the beginning, and external scrutinising assessment of designs after they are judged complete by designers. This distinction reflects PIA's single IA heritage and PA's dual PE/IA heritage.

Any notion that design-shaping IA—by designers and/or externals—can be so effective that additional post-design scrutiny is not needed, seems untenable on the grounds that designers have vested interests and necessarily limited knowledge, while full continuous external, and particularly public IA would be impossibly slow. Regardless of how well IA is integrated into design from the beginning, there will always be a need for some IA at the last design stage. As the final checkpoint, this stage should be taken most seriously, and from a democratic ideal, be fully public. The solution to IA's relative impotence as a post-design external add-on, is not to abandon this role but to supplement it with internal IA simultaneous with design.

Table 2. **Summary of differences between policy impact assessment (PIA) and policy assessment (PA)**

	PIA	PA
Outcomes assessed	Unplanned-for only	All
Policies assessed	Explicitly proposed only	All
Design role	External scrutiny after design	Shaping during, plus scrutiny after design

The breadth of PA requires not just breaking new conceptual ground but also the development of appropriate procedures and methods. Just what procedures and methods are appropriate is a function of the specific conception of PA, i.e., how formalised and positivistic it is conceived.

FORMS OF POLICY ASSESSMENT

Tensions Creating PA Types

The potential debates in PA are over how to resolve the tensions deriving from its tributary traditions of PE and IA. Whereas PIA comes in two forms, each of which reflects one of the two traditions, PA as a synthesis of PE and IA could be shaped by four possible polar resolutions of the methodological tension in PE (there being little procedural tension) and the procedural tension in IA (there being little methodological tension). On one end of the procedural tension would be those, often fully within government, who would favour informal PA, which is more in tune with the PE tradition. Informal PA—proponents would say 'pragmatic' PA—would emphasise flexibility over rules, trust over accountability, and education over sanction. At the extreme, informality would mean no rules for either inhouse procedures or public involvement: the proponent agency would retain total control of the process and content of assessment. At the other end would be many IA theorists who, focusing on the assessment component of policy assessment (the *A* in PA), would see PA, like project EIA, as best governed by formal procedures which include provision for public involvement. On one end of the methodological tension would be those who see PA as ideally technical and positivistic in the CBA and EIA traditions; on the other, would be those who look to more heuristic approaches that trade off precision for insight.

Both formality and informality could be associated with either positivism or heuristics. Both inhouse and participatory assessments can aim for positivistic certainty through exhaustive study, or for heuristic assessment sufficient for reasonable decisions to be made. Conversely, informal processes and formal procedures for getting information to publics and hearing from them could both be determined by whether they are to serve positivistic or heuristic inquiry. (Methodology has implications not only for the content of information exchanges among experts and between experts and publics, but also for the format of opportunities for exchange.)

Combined, the procedural and methodological tensions conceptually give rise to four possible forms (see Table 3). Only three of these forms are currently evolving owing to the fact that PA is being derived from either PE with its technocratic procedural informality and tradition of positivism but openness to heuristics, or from EIA with its more democratic procedural formality and positivistic practice. The three forms are: (1) informal positivistic PA which is evolving from CBA; (2) informal heuristic PA from PV; and (3) formal positivistic PA from SEA. In any given jurisdiction, actual PA is some combination of these types.

Table 3. **Types of policy assessment**

	Informal (as in PE)	**Formalised** (as in EIA)
Positivistic	Based on CBA	Based on SEA
Heuristic	Based on PV	No model yet

All three of these forms go beyond their respective practice bases by combining the substantive concerns of both IA and PE—i.e., objectives and externalities—to produce comprehensive assessment as part of policy design and as scrutiny of near-finished and existing policies. Informal positivistic PA broadens CBA's focus to incorporate as criteria hitherto externalities; informal heuristic PA broadens PV to address a wider range of externalities and to incorporate these as goals in design; formal positivistic PA broadens SEA's application to a wider range of policies and goals.

The fourth possible form, formal heuristic PA, has no basis in existing practice. This is because formality was introduced to policy planning by EIA—PE has no such tradition—and EIA is strongly positivistic because of its historical project-orientation, sponsorship, mandate, timing, and response to formality itself. But formal heuristic PA could have the greatest potential to be the continuous, comprehensive, influential, educational, efficient and democratic information support needed for good decision making in a complex society. To explain the potential of, and opportunity for, formal heuristic PA, it is first necessary to describe the characteristics of the other approaches (see Table 4).

Table 4. **Characteristics of policy assessment types**

What is PA. . .	Informal positivist	Informal heuristic	Formal positivist	Formal heuristic
genesis?	PE→CBA→PA	PE→PV→PA	EIA→SEA→PA	no precedent
formalised part?	nothing	product	process	product/ process
procedure?	understood	defined	regulated	transparent
activity start?	design	referral	screening	evaluation
staging?	cyclic	fluid	linear	continuous
controlled by?	designers	agencies	courts	networks
done by?	technicians	analysts	assessors	all
accountability?	technocratic	bureaucratic	legalistic	democratic
issue of concern?	improvement	management	mitigation	development
role in decision?	clarifying	enlightening	informing	suggesting
information goal?	certainty	sufficiency	completeness	insightfulness
learning goal?	technical	institutional	political	societal
worldview?	reductionist	systemic	legalistic	holistic
key concept?	trending	feedback	accumulation	contradiction
analytical focus?	facts	relations	significance	values, assumptions
data source?	statistics	literature	empirical research	collective knowledge
analytical tools?	indicators	checklists	criteria	scenarios
products?	studies	memoranda	statements	briefs
role for publics?	informed	excluded	consulted	participating
cost?	high	low	high	contingent

Informal Positivistic PA

The result of PA being derived from PE with its ideal of technocratic procedural informality and positivism, is the extension of CBA to address not just all outcomes of policies through full-cost accounting, but also the complete spectrum of policies throughout the planning and decision-making process. According to Sadler (1994:7), such an approach is supported by the United Kingdom's recent report, *Policy Appraisal and the Environment* (1991), which recommends "the application of extended benefit-cost analysis. . . . Environmental effects are to be considered with other economic and social factors during the design of policy options—with a view to clarifying tradeoffs and identifying a preferred alternative."

There are two possible approaches to extending CBA. One is the multiple accounts model that was adumbrated by the four accounts introduced to American water resources planning two decades ago (United States 1973) and that continues to be experimented with in many jurisdictions. Multiple accounts permit different methods and metrics. The other approach is to incorporate more into the utilitarian CBA framework by reducing more outcomes to its calculus. The second approach most straightforwardly enables economists to play the dominant role in shaping policy debates, but not even multiple accounts will create equality among experts: under either approach, the culturally and politically embedded utilitarian perspective will likely be increasingly comprehensive as quantitative techniques of modelling, risk analysis, and weighting become more sophisticated.

An important ongoing application of informal positivism will continue to be econometric assessment of fiscal and monetary policies. Fixated on transactional growth, economists within the now-worldwide neoclassical paradigm interpret quality of life and poverty alleviation in terms of quantifiable growth indicators, ignoring externalities such as global warming. Though the decisions informed by such PA have momentous consequences, few procedures govern this arcane art, and there is virtually no involvement by non-economists. Positivistic techniques inherently serve to exclude not just people, but also impacts not readily quantified by the dollar metric. Informality exacerbates the exclusion.

Informal Heuristic PA

Because this PA form is essentially methodologically enriched PV extended in focus, application, and timing, it reflects PV's strengths and weaknesses. Its particular strength is that it encourages qualitative, incisive judgements through flexible and efficient processes. It is as effective in producing useful information on the relationship between policies and societal goals as can be expected from inhouse assessment not subject to public scrutiny.

Canada has already produced a heuristically sensitive 'sourcebook of helpful concepts and ideas' for policy and program EA (Canada 1992a) which could also be applied to more holistic PA. It encourages early consideration of externalities in design, recognises that assessments "should be rigorous but not necessarily laborious. . . [and] *just sufficient* for informed decision making," that policy may be "relatively informal—even implicit. . . [and that] there are many [policy] areas where careful examination reveals very significant (though often subtle and/or indirect) environmental effects. . . [for example] loss of prime agricultural land due to urban sprawl accelerated by housing and mortgage policies" (Canada 1992a: 4-9). It offers a menu of heuristic techniques, criteria for selecting them in practice, substantive insights into specific impact categories, and key indicators. Summing up its environment-centred but comprehensive perspective of PA, the sourcebook states (p.16): "Integration is the new norm: separation of social and economic analysis from environmental analysis is becoming an outmoded approach to public policy and program development." At the same time, it ignores formality, and, in fact, gives very little attention to public involvement in its nine step iterative process model which begins with internal scoping and ends with public announcement "subject to any strictures of confidentiality."

Despite such advances as Canada's sourcebook, most current practice of PV and such extensions to PA as there have been, do not often appear systemically sophisticated. This may be due to: (1) limited systems awareness by policy analysts; (2) a policymaking culture that discourages abstract, theoretical or complex writing and conceptualisation; (3) the checklist structure of the PV process which militates against holism; and (4) the informality which provides for no check on slipshod work.

More significant than, and also because of, the potential methodological short-comings resulting from informality are the substantive dangers. Inhouse PA provides no external check on arbitrary government. It does not even help those weak departments and perspectives in government ostensibly established to protect weaker public interests.

Canada's peremptory and ideologically driven NAFTA PA process seems, from what can be learned about it, to be a good example of informality's dangers for heuristics. Not only was the environmental assessment suspect (as discussed above), but it seemingly ignored equity externalities and whether unregulated trade would, in fact, meet the objective of making Canadians better off economically, if stability, reduced dependency, and long-term productivity are considered. (Daly and Goodland's 1994 incisive assessment of free trade shows the shape good heuristic assessment could take.)

An example of informal heuristic PA is provided by the process being developed by the provincial government in British Columbia, Canada. Its Cabinet Planning Secretariat has initiated the development of procedures for departments to follow in preparing Cabinet submissions. The procedures require proponents to present 'real alternatives', to assess outcomes in terms of a long list of objectives and externalities (e.g., environment, human health, equity, women, indigenous people). There are no procedures for screening, scoping, or indeed for public involvement in any form.

For each Cabinet document, the process for developing and assessing the options will be idiosyncratic. Indeed, as the procedures have been developed, reaction from departmental policy analysts has resulted in the procedures having even less inhouse formality than originally conceived. Various ministries have been designated to take the lead in developing assessment tool kits that offer, rather than prescribe, salient concepts to help policy analysts ask the right questions. It is not expected that every ministry will have an impact specialist for each impact category, or that consultants will be routinely hired to produce impact statements. Internal PA processes such as this, should be monitored (to the extent legally and practically possible) to determine if their heuristic goals become compromised by their informality.

Formal Positivistic PA

Like SEA, this form of PA would explicitly extend to the policy level EIA principles and procedures governing screening, scoping, research, impact statements, decision-reporting, and especially public involvement. But it would consider all goals (not just environment) throughout the design/assessment processes (not just at designated scrutinising stages) and would be applied to all types of policy (not just infrastructure-oriented). Moreover, it would adapt EIA/SEA procedures not only to proposals but also to formal reviews of existing implicit and explicit policies. Being rooted in EIA, its methodological approach would tend to the positivistic.

This form of PA would likely be favoured by most IA theorists, explicitly and in the first place for its formality, implicitly and secondly for the technical, positivistic rigour it promises. Examples of IA theory headed in this direction are provided by Finsterbusch (1989), Sadler (1994), and Partidario (1993). Finsterbusch (1989: 17,21) answers his own generic question, 'how should policy decisions be made?', by rejecting not only utilitarianism and other one-goal perspectives, but also full democracy in which 'valuation would be achieved through discourse rather than by technical weighting procedures'. He proposes instead a 6-point 'ethical pluralism model' that 'balances' values in conflict by applying political processes to completed technical design and, most importantly from the standpoint of this paper, technical assessment.

Partidario (1993) and Sadler (1994) approach PA from the environmental perspective. Sadler (1994:3), still using the term, SEA, broadens its meaning.

> By most measures, conventional (that is, project-oriented) EA is a self-limiting and ineffective response to current scales and rates of ecological deterioration. More proactive, integrated approaches are required—in effect, a second-generation EA process that moves beyond the 'impact fixation' to address the causes of unsustainable development. These are located at the 'upstream' phase of the decision-making cycle, in the macro-economic policies and development programs pursued by governments of all political stripes. What is termed strategic environmental assessment (SEA) is a promising approach to ensure that policymaking takes account of sustainability principles.

While Sadler sees the need for procedural integration of IA with design and assessment of economic as well as infrastructure policies, he still sees environmental assessment as a process with its own identity "linked (with risk assessment, SIA, and cost-benefit analysis), into comprehensive project review to ensure sustainability" (1994:5-6). Sadler (1994:10) recognises that "the policy-making process does not necessarily correspond to. . . [an] idealised, hierarchical sequence. . . . It is more realistic to think of policymaking as a complex filtration process." Nevertheless, he (1994:11) suggests that "SEA procedures can be adapted from those established for project EA":

- Initiation and screening establishes the level of analysis required indicating whether no action is needed or whether a full or preliminary EA should be undertaken.
- Analysis and reporting covers the main steps involved in conducting full or preliminary assessment and preparing a statement of environmental effects (a succinct, policy-relevant interpretation of findings). . . .
- Monitoring and follow-up includes overseeing the implementation of decisions (Sadler 1994: 11).

Recognising that "scientific understanding is insufficient to make accurate predictions of the environmental impacts of development activity," Sadler (1994:12), like many others, calls for adaptive management. But he also adumbrates the need for heuristic PA, by pointing out that "the extended time and space scales of an SEA mean that even more factors will affect the results of policy review, often in unforeseen ways. Under these circumstances, relatively simple methods that help clarify tradeoffs and conflicts will be most effective for initial analysis, rather than the application of more complicated models" (Sadler 1994:13).

As Partidario (1993:1) sees it, "EIA could change to become a new form of environmental assessment and review (EAR) tool to help integrated policy and planning processes move towards more sustainable approaches." EAR would be a 'dynamic process' similar to adaptive management, involving both shaping and scrutinising of design. Partidario's EAR "encourages public involvement at three specific stages: identification of existing problems and impacts of evolution, assessment of alternatives [options], and post-evaluation or process review" (1993:4). Her conception goes well beyond current SEA practice, though notably absent from her list of involvement stages is that of formulating options. This is where design and assessment power lies and where public involvement could make a significant difference. (Asplund and Hilding-Rydevik (1993) concluded from their study of Swedish SEA that effectiveness depends on two crucial factors: the constellation of the planning group which determines perceptions and interests, and how early the alternatives and environmental issues are raised.)

Partidario's and Sadler's conceptions of EAR and SEA more clearly integrate environmental considerations into design (as well as being a check on design products) than do Wood and Dejeddour's. Still, there are three problems. First, by the very names, their conceptions are restricted to environmental goals and therefore could partialise what should be a holistic process of considering all societal goals simultaneously. The goal of sustainability is best met, not by giving it separate attention, but by seeking means to achieve it while enhancing equity, quality of life, and efficiency.

Second, their proposed extensions of EIA procedures seem still too linear. Their proposals do not fully accommodate the need for effective public involvement in the fluidity, diffuseness, indeed chaos and sometimes subterfuge of policymaking (though Partidario suggests new roles for publics). In developing PA, it would be better to begin with the reasons for formalising public involvement rather than tailoring the procedures that have evolved through EIA.

Third, their conceptions do not identify the connections between method and procedure (though Sadler identifies the need for heuristics). They leave open the possibility that positivism could be induced in PA by formality of the EIA-type, as it has at the project level. Ideal PA would eliminate these three problems.

IDEAL POLICY ASSESSMENT: HEURISTIC AND APPROPRIATELY FORMAL

The Need for Formal Heuristic PA

Considering the preceding forms, it can be seen that the potential of PA can founder on the shoals of positivism and its dangerous promise of certainty, and/or the shoals of informality and its dangerous promise of flexibility. It can also be seen that just avoiding these shoals is insufficient. PA must aim for powerful heuristics, not nonchalance, and for formality that supports meaningful public involvement, not ritualism.

The starting point to the ideal is to break away from the limitations posed by simply extending PA from either PE or IA alone. Starting afresh, PA could synergise SEA's concept of formality (not its specific procedures) with PV's heuristic potential (not its checklists). This would create PA formalised enough to safeguard against its becoming window-dressing for done-deals or dead paradigms, and heuristic enough that the insights it yields are widely understood and useful in making crucial decisions. At the same time, ideal PA, while formalised, would be efficient and fluid enough to meet the political and administrative realities of policymaking, and while heuristic, empirical enough to reflect reality rather than cant.

The role of appropriately formal heuristic PA would be to promote wide comprehensive discussion on fundamental issues in policy realms ranging from trade to transportation to taxation. It would be integrative: substantively, by addressing the systemic links among goals, policy realms, and decision-levels from implicit policy to day-to-day work planning; methodologically, by combining empiricism with systems theory at once sophisticated and accessible; and procedurally, by informationally linking publics and governments throughout the flux of policymaking and review.

Formal Heuristic PA Is Not Only Adaptive Management

Formal heuristic PA would be quite different than adaptive management (Holling 1978), despite some seeming similarities. Recognising the by now obvious fact that uncertainty increases with complexity, adaptive management adds to planning in complex situations a capacity for monitoring action outcomes and for responding quickly to surprises. It confronts positivistic science's inability to provide accurate predictions, but it is an inadequate solution. It does not obviate the need to reduce uncertainty before making policy decisions, nor the need for adopting the Precautionary Principle[12] (Dethlefsen et al. 1993).

Reduction of uncertainty is especially important in situations where bad decisions can have serious irreversible consequences. Putting lifeboats on ships is not a substitute for building safer ships and setting safer courses. When interventions into fragile and/or self-regulating and/or self-making systems such as ecosystems or cultures are being considered, the proper response to uncertainty is not to wade in, intending to deal with problems as they arise, which adaptive management purists propose, but to hold back and get better information and understanding. Thus building adaptive management (or action-research) capacity into government is necessary but far from sufficient. The limits of positivism need to be addressed also by improving predictive methods. The adaptive management capacity that is needed can be provided through PA procedures that formally require monitoring and policy audits, and that suggest heuristically powerful concepts for framing questions and interpreting answers.

Participatory Heuristic Methods

A heuristic approach to prediction would focus on: the big picture over details; direction of change over quantity of change; systems processes over systems states; and would favour insight over rigour. It would use soft systems concepts—e.g., of feedback, entropy, variety, and recursion—to model social, cultural, institutional, biophysical, and ecological systems affected by policy. Comparison of a system assuming the policy being assessed is in place (the with-policy scenario) with the system assuming the policy is not in place (the without-policy scenario) would acknowledge that in both cases the system is dynamic—even the without-policy system will not be the same tomorrow as today—and that this dynamism is often a function of human agency and choice; that making and comparing 'movies' of processes is better than comparing snapshots.

The widespread concern that a systems approach is necessarily authoritarian is understandable but best dealt with, not by abandoning systems thought, but by emphasising the heuristic potential of systems modelling for aiding thought and dialogue among policy analysts, decision makers, and publics. To realise this potential, procedures would have to be designed to ensure that citizens participate in building the models, rather than simply being possessors of characteristics to be studied through the model (as they are in the positivistic conception), to ensure, in short, that citizens are the predictors as well as the predicted. Participation in model-building would include identifying key variables and their relationships as well as the intervention options whose if–then consequences are to be explored. This goes far beyond the scoping of EIA. Computers could aid analysis of consequences,

---

[12] The Precautionary Principle states that it is better to be safe than sorry by reducing potentially harmful activity in situations where there is uncertainty about the harm yet still reason, e.g., because of systemic analysis, to worry.

but they also pose dangers in distorting communication, and by appearing to offer more clarity, precision, and certainty than is warranted. Electronic information networks could be used for information exchange and dialogue, again with caution. Felt pens and sheet paper, whiteboards and markers, could be the most useful hardware for heuristic PA, analogies and meditations the most useful software.

Policy Assessment Procedures

A growing number of IA theorists and practitioners are coming to the conclusion that PA procedures should be developed from scratch rather than be adapted from those created for EIA (Bregha et al. 1990; Boothroyd 1993). Formal heuristic PA, in contrast to PA based on SEA, would follow these admonishments. Rather than linear EIA-like stages starting with screening by the initiating agency, formal heuristic PA would be a continuous part of ongoing policy development. Discrete exercises would typically start with evaluation of existing policy (or, what amounts to the same thing, with identification of problems or policy needs). Unlike SEA which assumes hierarchical deductive decision making through logical decision-trees from policy to project, formal heuristic PA would recognise the fluidity of decision making and the complexity of relations among decisions by encouraging assessment at any time, at any policy level from the general to the specific, at any point in the stream of causes and effects from policy to activities, in any policy sphere, by any party.

Formal heuristic PA procedures—collectively indicating what outcomes of what policies should be assessed how, when, with what products for whom—fall into two categories: (1) mechanisms to ensure meaningful public involvement in all aspects of PA; and (2) guidelines on PA methods and products.

Mechanisms to Facilitate Public Involvement in Policy Assessment

Rather than the EIA/SEA procedures—which only give publics the roles of attempting to make sure scientists do their job satisfactorily and of putting values on their findings—formal heuristic PA would incorporate citizen values and knowledge throughout. Procedural mechanisms would ensure publics receive policy-relevant information in useful forms and have structured opportunities to use it. Though PA would be continuously open to public input, specific mechanisms would mandate proactive involvement of publics in identifying needs for policy change, and in design as well as scrutiny of options. These could vary from public audits of existing policies, to structured participatory processes for policy design, to provisions for public systematic monitoring of day-to-day decision making as a revelation of implicit policy (see Table 5).

Existing procedures. EIA procedures relating to screening, scoping, impact statements, public review, etc. would be rigorously applied to the design scrutiny component of PA, but the content of assessment pursuant to these procedures would be expanded to become more holistic in scope and systemic in method. Procedures governing IA of projects will also continue to be needed in order to monitor their consistency with explicit policies and to reveal the unstated, unknown, or lingering policies that initiate and encourage projects contradictory to current rhetorical goals.

Table 5. **Possible procedural mechanisms for involving publics in heuristic policy assessment**

Public needs to be informed of:	Agencies could be required to:
Explicit policies in place & outcomes relative to societal goals	Explicate & assess existing policies
Consistency of policies with each major goal	Conduct policy audits; initiate other & participatory policy reviews
Implicit policies in place & outcomes relative to major societal goals	Do class audits of project EISs to show cumulative effects ignored; adhere to freedom-of-information laws
Major emerging or latent societal problems	Sponsor & disseminate state-of-society reports comparing trends
Policies being reviewed or generated, new issues being worked on	Issue disclosure statements (as for conflict of interest)
New policies being considered	Prepare PA statements
Thinking behind policy decisions taken	Trace PA history of decisions
Policymaking process	Publicise internal procedures
Consistency of plans and day-to-day decisions with explicit policy	Report decisions/plans in useful format
Actual outcomes of policies	Conduct research, disseminate findings
Publics need to be able to:	**Agencies could be required to:**
Initiate policy audits	Respond to petitions
Conduct policy audits	Adhere to information freedom laws
Initiate policy design/redesign	Respond to petitions
Contribute to policy design (including policy on policymaking)	Facilitate public input to framing problems & options; allow public servants to consult with publics; devolve power to local governments; mediate conflict to get win–win
Scrutinise policy proposals	Enable public assessments before final decisions; fund intervenors

Not referendums. State-initiated referendums are not suggested as a possible mechanism for involving the public in policy-decision making. Referendums (including electronic voting) are promoted by the utilitarian theory of 'direct democracy' that assumes all questions can be simple and all votes (like dollars) equal, regardless of the knowledge or morality of the voter, or the influence of media advertising. Use of referendums can promote debate but also can be destructive of sustainable development, equity, and quality of life by encouraging short-sighted individualistic protection and thus Hardin's (1968) 'tragedy of the commons'. In any event, the referendum framer retains more power than given to the voter. The heuristic ideal is best served by making democracy increasingly participatory, i.e., by linking citizens, their representatives, and public servants in thoughtful, non-sloganising dialogue about crucial issues. This leads to better decisions based on wider information, ideas, and scrutiny from an increasingly knowledgeable and thoughtful citizenry, bureaucracy and legislature.

A transportation example. Consider the possibilities for an annual public and participatory transportation policy audit. Planners and policy analysts describe policy contexts now when providing background to recommendations to decision makers. The new procedural possibility would regularise the timing of this work, bring it from the background to the foreground, enrich it by analysing compatibility of explicit policy with societal goals (e.g., sustainable development) and with implicit policy as revealed by day-to-day decisions, and make it accountable to publics by facilitating public scrutiny. Selected information and insight-rich literature on key transportation issues (e.g., Goodland et al. 1993) would aid analysis and scrutiny. Such audits would replace futile exercises in simplistically surveying people to identify their wishlists without regard for tradeoffs, systems dynamics, or radical options.

Compliance. Compliance with requirements for audits could be ensured through legislation, thus offering recourse to the courts, or through watch-dog agencies such as the Parliamentary Commissioner for the Environment established under New Zealand's Environment Act 1986 (Bregha et al. 1990). Introducing audit requirements could increase democratic expectations, which in turn, could lead to further formalisation.[13]

Guidelines on the Substance and Methods of Policy Assessment
In addition to mandating public access to information and community input to design and decision making, procedures for formal heuristic PA could also take the form of guidelines on (a) substantive issues to be addressed; and (b) conceptual and technical methods for answering substantive questions. Guidelines could suggest that assessors:

---

[13] In a Canadian workshop on environmental assessment of policy, there was "general agreement. . . that the single most effective procedure to ensuring the success of policy assessment is to create mechanisms which promote ministerial accountability. . . [and that] governments need to define environmental objectives for which they can be held accountable . . . Very quickly, the definition of environmental objectives will move beyond technical matters to the question of what kind of society Canadians want" (Bregha 1990: 2-5).

- Clarify the nature of policies being assessed, their context (relation to other policies), objectives, and tradeoffs intrinsic to the policy area
- Identify the nature of systems (mechanical, natural, ecological, cultural, social, human) impacted by the policy being assessed
- Analyse systemic processes and directions of change impacted—using such heuristically rich but easily explained and applied concepts as recursion, requisite variety, negative and positive feedback, entropy, throughput, and robustness
- Consider critical sustainable development variables—e.g., local and global natural capital, biodiversity, appropriated carrying capacity, social carrying capacity, cultural survival, gender relations, personal and community health, aesthetic pleasure, spiritual development—and their relationships
- Consider the usual EIA list of impact qualities—duration, intensity, effectiveness, timing, significance, reversibility, mitigatability, positiveness or negativeness, etc.

The guidelines could offer information on a range of predictive or explanatory methods providing various mixes of heuristic power, accuracy, completeness, and efficiency. The methods could be positioned along a continuum from the simple to the highly sophisticated, with the choices of where to enter the continuum, how often to iterate, and what tradeoffs to make among assessment transparency, efficiency, accuracy, and comprehensiveness, being left to all the assessors and their case-by-case appraisals of their policymaking contexts (in terms of such categories as urgency, apparent significance of impacts, open-mindedness of superiors, integrity of assessor, workload, etc).

Menus of interpersonal and individual problem-solving techniques could be presented, including holistic methods such as scenario building, iterative methods such as Delphi, and systematising methods such as goal achievement matrices. Guidelines could indicate the kinds of people to be considered for involvement, and how, in various kinds of PA exercises. They could suggest the format and substance of reports, e.g., descriptions of methods used. Guidelines for doing PA through project IA could be developed. These would show how to analyse project decisions, plans and options presented by proponents to discover implicit policies and their consistency with explicit policies. Conversely, project IA could be enriched with guidelines calling for analysis of project-policy consistency.

Education

As discussion groups at the 1992 and 1993 IAIA annual meetings concluded, procedural steps will not be sufficient to produce good PA of any kind. Popular education on policy issues and processes is necessary for tough collective choices to be popularly supported and promoted. If participatory heuristic PA in particular is to thrive, it is necessary that all those who practice it, including citizens, be versed in heuristic methods and be thoughtful about development paradigms. Education to develop PA motivation, knowledge and skills can occur in the schools, can result from meta-assessment of current PA practice, and can be built into future practice. Continuing education programs could introduce appropriate systems theory to elected and permanent officials, planners, analysts, technical experts, and active citizens.

Good PA education and practice would reinforce each other. The more that is learnt substantively and procedurally from following procedures, the more effective we will become in working within those procedures and in designing better ones. The more we learn about systems theory and democratic principles through schooling, the better we are in practice and in learning from our experience. Most broadly put, formal heuristic PA would facilitate wide social learning about paradigms and actual human and ecological systems.

Professional associations such as the International Association for Impact Assessment (IAIA) could play an important role in promoting mutual learning about PA concepts, processes, tools, and institutions. In particular, they could—

- Work with teachers to develop heuristic PA curricula which could include real, proactive activities. Pilot exercises could test the hypothesis that children quickly grasp general systems theory concepts and usefully apply them to unravelling 'tragedy of open access' conundrums manifested in dynamics ranging from local littering to global warming.
- Show that community and regional public planning processes can usefully involve heuristic PA. Instead of publics being asked to generate wish-lists or express opinions about pre-established policy options, they could be helped to analyze the dynamics of unsustainable social and ecological systems and to evolve good solutions.
- Sponsor national or international forums in which macro-economic and social policies are not debated but rather analysed and developed. The emphasis would be on sincere effective communication (Forester 1989) about systems dynamics, with technical information from experts available on request. Existing or potential policies would not just be espoused or denounced on the basis of different single goals, but rather comprehensively and paradigmatically assessed in terms of agreed-on sets of goals. Bases for agreement and unresolvable differences would be clarified.
- Provide continuing education on systems theory and participatory heuristics methods for members and others.
- Host workshops on formal heuristic PA perspectives and procedures for government officials and NGOs.

OBJECTIONS TO FORMAL HEURISTIC POLICY ASSESSMENT, WITH REJOINDERS

Some or all of the procedural mechanisms suggested for enhancing public involvement in heuristic PA will dismay those who believe closed representative government equals efficient government which equals good government. The political and bureaucratic constraints to public pre-decision PA are well identified by Bregha (1990) and Bregha et al. (1990). But the benefits of formalisation and therefore democratisation of policy planning are shown by the social learning and design benefits already resulting from EIA (despite its many shortcomings) and, increasingly, from freedom-of-information laws. The possible formal public involvement mechanisms identified above for heuristic PA are intended to show there is room for further democratisation. Although they, like all democratisation initiatives, could make policymaking more cumbersome, they could also improve the overall efficiency of governance by producing better upstream decisions and downstream implementation in the short and long terms.

Specific objections to formal heuristic PA, with rejoinders in *italics*, could be the following:

1. Ultimately, policy is about beliefs. We cannot resolve fundamental paradigmatic issues.
   Beliefs change on the basis of new information and careful logic, both of which would be fostered by formal heuristic PA.
2. People avoid fundamental issues. They fear conflict, thinking difficulty, or the unbearability of important conclusions.
   Procedural mechanisms and guidelines can encourage assessment processes that are fun, satisfying, and redemptive.
3. Paradigmatic thought cannot be forced by procedures.
   People do not have to be forced to think about and discuss major problems; they need the opportunity, useful information, concepts and perhaps facilitation, all of which procedures can promote.
4. Paradigmatic thought cannot be directly applied to policy decisions; policies do not effect fundamental social change.
   Policy reflects assumptions; fundamental change to equitable, high quality-of-life, sustainable development can happen through policy-by-policy decision making.
5. Policy is made continuously and fluidly by many decision-centres; to formalise PA is to ossify it.
   Formalisation could enliven PA by increasing the variety of opportunities for initiating, applying and conducting PA.
6. Formal PA is too slow; we need quick processes responsive to urgent issues, not more public consultation. Governments do not have enough time in policy-making to engage in dialogue with publics about fundamental systems.
   Formal heuristic PA may slow some decisions but speed up the initiation and resolution of others; in any event, it is better to make a good decision slowly than a bad one quickly. Choosing not to dialogue can result in time lost to cleaning up problems and conflicts that dialogue could have prevented. Some of the time now spent pushing partial agendas could thus be redirected more usefully to dialogue. Procedures could improve both the timing and quality of information flows. Formal heuristic PA need not be slowed by research needs because policy-relevant data collection and proactive systematising of knowledge into policy-relevant forms should be ongoing rather than exercise-specific.

7. Social learning and better decision making are already being provided continuously, and more effectively and efficiently, through the mass media and through pluralistic politics.
    Existing policymaking processes are not leading to sustainability, equity, and quality of life. The trickle-down growth paradigm—which is at the root of the problem—needs to be explicitly revealed and confronted, policy by policy, decision by decision.

8. Formalised processes should be saved for big issues.
    PA has to be widespread and continuous because little decisions reflecting implicit policy produce cumulative effects.

9. Governments will only use public PA to defuse hot issues, in the same way they use costly inconclusive commissions of inquiry.
    Rhetoric-promoting public hearings can be replaced with processes designed to encourage mutual learning and collective problem solving.

10. Formalising PA may result in attention to procedures overwhelming consideration of substantive issues; PA could become ritualised but trivial.
    The solution is to delink the formality introduced to planning by EIA from EIA's technicality, to combine public involvement mechanisms with heuristic guidelines.

11. Formal IA should be confined to technical EIA of policies governing physical projects. Assessment of other kinds of policy and other kinds of impacts should be left to ongoing political, social and cultural processes. Assessing the social impacts of policy is fundamentally and unavoidably a political act.
    EIA is better than nothing but it will not lead to sustainable development so long as it operates within the trickle-down growth development paradigm. Paradigm change can only be achieved through systemic analysis of the links among policy realms and levels of decision. Formal heuristic PA introduces such analysis to the ongoing political process; values still guide thought, heuristics explore their implications.

12. Heuristics equals soft-headedness; outcomes should be assessed using predictively powerful and precise quantitative techniques and detailed empirical data.
    Technical scientific studies are necessary but not sufficient for good assessment. For whole pictures to be kept in sight, technical findings must be contextuated by systems concepts and knowledge of real systems. Some assessment resources need to be diverted from learning more and more about less and less to understanding the total implications of what is already known.

13. The interest of those who hold the ultimate power in society (the rich, their politicians and their media) is not served by enactment and enforcement of procedures to ensure public assessment of important issues. They have the most to lose from public understanding, and therefore possible rejection of trickle-down growth development policies. Because they control politics now, they will not implement formal heuristic PA.

Heuristic PA can start from the bottom. Increments of systemic awareness by publics and elites can change conceptions of what constitutes self-interest, which in turn can lead to stronger procedures and heuristics.

14. Participatory PA will not work because the average citizen is ignorant, mean-spirited and/or apathetic, which means special interest groups dominate participatory processes.

PA can be designed to be broadly participatory and learningful. Interest groups are part of the solution—even parochial groups facilitate learning and usually, at least, internal social responsibility. Isolated individuals are culturally bereft and manipulable.

CONCLUSION

This chapter has argued that introducing impact assessment (IA) to the policy level to create policy impact assessment (PIA), broadening PIA to comprehensive policy assessment (PA), then shaping PA into a formal heuristic form, is a necessary progression if we are to have equitable sustainable development with high quality of life. It is a progression already underway. PIA in its policy-vetting (PV) form is in place and there is observable movement to formalising environmental impact assessment (EIA) of policy under the name of strategic environmental assessment (SEA). Policy analysis and evaluation (PE) and IA theorists, and meta-policy-makers, are grappling with the issues involved in creating a practicable, effective PA. Many of them look forward to PA that formalises opportunities for meaningful public involvement and creates wide social learning of useful systemic knowledge. Some of the heuristic and procedural pieces of this ideal are already in development. Resistance can be expected, but countered.

REFERENCES

Asplund, E. and T. Hilding-Rydevik. 1993. "Strategic Environmental Assessment in Swedish Municipal Comprehensive Planning: A Research Project." Paper presented at the International Association for Impact Assessment 13th Annual Meeting, Shanghai, 11–15 June 1993.

Beer, S. 1975. *Platform for Change*. New York: John Wiley

Bennett, J.W. 1976. *The Ecological Transition: Cultural Anthropology and Human Adaptation*. Englewood Cliffs: Pergamon Press.

Berger, T.R. 1977. *Northern Frontier, Northern Homeland: Report of the Mackenzie Valley Pipeline Inquiry*. Ottawa: Supply and Services Canada.

Berkes, F. 1988. "The Intrinsic Difficulty of Predicting Impacts: Lessons from the James Bay Hydro Project." *Environmental Impact Assessment Review* 8(3): 201–220.

Boothroyd, P. 1992. "Principles for Policy Impact Assessment: Report of an Experts' Workshop Conducted by IAIA for Indonesia Ministry for Population and Environment, Washington DC 19 August 1992." In *Policy Impact Assessment for Sustainable*

Development. David Marshall, ed. Unpublished report prepared by the International Association for Impact Assessment for the Indonesian Ministry of State for Population and Environment, Appendix 6.

Boothroyd, P. 1993. Shanghai IAIA Policy Impact Assessment Workshop Notes. Unpublished.

Boothroyd, P. 1994. "Managing population–environment linkages: A general systems theory perspective." In *Population–Environment Linkages: Toward a Conceptual Framework.* Peter Boothroyd, ed. Halifax, Jakarta: Environment Management Development in Indonesia Project, 141–160.

Bregha, F. (rapporteur). 1990. "Report of the Workshop on Strengthening the Environmental Assessment of Policy." Ottawa: Canadian Environmental Research Assessment Council and National Roundtable on Environment and Economy.

Bregha, F. et al. 1990. *The Integration of Environmental Considerations into Government Policy.* Ottawa: Canadian Environmental Assessment Research Council.

Canada. 1992a. "Developing Environmentally Responsible Policies and Programs: A sourcebook on environmental assessment." Final Draft. Ottawa: Federal Environmental Assessment Review Office.

Canada. 1992b. North American Free Trade Agreement: Canadian Environmental Review. Ottawa: Federal Environmental Assessment Review Office.

Canada. 1993. The Environmental Assessment Process for Policy and Program Proposals. Ottawa: Federal Environmental Assessment Review Office.

Canadian Environmental Assessment Research Council (CEARC) and United States National Research Council (USNRC) Board on Basic Biology, Committee on Applications of Ecological Theory to Environmental Problems. 1986. Proceedings of the Workshop on Cumulative Environmental Effects: A Binational Perspective. Ottawa: Supply and Services.

Carson, R. 1962. *Silent Spring.* Boston: Houghton Mifflin.

Centre for Environmental Management and Planning (CEMP). 1990. *Environmental Policy and Management: An International Forum.* Alfriston: The British Council.

Churchman, C.W. 1979. *The Systems Theory and Its Enemies.* New York: Basic.

Daly, H. and R. Goodland. 1994. "An Ecological–economic assessment of deregulation of international commerce under GATT. *Ecological Economics* 9(1): 73–92.

Dethlefsen, V., T. Jackson, and P. Taylor. 1993. "The precautionary principle: Towards anticipatory environmental management." In *Pollution Prevention: A Practical Guide for State and Local Government.* D.T. Wigglesworth, ed. Boca Raton: Lewis, 42–54.

Economic Commission for Europe (ECE). 1992. *Application of Environmental Impact Assessment Principles to Policies, Plans and Programmes.* New York: United Nations

Finsterbusch, K. 1989. "How should policy decisions be made?" *Impact Assessment Bulletin* 7(4): 17–24.

Forester, J. 1989. *Planning in the Face of Power.* Berkeley: University of California Press.

Geddes, P. 1915. *Cities in Evolution.* London: Williams and Norgate (1949).

Georgescu-Rogen, N. 1971. "The entropy law and the economic problem." In *Toward a Steady State Economy* (1973). H.E. Daly, ed. San Francisco: Freeman, 37–49.

Goodland, R., P. Guitink, and M. Phillips. 1993. "Environmental Priorities in Transport Policy." Informal discussion draft provided at Policy Impact Assessment Workshop, International Association for Impact Assessment 13th Anual Meeting, Shanghai, 11–15 June 1993.

Hardin, G. 1968. "The Tragedy of the Commons." *Science* 162: 1243–1248

Heylighen, F., E. Rosseel, and F. Demeyere, eds. 1990. *Self Steering and Cognition in Complex Systems: Toward a New Cybernetics.* New York: Gordon and Breach.

Hill, M. 1968. "A goals-achievement matrix for evaluating alternative plans." *Journal of the American Institute of Planners* 34(1): 19–28.

Holling, C.S., ed. 1978. *Adaptive Environmental Assessment and Management.* New York: Wiley.

Jacobs, J. 1961. *The Death and Life of Great American Cities.* New York: Random House.

Laszlo, E. 1974. *A Strategy for the Future: The Systems Approach to World Order.* New York: Braziller.

Lee, N. and F. Walsh. 1992. "Strategic environmental assessment: An overview." *Project Appraisal* 7(3): 126–136.

Lichfield, N., P. Kettle, and M. Whitbread. 1975. *Evaluation in the Planning Process.* Oxford: Pergamon.

Marshall, D., ed. 1992. Policy Impact Assessment for Sustainable Development. Unpublished report prepared by the International Association for Impact Assessment for the Indonesian Ministry of State for Population and Environment.

Morse, B. and T. Berger. 1992. *Sardar Sarovar: Report of the Independent Review.* Ottawa: Resource Futures International.

Odum, H.T. 1971. *Environment, Power, and Society.* New York: Wiley.

O'Riordan, T. and W.R.D. Sewell. 1981. "From project appraisal to policy review." In *Project Appraisal and Policy Review.* T. O'Riordan and W.R.D. Sewell, eds. Chichester: Wiley, 1–28.

Partidario, Maria do Rosario. 1993. "Proposal for an EA Procedure in Comprehensive Land-Use Planning: A Contribution for Discussion." Paper presented at the International Association for Impact Assessment 13th Annual Meeting, Shanghai, 11–15 June 1993.

Prest, A.R. and R. Turvey. 1965. "Cost–benefit analysis: A survey." *Economic Journal.* 75(300): 685–705.

Rees, W.E. 1992. "Ecological footprints and the appropriated carrying capacity: What urban economics leaves out." *Environment and Urbanization* 4(2): 121–130.

Rees, W.E. 1995. Cumulative Environmental Assessment and Global Change. Policy Issues and Planning Responses Working Paper No 8, Vancouver: University of British Columbia, Centre for Human Settlements.

Sadler, Barry. 1994. "Environmental assessment and development policymaking." In *Environmental Assessment and Development.* R. Goodland and V. Edmundson, eds. Washington DC: World Bank, 3–19.

Sonntag, N.C. et al. 1987. *Cumulative Effects Assessment: A Context for Further Research and Development.* Ottawa: Canadian Environmental Assessment Research Council.

Therivel, R. 1993. "Systems of strategic environmental assessment." *Environmental Impact Assessment Review* 13(3): 145–168.

United Kingdom, Department of the Environment. 1991. *Policy Appraisal and the Environment.* London: HMSO.

United States. 1973. "Principles and standards for planning water and related land resources." *Federal Register* 38(174): 24777–24869.

United States, The Interorganizational Committee on Guidelines and Principles for Social Impact Assessment. 1994. *Guidelines and Principles for Social Impact Assessment.* Washington DC: Dept of Commerce, National Oceanic and Atmospheric Administration, National Marine Fisheries Service.

van Eck, M. 1993. "Environmental Impact Assessment for Policy Plans and Programmes in the Netherlands." Paper presented at the International Association for Impact Assessment 13th Annual Meeting, Shanghai, 11–15 June 1993.

Wackernagel, M. et al. 1993. *How Big Is Our Ecological Footprint?: A Handbook for Estimating a Community's Appropriated Carrying Capacity*. Vancouver: Task Force on Planning Healthy and Sustainable Communities, University of British Columbia.

Wolf, C.P. 1974. *Social Impact Assessment: The State of the Art*. Milwaukee: Environmental Design Research Association.

Wolf, C.P. 1990. "A systems approach to impact assessment." *Social Impact Assessment* 14(1): 3–16.

Wood, C. and M. Dejeddour. 1992. "Strategic environmental assessment: EA of policies plans and programmes." *Impact Assessment Bulletin* 10(1): 3–23.

World Commission on Economy and Development (WCED). 1987. *Our Common Future*. Oxford: Oxford University Press.

Part II:

TOOLS OF IMPACT ASSESSMENT

Chapter 5

Economic and Fiscal Impact Assessment[1]

F. LARRY LEISTRITZ
North Dakota State University, USA

When major resource or industrial development projects are proposed or when policy changes that will substantially affect patterns of economic activity and resource use are considered, decision makers are increasingly requesting analyses of the socioeconomic impacts that may result. The *socioeconomic impacts* of development projects and programs have been categorised in a number of ways. One classification of such impacts identifies: (1) *economic* impacts (including changes in local employment, business activity, earnings, and income); (2) *demographic* impacts (changes in the size, distribution, and composition of the population); (3) *public service* impacts (changes in the demand for, and availability of, public services and facilities); (4) *fiscal* impacts (changes in revenues and costs among local government jurisdictions); and (5) *social* impacts (changes in the patterns of interaction, the formal and informal relationships resulting from such interactions, and the perceptions of such relationships among various groups in a social setting) (Leistritz and Murdock 1981; Leistritz and Ekstrom 1986; Murdock et al. 1986). This paper provides a brief overview of the conceptual bases, methodological alternatives, and assessment techniques that are commonly utilised in assessing two of these categories of impacts (i.e., economic and fiscal impacts). In addition, the importance of economic and fiscal impact assessment to policy making and impact management is discussed, and the likely future of the field is described.

ECONOMIC IMPACT ASSESSMENT

The purpose of an economic impact assessment is to estimate changes in employment, income, and levels of business activity (typically measured by gross receipts or value added) that may result from a proposed project or program. As with the assessment of other categories of impacts, the general approach involves projecting the levels of economic activity that would be expected to prevail in the study area with and, alternatively, without the project. The difference between the two projections measures the impact of the project.

[1]*Environmental and Social Impact Assessment* - Edited by F. Vanclay and D.A. Bronstein. Copyright © 1995 by the International Association of Impact Assessment. Published in 1995 by John Wiley & Sons Ltd. A version of this chapter will appear in *Impact Assessment*, **12**(3) the quarterly journal of IAIA.

Conceptual Bases

Export base theory (also termed economic base theory) provides the conceptual foundation for all operational economic impact assessment models. A fundamental concept of export base theory is that an area's economy can be divided into two general types of economic units. The basic sector is defined as those firms which sell goods and services primarily to markets outside the area. The revenue received by basic sector firms for their exports of goods and services is termed basic income. The remainder of the area's economy consists of those firms which supply goods and services primarily to customers within the area. These firms are referred to as the nonbasic sector or sometimes as residentiary or local trade and service activities.

A second key concept in export base theory is that the level of nonbasic activity in an area is uniquely determined by the level of basic activity, and a given change in the level of basic activity will bring about a predictable change in the level of nonbasic activity. This relationship is known as the multiplier effect. Thus, export base theory emphasises external demand for the products of the basic sector as the principal force determining change in an area's level of economic activity.

The basis for the multiplier effect is the interdependence (or linkages) of the basic and nonbasic sectors of an area's economy. As the basic sector expands, it requires more inputs (e.g., labour and supplies). Some of these inputs are purchased from local firms and households. As the firms in the nonbasic sector expand their sales to the basic sector, they too must purchase more inputs, and so on. Increased wages and salaries paid to labour and management by the basic sector, together with similar payments by the nonbasic sector, lead to increases in the incomes of area households. Some of this additional income is spent locally for goods and services, some is saved, and some leaves the area as payments for imported goods and services (or as additional tax payments to government). To the extent that additional income is spent locally for goods and services, the output of local firms is increased and additional cycles of input purchases and expenditures result. This cycle of spending and respending within the local economy is the basis for the multiplier effect (Leistritz and Murdock 1981).

The magnitude of the multiplier effect is determined by the proportion of a given dollar of additional income that is spent locally. High multiplier values are associated with high levels of local spending, which in turn imply a diversified, relatively self-sufficient economy. Larger regions tend to have higher multiplier values.

Assessment Methods

When estimating the magnitude of secondary economic effects (resulting from the multiplier process) for a specific project in a given area, most analysts employ either an *export base model* (employment or income multipliers) or an *input–output (I–O) model*. In recent years, input–output models have been applied with increased frequency in impact assessment. Some reasons for the increasing use of I–O models are: (1) this technique provides more detailed impact estimates (e.g., business volume and employment by sector) than other approaches and can better reflect differences in expenditure patterns among projects; and (2) data bases and data management systems are now available that enable development of I–O models tailored to local conditions but based largely or totally on secondary data sources. Commonly used I–O models of this type include REMI (Treyz et al. 1977, 1992), RIMS (US Department of Commerce 1992), and IMPLAN (Alward et al. 1989). Recent evaluations of these and similar I–O models are provided by Crihfield and Campbell (1991), Brucker et al. (1987), and Brucker et al. (1992). For examples of studies that apply input–output models in the analysis of various projects and programs, see Mortensen et al. (1990), Mulkey and Clouser (1991), Bangsund and Leistritz (1992), Siegel and Leuthold (1993), and Parsons and Johnson (1994).

Practical Problems

Whatever modeling system is used, the analyst will need specific information about the proposed project in order to prepare an assessment of its economic impacts. The magnitude and distribution of impacts from any project are dependent on many factors, but among the most important of these are: (1) work force requirements, including temporary vs. permanent workers, timing of employment patterns (e.g., duration of construction periods), earnings, and skill requirements; (2) capital investment; (3) local input purchase patterns; (4) output; and (5) resource requirements (Murdock and Leistritz 1979; Leistritz et al. 1982). Obtaining reliable information on these topics can be a major task and may require not only extensive consultation with project officials but also examination of experience in developing analogous projects in similar areas. On the other hand, much of the information is useful in assessing other impact dimensions as well.

A factor that may add complexity to the impact assessment process is the potential interaction between the proposed project and other basic industries. For example, a surface mining project may displace existing agricultural production, either temporarily or permanently, or the access roads associated with a resource development project may enhance tourism. When such effects can be reasonably anticipated, they should be incorporated in the assessment. However, developing quantitative estimates is often challenging. Experiences in developing analogous projects often may be the best guide.

History

Interest in the economic base concept goes back more than 60 years (Haig 1926; Hoyt 1933), while the input–output model concept can be traced to the works of Leontief (1936, 1941). Early work on the export base concept emphasised alterative methods for estimating employment and/or income multipliers (Gillies and Grigsby 1956; Levan 1956; Tiebout 1962; Ullman and Dacey 1960), while more recent work has included estimation of disaggregated multipliers (i.e., separate multipliers for each basic industry) (Weiss and Gooding 1968; Braschler 1972; Bender 1975). The initial development of I–O models was at the national level (Leontief 1941), and subsequent work was directed at estimating I–O models for states and regions (Isard 1951; Miernyk 1965; Roesler et al. 1968). State and regional I–O models have been developed both from primary data (from surveys of firms and households in the study area) and from secondary data (by adjusting national coefficients), as well as by methods using a combination of primary and secondary data (e.g., Henry et al. 1980).

Another trend in the development of economic impact assessment methods and models has been the development of integrated assessment models that incorporate multiple impact dimensions (e.g., economic and demographic). The initial models of this type were developed during the 1960s, primarily for use as regional planning tools (Hamilton et al. 1969). Development of such integrated assessment models was rapid during the 1970s and early 1980s, largely in response to needs related to assessing the impacts of large-scale development projects (Leistritz et al. 1986). These models often incorporated economic, demographic, public service, and fiscal impact dimensions and allowed for rapid analysis of alternative scenarios. More recently, demands for assessments of large-scale projects have been less frequent (at least in North America), and so the degree of interest in large integrated assessment systems appears to have lessened.

FISCAL IMPACT ASSESSMENT

The purpose of fiscal impact assessment is to project the changes in costs and revenues of governmental units that are likely to occur as a result of a development project. The government units of primary interest are those local jurisdictions that may experience substantial changes in population and/or service demands as a result of the project. The fiscal implications of a new project are determined by the interactions of a number of factors, including project characteristics (e.g., the magnitude of investment, the size and scheduling of the work force) and site area characteristics (e.g., state and local tax structure, the capacity of existing service delivery systems) and by the nature of the economic and demographic effects resulting from the project. Further, because the fiscal impacts of a project are of considerable interest to local officials and their constituents and to developers, the fiscal impact assessment should be designed to produce information in a form that is most useful to policymakers (Leistritz and Murdock 1988).

Fiscal impact analysis is sometimes confused with other evaluative techniques, particularly cost-benefit and cost effectiveness analysis. Fiscal impact analysis, also sometimes termed cost-revenue analysis (Burchell and Listokin 1978), focuses exclusively on the public sector costs and revenues associated with a project or program. The key feature of fiscal impact analysis is that it focuses exclusively on the revenues and costs incurred by governmental units. Another important feature is that the distribution of revenues and costs through time is frequently included in the analysis. Cost effectiveness analysis focuses on the cost of providing selected services, or more broadly, of achieving selected objectives. Cost-benefit analysis is the broadest of the three techniques and involves comparison of both tangible and intangible costs and benefits of a project. The costs and benefits considered include not only the expenditures and revenues of public sector entities but also benefits and costs experienced by private businesses and individuals, including both tangible and intangible benefits and costs (Randall 1987).

Issues Related to Fiscal Impact Assessment

Some issues that frequently concern policymakers relate to the distribution of project-related costs and revenues, both over time and among jurisdictions, and the risks to which the local government may be exposed because of uncertainty regarding the future of the project and/or the nature of its impacts. The problem of cost and revenue timing, frequently referred to as 'the front-end financing problem', arises because during the early years of a project, local public sector costs frequently increase more rapidly than project-induced revenues. While project-related revenues may exceed project-related costs over the life of the project, local jurisdictions may face short-run cash flow problems. These problems can be exacerbated if local governments are unable to obtain funds to offset revenue shortfalls through borrowing. Uncertainty associated with a proposed project also may discourage local officials from incurring financial obligations, even though borrowing might seem a logical approach to financing new infrastructure. Questions concerning: (1) whether a project will actually be developed; (2) whether it may be abandoned prematurely; and (3) what the actual magnitude and distribution of project-related growth will be, may make local officials reluctant to make commitments.

The interjurisdictional distribution problems may be as severe as those associated with cost and revenue timing. The project facilities that generate most of the new public sector revenues may be located in one local jurisdiction, while most of the project-related population may live in a different school district, local jurisdiction, or even a different state. Fiscal impact assessments should be designed to identify these intertemporal and interjurisdictional distribution problems in advance so that decision makers can have the opportunity to devise strategies for coping with them. (For more detailed discussions of these coping strategies, see Leistritz et al. 1983; Leistritz and Murdock 1988.)

Fiscal Impact Assessment Techniques

Specific techniques employed to estimate the fiscal impacts of new projects or programs differ somewhat in the details of the estimation procedure, and assessments differ substantially in the scope of costs and revenues addressed. In general, the revenues of local governments can be broadly classified as own-source revenues (taxes and charges assessed and collected directly by the local jurisdictions) and intergovernmental transfers (funds received from state and federal levels). Own-source revenues can be further classified according to their primary determinants into those based on property valuation, those based on income or sales, those based on the level of production of some industry, and those based largely on changes in population. The most appropriate techniques for estimating revenues from these sources will differ depending on the revenue source (Burchell and Listokin 1978).

Intergovernmental revenues are often more difficult to project than own-source funds. These difficulties arise because the allocation formulas are frequently complicated, eligibility for certain forms of assistance changes as local wealth or other indicators change, and overall community effects often must be considered. They are also subject to political manipulation and intervention. In the United States, for example, state school aid often is inversely related to local wealth, and so a new project that significantly affects the local tax base could affect the level of state assistance not only for the new students associated with the specific project, but also for all other students in the locality. In such situations, the analyst must take account of this overall net change in order to obtain a realistic estimate of the effect of the project on the community.

A number of approaches can be employed in estimating the community service costs associated with growth. Methods for estimating service costs are, of course, closely related to those used in projecting service requirements. The major difference is the nature of estimates developed. Whereas the objective of public service analysis is to evaluate changes in requirements for service facilities and personnel, fiscal impact analysis involves estimating the capital and operating costs of these services. Cost estimation methods can be categorised into average cost and marginal cost approaches by the nature of the cost estimates they provide. The average cost approaches include the per capita expenditures method, the service standard method, and the use of cost functions derived from cross-section regression analyses. Marginal cost approaches include the case study approach, comparable city analysis, and economic-engineering methods. (For a detailed discussion of fiscal impact assessment techniques, see Burchell et al. 1985; Leistritz and Murdock 1981; Burchell and Listokin 1978.)

History

Fiscal impact analyses have been part of the planning profession since the 1930s (Mace 1961). Planners first employed this type of analysis in connection with public housing projects, seeking to justify replacement of deteriorated housing due to its negative local fiscal effects. In the 1940s, fiscal impact analysis was used in the urban renewal process to demonstrate the advantages of the new land use over the old (Burchell and Listokin 1978). During the 1950s, it was employed during the suburbanisation movement to gauge the impact of single family homes on local school districts. In the 1960s, supported by local planning assistance funding (primarily provided by the HUD 701 program), it was used to evaluate the fiscal effects of the master plan (Burchell and Listokin 1978).

During the 1960s, fiscal impact techniques also were applied to evaluate the effects of industrialisation on local governments (Hirsch 1964; Kee 1968). In the 1970s, fiscal impact analysis emerged as an almost universal accompaniment to large-scale development proposals, either volunteered by the developer or required by local governments or state regulatory bodies.

IMPORTANCE TO POLICY MAKING AND MANAGEMENT

Economic and fiscal impact assessments are increasingly demanded by policymakers and resource managers because they address issues that are key to a wide variety of decisions. For example, in determining whether to designate certain public lands as wilderness areas, land managers may feel a need to consider the economic and fiscal impacts of alternative land uses (e.g., wilderness vs. ranching or mining). When large-scale mining and resource development projects have been proposed, the local economic and fiscal impacts often have been one of the principal topics of debate (Leistritz and Murdock 1988), and special taxes and/or impact payments have sometimes been imposed to mitigate potential fiscal problems for local governments. On the other hand, the economic impacts of proposed resource and industrial development projects are often seen as among the most positive (Murdock et al. 1986), and project proponents frequently volunteer estimates of secondary employment and income effects as part of their applications for required permits.

As state and local governments become more heavily involved in economic development efforts, economic and fiscal impact analysis tools can be useful in helping to establish priorities for incentive programs. While a number of authorities are now using selected measures of direct economic impact (generally the number of jobs created) as criteria in awarding financial support (Leistritz and Hamm 1994), the total economic impact (including secondary effects) would appear to be a more meaningful criterion. Similarly, local governments have long been involved in providing tax abatements and other incentives to new firms. In an era of budget stringencies, local units may feel an increasing need to examine secondary as well as direct benefits and costs in determining the use of scarce resources for incentive programs. Economic and fiscal impact analysis offers tools that can be useful in guiding such decisions (Lansford and Jones 1991; Coon et al. 1993).

THE LIKELY FUTURE OF THE FIELD

Economic and fiscal impact analyses have developed as areas of applied research largely in response to demands of clientele. Over the past several decades, decision makers have increasingly been demanding information regarding the economic base of their community or region, the likely effect of a specific project or program on the area's economy, and the effect of specific projects or programs on the costs and revenues of local governments. Researchers and analysts have responded using the principles and methods of regional economics and public finance. Over time, the ability of economic and fiscal impact assessment practitioners to provide timely and reliable information in response to such requests has improved, but the field continues to be one that is largely driven by clientele demands.

Looking to the future, the development of economic and fiscal impact assessment likely will continue to be heavily influenced by the demands of decision makers. Recent emphases on assessing potential impacts at the level of policies and programs, rather than specific projects, will probably lead to the development of analytical tools better suited to such applications (whereas most previous applications have been at the project level). In particular, there likely will be increased demand for economic and fiscal impact analysis at the national and state levels. The emphasis on moving impact assessment to higher levels in the decision-making process also will likely lead to a more proactive approach to assessments.

In keeping with the tradition of development and evaluation in response to user needs, economic and fiscal impact assessment will likely continue to be applied to problems and issues of priority concern to decision makers, which may be somewhat different from those which have been the focus of such assessments in the recent past. For example, with economic development becoming a high priority issue for many states and regions, economic and fiscal impact analysis likely will see increased applications in economic development planning and analysis. For economic impact analysis, this represents something of a full circle, as many of the applications of economic impact tools in the 1960s and early 1970s were in conjunction with regional development and planning efforts (Hamilton et al. 1969; Bohm and Lord 1972; Battelle Columbus Laboratories 1973). On the other hand, fiscal impact assessment has generally been undertaken largely at the level of specific projects with local governments as the major focus of concern. In the future, we may anticipate increased demands for development and application of methods suitable for analysing the fiscal impacts of projects, policies, and programs of state and national governments.

To summarise: economic and fiscal impact assessment have developed as pragmatic approaches attempting to bring appropriate concepts and tools from regional economics and public finance to bear on problems of concern to policymakers. Their development as areas of applied research has been enhanced by developments in their parent disciplines; in turn, pressures to improve economic and fiscal impact analyses have stimulated advances in those disciplines. Further development of this field can be expected, and the nature of that development will be heavily influenced by the expressed needs of clientele groups, as these needs evolve over time.

REFERENCES

Alward, G.S. et al. 1989. Micro IMPLAN Software Manual. St. Paul: University of Minnesota.

Bangsund, D.A. and F.L. Leistritz. 1992. Contribution of Public Land Grazing to the North Dakota Economy. Agricultural Economics Report No. 283. Fargo ND: NDSU, Dept of Agricultural Economics.

Battelle Columbus Laboratories. 1973. Final Report of the Arizona Environmental Economic Trade-off Model. Phoenix: Arizona Office of Planning and Development.

Bender, L.D. 1975. Predicting Employment in Four Regions of the Western United States. USDA Technical Bulletin No. 1529. Washington DC: USDA, ERS.

Bohm, R.A. and J.H. Lord. 1972. *Regional Economic Modeling—The TVA Experience*. Knoxville TN: Tennessee Valley Authority.

Braschler, C. 1972. "A comparison of least squares estimates of regional employment multipliers with other methods." *Journal of Regional Science* 12(3):457–468.

Brucker, S.M., S.E. Hastings, and W.R. Latham. 1987. "Regional input–output analysis: A comparison of five ready-made model systems." *Review of Regional Studies* 17(2): 1–16.

Brucker, S.M., S.E. Hastings, and W.R. Latham. 1990. "The variation of estimated impacts from five regional input–output models." *International Regional Science Review* 13(1–2): 119–139.

Burchell, R.W. and D. Listokin. 1978. *The Fiscal Impact Handbook.* Piscataway NJ: Rutgers University, Center for Urban Policy Research.

Burchell, R.W., D. Listokin, and W.R. Dolphin. 1985. *The New Practitioner's Guide to Fiscal Impact Analysis.* Piscataway NJ: Rutgers University, CUPR.

Coon, R.C., R.R. Hamm, and F.L. Leistritz. 1993. North Dakota Microcomputer Economic–Demographic Assessment Model (MEDAM): User's Guide and Technical Description. Agricultural Economics Software Series No. 8. Fargo ND: NDSU, Department of Agricultural Economics.

Crihfield, J.B. and H.S. Campbell, Jr. 1991. "Evaluation of alternative regional planning models." *Growth and Change* 22(2): 1–16.

Gillies, L. and W. Grigsby. 1956. "Classification errors in base-ratio analysis." *Journal of the American Institute of Planners* 22: 17–23.

Haig, R. 1926. "Toward an understanding of the metropolis: Some speculations regarding the economic basis of urban concentration." *Quarterly Journal of Economics* 40: 179–208.

Hamilton, H.R. et al. 1969. *Systems Simulation for Regional Analysis: An Application to River Basin Planning.* Cambridge MA: MIT Press.

Henry, M.S. et al. 1980. "A semi-survey approach to building regional input–output models: An application to western North Dakota." *North Central Journal of Agricultural Economics* 2(1): 17–24.

Hirsch, W.Z. 1964. "Fiscal impact of industrialization on local schools." *Review of Economics and Statistics* 46: 191–199.

Hoyt, H. 1933. *One Hundred Years of Land Values in Chicago.* Chicago: University of Chicago Press.

Isard, W. 1951. "Interregional and regional input–output analysis: A model of a space economy." *Review of Economics and Statistics* 33: 318–328.

Kee, W.S. 1968. "Industrial development and its impact on local finance." *Quarterly Review of Economics and Business* 8: 19–24.

Lansford, N.H. and L.L. Jones. 1991. "Tax abatement as a development incentive: Economic impact on rural communities." *Impact Assessment Bulletin* 9(3): 31–42.

Leistritz, F.L. and B.L. Ekstrom. 1986. *Socioeconomic Impact Assessment and Management: An Annotated Bibliography.* New York: Garland Publishing.

Leistritz, F.L. and R.R. Hamm. 1994. *Rural Economic Development 1975–1993: An Annotated Bibliography.* Greenwich CT: Greenwood Publishing Group.

Leistritz, F.L. and S.H. Murdock. 1981. *The Socioeconomic Impact of Resource Development: Methods for Assessment.* Boulder CO: Westview Press.

Leistritz, F.L. and S.H. Murdock. 1988. "Financing infrastructure in rapid growth communities: The North Dakota experience," pp. 141–154 in *Local Infrastructure Investment in Rural America.* T.G. Johnson, B.J. Denton, and E. Segarra, eds. Boulder, CO: Westview Press.

Leistritz, F.L., S.H. Murdock, and A.G. Leholm. 1982. "Local economic changes associated with rapid growth," pp. 25–61 in *Coping With Rapid Growth in Rural Communities*. B. Weber and R. Howell, eds. Boulder CO: Westview Press.

Leistritz, F.L. et al. 1983. "Socioeconomic impact management: Program design and Implementation considerations." *Minerals and the Environment* 4: 141–150.

Leistritz, F.L., R.A. Chase, and S.H. Murdock. 1986. "Socioeconomic impact models: A review of analytical methods and policy implications," pp. 148–166 in *Integrated Analysis of Regional Systems*. P. Batey and M. Madden, eds. London: Pion.

Leontief, W. 1936. "Quantitative input and output relations in the economic system of the United States." *Review of Economics and Statistics* 18: 105–125.

Leontief, W. 1941. *The Structure of the United States Economy 1919–1939*. Cambridge MA: Harvard University Press.

Levan, C.L. 1956. "Measuring the Economic Base." Papers of the Regional Science Association 2: 250–258.

Mace, R.A. 1961. *Municipal Cost-Revenue Research in the United States*. Chapel Hill NC: University of North Carolina.

Miernyk, W.H. 1965. *Elements of Input-Output Economics*. New York: Random House.

Mortensen, T.L. et al. 1990. "Socioeconomic impact of the conservation reserve program in North Dakota." *Society and Natural Resources* 3: 53–61.

Mulkey, D. and R.L. Clouser. 1991. The Economic Impact of the Dairy Industry in Okeechobee County, Florida. SP #91-40. Gainesville FL: University of Florida, Institute of Food and Agricultural Sciences.

Murdock, S.H. and F.L. Leistritz. 1979. *Energy Development in the Western United States*. New York: Praeger.

Murdock, S.H., F.L. Leistritz, and R.R. Hamm. 1986. "The state of socioeconomic impact analysis in the United States: Limitations and opportunities for alternative futures." *Journal of Environmental Management* 23: 99–117.

Parsons, R.L. and T.G. Johnson. 1994. "The potential economic impacts of bovine somatrophics use on regional and state economics in Virginia." *Review of Agricultural Economics* 16(2): 175–186.

Randall, A. 1987. *Resource Economics*. New York: John Wiley and Sons.

Roesler, T.W., F.C. Lamphear, and M.D. Beveridge. 1968. *The Economic Impact of Irrigated Agriculture on the Economy of Nebraska*. Lincoln: University of Nebraska, Bureau of Business Research.

Siegel, P.B. and F.O. Leuthold. 1993. "Economic and fiscal impacts of a retirement/recreation community: A study of Tellico Village, Tennessee." *Journal of Agricultural and Applied Economics* 25(2): 134–147.

Tiebout, C.M. 1962. The Community Economic Base Study. Supplementary Paper No. 16. New York: Committee for Economic Development.

Treyz, G. et al. 1977. *The Massachusetts Economic Policy Analysis (MEPA) Model*. Amherst: University of Massachusetts.

Treyz, G.I., D.S. Rickman, and G. Shao. 1992. "The REMI economic-demographic forecasting and simulation model." *International Regional Science Review* 14(3): 221–253.

Ullman, E.L. and M.F. Dacey. 1960. "The Minimum Requirements Approach to the Urban Economic Base." Papers of the Regional Science Association 6: 175–194.

US Department of Commerce, Bureau of Economic Analysis. 1992. *Regional Multipliers: A User Handbook for the Regional Input-Output Modeling System* (RIMS II), 2nd ed. Washington DC: US Government Printing Office.

Weiss, S.J. and E.C. Gooding. 1968. "Estimation of differential employment multipliers in a small regional economy." *Land Economics* 44(2): 235–244.

Chapter 6

Demographic Impact Assessment[1]

HENK A. BECKER
Utrecht University, The Netherlands

INTRODUCTION

Impact assessment (IA) can be defined as the process of identifying the future consequences of a current or proposed action. Thus, demographic impact assessment (DIA) is the process of identifying the future consequences of a current or proposed action that may have an impact on demographic processes, or on issues related to demographic processes. Demography as a scientific discipline was once restricted to the analysis of fertility, mortality, family formation, and related events in the life course of individuals. Recently, however, demographers have expanded their field to also look at the level of education, employment, income, housing, retirement, and many aspects of the populations involved. The broad definition of demography implies that the boundaries between demography and related social sciences (e.g., sociology, political science) have become less clear.

Two types of DIA can be conceived:

1. Processes in which demographic characteristics are dependent variables. The following examples will elaborate this. The diffusion of contraceptive devices and practices may have an impact on the birth rate of the 'target population' of the intervention; the building of a large reservoir may have an impact on the death rate of the original population of the area to be flooded because this population will have to be resettled elsewhere. Resettlement means uprooting the very young and the feeble elderly. The raising of the level of education of women in a specific country may have an impact on the birth rate as well as the occupational opportunities and level of income of the women concerned.

2. Processes in which demographic characteristics are independent variables. In western countries, there is a relatively high consumption pattern amongst people in the middle age categories, especially 45 to 54 (Becker 1994a). Because of dramatic increases in the number of births following catastrophic events such as the world wars (the baby booms), an increase in the proportion of the population will enter this age category in a number of Western countries in the near future. The extra numbers in the baby boom cohorts will

[1]*Environmental and Social Impact Assessment* - Edited by F. Vanclay and D.A. Bronstein. Copyright © 1995 by the International Association of Impact Assessment. Published in 1995 by John Wiley & Sons Ltd. A version of this chapter will appear in *Impact Assessment,*the quarterly journal of IAIA.

add considerably to the volume of consumption in many western countries. As a consequence, the economy in these countries will be stimulated, and there will be an increased burden on the environment.

At first consideration, the second type of DIA does not fit into the concept of IA as given, but it does deal with analysing consequences of current or proposed actions. For example, this is evident when national policy is considered: fiscal policy could be used to *reduce* the spending power of the dual income baby boom families in the years to come; social policy could be used to *increase* the spending power of single income (and single parent) families in the future. Another example of Type 2 DIA is the $I = P \cdot A \cdot T$ model which analyses the impact of population on (and in conjunction with) the level of affluence and the level of technology (see chapter by Goodland and Daly).

Both types of DIA deal with developments in contemporary society that are crucial to human welfare and wellbeing, even to the survival of humanity! An understanding of the processes in which demographic characteristics are dependent or outcome variables may help to curb the population explosion. When demographic characteristics are the independent variables, they often constitute boundary or limiting conditions for future interventions. For example, the volume of 'social capital' in a country (i.e., the number of trained and experienced members of the workforce) is a boundary condition to planned economic growth. DIA can be applied to developments at the international and national policy level. If specific interventions are considered, the relevance of DIA is also obvious, such as building a large reservoir, the reorganisation of the social security system in a country; or planned strategies of business corporations that have consequences for the life courses of consumers.

In many countries IA is required by law. Environmental impact assessment (EIA) is an example of a legally prescribed analysis of the consequences of proposed actions. In a number of countries, legal requirements for IA are being expanded or have already been expanded to include, amongst other things, DIA. In The Netherlands, each government department is required to research the efficiency and effectiveness of its major activities. The General Court of Audits is responsible for enforcing this law. Each year, the Court of Audits succeeds in getting more government departments to apply IA to their activities and to present to Parliament documents showing the outcomes of their analyses. It is presumed that the institutionalisation of IA by government departments will result in an increase in efficiency and effectiveness of government activities. This kind of IA always includes DIA. It is likely that many countries will follow the example of The Netherlands.

Until recently, DIA was part of social impact assessment (SIA), and in many cases DIA was the only form of SIA undertaken in an EIA (see chapter by Burdge and Vanclay). Nowadays SIA, EIA, and technology assessment (TA) (see chapter by Porter) have expanded to such an extent that the taxonomy of IA is ripe for a major revision, as demonstrated by this book.

A BRIEF HISTORY OF DIA

Malthus warned that population growth was going to reach the limits set by food production. He also warned that, ultimately, a shortage of food would lead to the death of the number of people who had been born in excess of the food available. Thus Malthus used the first and the second type of DIA as identified above. We know now that Malthus did not foresee the improvements in agricultural technology that made an increase in food production possible. Malthus forgot TA.

The Marquis de Condorcet was the first to conduct a DIA as part of a SIA. In 1775–76, he collected birth and death records and related socioeconomic information for a region, part of the Somme valley north of Paris, that was the site of a proposed canal. He put this information into a predictive model of the effects of canal seepage on mortality rates and made a number of recommendations. Condorcet's inquiry differed fundamentally from the informal forecasts that advisers had historically provided the court. The selection of independent and dependent variables, the attempt to isolate the effects of canal seepage from other (largely socioeconomic) causes of mortality, and the use of quantification marked a new application of scientific method to problems of public administration (Pendergast 1989).

Following the pioneering approach of Condorcet, DIA was applied only on an incidental basis. Regular DIA began with the passage of the National Environmental Policy Act (NEPA) in the United States in 1969. Generally, EIAs do have a demographic submodel, although at first these submodels were very restricted in their coverage.

Jay Forrester introduced a substantial improvement of DIA by using demographic variables as a major component of computer simulations. In *World Dynamics* (1971), he explored the consequences of current and future actions related to population growth, food production and pollution. In *The Limits to Growth* (Meadows et al. 1972), Forrester's model was further developed, and their computer simulation of the world system can be considered as an analysis of consequences elaborating the model of Malthus. Like Malthus, Meadows et al. underrated the impact of technological innovation on the dependent variables in their model.

Forrester's pioneering activities have been followed by a large number of world models in which the demographic submodel gradually acquired a dominating position, illustrated by the Brundtland Report (WCED 1987) that tried to demonstrate that 'sustainable development' was feasible only if population growth was curbed, the devastation of the environment was stopped, and 'sustainable technology' was introduced.

Since the early 1960s, demographic trends in Western Europe and the United States have been analysed on a regular basis (OECD 1970). Gradually, demographic trends in the whole world were assessed periodically. These statistics provided a basis for DIA on a world scale, and at the national, regional and local scale. However, these statistics have a major drawback: they tend to provide transversal data only. Transversal data only provide snapshots of isolated situations in a process. In most DIAs, longitudinal data are required, especially cohort-related data on the life courses of individuals.

In 1957, the American economist Paul Samuelson observed that the baby boom that had appeared in most western countries after World War II would lead to severe social problems. It was thought to be unlikely that the small cohorts that followed the booms would be willing to pay for the old age pensions of large cohorts. Shortly after the Second World War, most western countries experienced a baby boom. Between 1965 and 1970, western countries experienced a baby bust. Social scientists pointed out that the combination of these demographic developments would lead to both a greening and a greying of the population and that it was unlikely that small cohorts would be willing to pay for the old age pensions of large cohorts (Baltes and Mittelstrass 1992; Gillis et al. 1970; van de Kaa 1987; Lutz 1991). DIA entered a boom too. In numerous assessments, the consequences of the changing demographic profile have been analysed for the educational system, the labour market, health care, and old age pensions. Type 2 DIAs were involved.

As a next step, the countries involved tried to find out how they could cope with these consequences of the demographic irregularities. Should they stimulate fertility (Hoehn 1991)? Should they stimulate immigration (Wils 1991)? Here, Type 1 DIAs were being considered. The 1960s increased moral concern about demographic developments and their consequences. Was there an implicit social contract that obliged one generation to support another generation in its old age? Problems about social justice related to controversies raised by demographic developments have been discussed, for example in Rawls' (1972) theory of social justice.

In the sixties, a moral debate developed over the right to privacy of individuals interviewed in a census or a demographic research project. Subsequently, in a number of countries, among them the former West Germany and The Netherlands, the national census was abolished; in other countries, it is undertaken on a reduced frequency (e.g., every ten years instead of five); in some countries there has been pressure to reduce the number of questions in the census, and there have been restrictions on the output that is available; and in some countries the national census is based on a sample, not the population. All of this has decreased the opportunities for, or extent of, demographic research in general, and DIA in particular. Research on the population as a whole was no longer possible and additional sample based research had to be utilised instead.

DIA prospered from the 1960s on, not only at the international and national level, but also at regional levels. Organisational and regional development projects all required some kind of impact assessment, and each time a demographic sub-model was included. Because demographic issues became of greater consequence, the submodels grew in size, complexity, and importance. Over time, demographic forecasting had more and more support from basic sciences. Keyfitz (1982) explored how knowledge could improve forecasts by, for example, *ex post* evaluation of 'old' forecasts. Stoto (1983) analysed the accuracy of population projections and provided techniques for measuring accuracy. The *ex post* evaluation of demographic forecasts has now become a substantial research field (see, e.g., Keilman 1990; Dewulf 1991). Demographic issues on a world scale will be important for a very long time in the future. Demographic fluctuations related to the baby boom and the baby bust will be felt in western countries until about the year 2040. Thus, that the history of DIA has only just begun!

SUBSTANTIVE ISSUES IN DIA

There is a debate about the survival of humanity. The Brundtland Report (WCED 1987) has stimulated this debate, taking the concern for the environment and the quest for sustainable development as a starting point. The historian Paul Kennedy (1993) has contributed substantially to this discussion by taking up the Malthusian model. He has explored how humanity could prepare itself for the 21st century, stressing that Darwin's concept of the survival of the fittest still applies. This issue does not lead to defeatism, however. Social policy could mitigate the risks involved. Thurow (1993) suggests that the growth of world population is only possible if deforestation is curbed. He advocates negotiations between developed and developing countries leading to contracts that induce developing countries to stop deforestation and practise reforestation and that oblige developed countries to pay the developing countries rent for the maintenance of the forest areas. Satellite imagery could establish whether the developing countries keep their promises.

Second, DIA at the level of organisations demands our attention. Organisations are forced by circumstances to become 'learning organisations'. A learning organisation has institutionalised social techniques to monitor its turbulent environment and to adapt its strategies to its changing context. Only if they adapt to changes in their economic, social, and cultural environment quickly enough will they be able to survive. Strategic learning requires a systems approach. In the systems model, a submodel has to deal with the dynamics of the labour force of the organisation. In strategic learning scenario analysis, *ex ante* evaluation (IA) and *ex post* evaluation are integrated. A model for strategic learning always has a submodel representing demographic variables. In this way, DIA is integrated into larger models. This does not imply, however, that IA incorporates *ex post* evaluation. The methodological requirements for *ex post* evaluation differ substantially from those for *ex ante* evaluation. *ex ante* evaluation has to work with data about events that have not taken place yet.

Third, DIA should be taken into account in development projects, such as in the feasibility studies that are required by the World Bank and other funding agencies in preparation of, for example, the subsidising of large reservoirs. DIA is a standard submodel of this type of *ex ante* evaluation models. One specific example is the feasibility study for the Three Gorges Dam in the People's Republic of China. In a critical review of the feasibility study, *Damming the Three Gorges* (Barber and Ryder 1990), the conclusions and premises of the feasibility study have been questioned. Have they policy analysts paid enough attention to the interests of the population of the area to be flooded?

DIA is a customer of basic science which provides new theories and new methods. How can DIA practitioners ensure that they get new knowledge in time?

Primarily this is the responsibility of professional associations in cooperation with universities and other bodies. A mature field of applied science should institutionalise the production of 'state of the art' reports on a regular basis. Also training seminars ought to be available periodically. DIA practitioners have to keep informed about achievements in demography, sociology, economics, political science, informatics, and a range of other disciplines.

DIA has to contribute to the growth of knowledge. The enhancement of theory and methodology of DIA primarily requires cooperation with scientists engaged in the *ex post* evaluation of forecasts and related statements about the future. Basic science has developed and tested 'error theories' to explain the degree of accuracy DIAs and other types of futures analysis accomplish (see, for example, Dewulf 1991). An error theory explains why statements about the future show a relatively low degree of accuracy. To give an example: sometimes social scientists are late in adjusting their models to new information.

The substantive issues listed above are not restricted to DIA. Most kinds of IA show similar areas of concern. Nevertheless the substantive issues in DIA have something special because demographic developments deal with human behaviour that is closely related to choices concerning major events in the life course: having children, deciding to leave the educational system, the age of retirement, and so on (Soerensen et al. 1986).

THE THEORETICAL POSITIONS OF DIA

In DIA nowadays we not only have to answer 'what' questions, but also 'why' questions. The core of the theoretical issues in DIA lies in the explanation of behavioural choices concerning major life course events. Two theories are used in demographic research in general and DIA in particular to solve this explanatory problem: the theory of individualisation and the theory of generations.

Individualisation theory takes the modernisation of Western society at the end of the 18th century as a starting point. At that time, rationalisation, secularisation, urbanisation, and industrialisation increased rapidly. As a result of those developments, individualisation theory predicted that individuals in modern society would follow life courses that would increasingly differ from standardised life courses, the values of the individuals will gradually become less standardised, and individual diversity would increase (Beck 1992; Ester et al. 1993). If research data were to show a new standardisation of life courses and values, individualisation theory would not be able to explain this new standardisation, and would not be supported.

Contemporary generations theory owes much to the work of Mannheim (1928–1929), a sociologist of culture. Mannheim argued that individuals have a formative period, approximately between the age of 10 and the age of 25. If individuals experience major events (wars, economic recessions, cultural shifts) during their formative period, these events will leave their marks on their life courses and on their values. Generations theory has been elaborated and tested on a large scale by Inglehart (1977; 1993), a sociologist and political scientist. Inglehart was looking for an explanation for the waves of social unrest that had inundated the West in the late 1960s and the early 1970s. In his model, two generations are represented. The first generation is that of materialists. Individuals who were born before the Second World War and grew up in years of poverty, unemployment, war, and political unrest, have as their main goal in life to achieve material prosperity and physical security. In opposition to this older generation stood the new one of the post-materialists. As youngsters, they grew up in a period of unprecedented prosperity and political stability. The members of this generation took prosperity for granted and thought the time right for far-reaching changes to society. Since the economic and physical needs were now largely met, it was time to give priority to nonmaterial matters. Inglehart called the new generation, the postmaterialist generation. Its members show traits of hedonism. Their life courses are nontraditional, such as living together without being married. The theory of Inglehart has been corroborated by research in the 1980s and early 1990s.

In 1985, Becker introduced the theory of the emergence of a four-generation pattern: the Prewar Generation, born between 1910 and 1930; the Silent Generation, born between 1930 and 1940; the Protest Generation, born between 1940 and 1955; and the Lost Generation, born between 1955 and 1970 (Becker 1985; 1990; 1992; 1993). From this theory of generations, predictions have been derived concerning life course events and values. Among other things, the transition from school to work has been studied as a major life event in a number of western countries (Sanders and Becker 1994). Individualisation theory has been used as a rival theory and predictions tested. In six research projects, the theory of generations led to better predictions than individualisation theory (Becker 1993). Of course this does not imply that individualisation theory has been falsified. The outcomes of the testings show that using rival theories in explanatory research projects can be productive. Testing of rival predictions shows which theory provides the best explanation under the given circumstances.

The theory of generations (in conjunction with the theory of individualisation) has been applied in DIA. In one Type 1 DIA, the consequences of proposed government policies regarding careers of scientists in universities have been simulated (Becker and Hermkens 1993). In another case, a Type 2 DIA, the consequences of the baby boom cohorts entering the age category of 45 to 54 have been simulated (Becker 1994b).

METHODOLOGICAL ISSUES IN DIA

DIA shows a number of methodological problems that have not yet been fully been resolved. In the first place, researchers have to come to terms with incomplete data. The appropriate management of data, especially when there is no, or limited information about the future, is a complex problem. In DIA, hazard rates are used to estimate data that are missing. Hazard rates and related techniques (for example, event history analysis) are adapted from biology and medical science. Plants and animals have life courses too. Estimating the timing of a major event in the life course of a tree or a cow does not differ methodologically from estimating the timing of a major event in the life course of the member of a birth cohort.

In the second place, some variables in cohorts show age, period, and cohort effects. Age effects relate to the impact of growing older on values and behaviour. Period effects relate to the *Zeitgeist*, 'the spirit of the time'. Cohort effects relate to the impact of the formative period in the life course on values and behaviour. To unravel these effects, most researchers try to specify one of the effects as to its historical setting. After the specification of one effect, the other two effects can be located (see, for example, Beekes 1990). This problem applies to variables that are subject to change over time, like values. Life course events like leaving school, taking a job or having a child are not subject to age-effects after the event.

In the third place, a number of methodological problems in DIA that relate to futures analysis in general are pertinent. Projections only extrapolate trends from the past. Their drawback is that they are not related to explanatory models. Forecasts are, by definition, related to explanatory models. Forecasting methodology is used when the processes to be predicted show a relatively high degree of predictability. If the processes to be analysed show a relatively low level of predicability, futures exploration methodology should be used. In this case, a number of scenarios will represent the turbulent environment of the actor involved. In demographic research in general, and in DIA in particular, the use of scenarios and related types of futures exploration is increasing.

In the fourth place, the relationship between futures analysis and theoretical explanation in DIA demands our attention. Systems theory provides a mode for discussion, distinguishing between relatively closed and relatively open systems. Boudon (1986) has elaborated a distinction between theories in a strict sense and formal theories. Theories in a strict sense can be used to explain processes in relatively closed systems. Formal theories are designed for explanations in relatively open systems. The future of social systems is relatively open in most cases. For these reasons, formal theories are preferred in this case. There are methods available for translating theories with relatively high levels of abstraction (like formal theories) into theoretical propositions, research hypotheses, and statements about observations (Bryant and Becker 1990).

PRACTICAL ISSUES IN DIA

In DIA, as in social research in general, it is desirable to keep costs low, but still to obtain a high response rate to surveys. Demographic data can often be gathered by telephone interviews. This implies that the advantages of computer-assisted telephone interviewing (CATI) are available for most DIAs. CATI usually yield a relatively high response and the costs of data processing are reduced. Besides CATI, indepth interviewing, such as of policymakers and other stakeholders, is often needed in DIA. The Delphi approach also has been applied with success (Kenis 1995).

New data may not always be necessary. Data archives are relatively well stocked with demographic data, particularly transversal data. Increasingly longitudinal data are included as well. In many cases international comparison is possible.

Public participation in DIA is feasible in a lot of cases. If the information is presented in the format of a simple cohort- or generations-replacement model, most audiences are capable and willing to discuss the issues that are relevant to the DIA. The 'visibility' of a pattern of generations is high in the eye of the general public. Simple models have been discussed even in primary schools.

THE LIKELY FUTURE OF DIA

A reflection on the future of DIA needs to consider scenarios that provide possible contexts for this type of applied social science research. For the sake of argument, I take a 'balanced growth' scenario as the frame of reference for my speculation about DIA in the next two to three decades. This scenario presumes market economies and vital cultures in the dominating parts of the world, and that all major parts of the world will experience economic and cultural growth (CPB 1992; Thurow 1993). It is likely that three major areas of the application of DIA will become increasingly important: the social setting for DIA, the contribution of DIA to science, and social networking related to DIA.

The first area of application of DIA that will become more important is migration and fertility. Governments and other actors will try to influence both issues. DIA can be used to get a preview of how various policy options will influence migration and fertility patterns (Type 1 DIA). They will also want to know how changing migration and fertility rates may put boundaries to their actions (Type 2 DIA).

The second area of growing application is social capital and economic vitality. Governments and related actors will try to find hidden resources of human capital, such as amongst the elderly. Delayed retirement policies already exist in a number of countries. Assessment of the effects of delayed retirement requires Type 1 DIA. On the other hand, the given social capital is a limit to economic growth. Can guided immigration be used to fill the gaps in the labour force that result from the baby bust? Such analysis requires Type 2 DIA.

The third area of an increasing demand for DIA is the relationship between generations. Contemporary elderly tend to be well-organised, such as in the USA with, among other groups, the American Association of Retired Persons. Contemporary young people are beginning to protest against the growing flow of financial resources going to the elderly. The young protest in an indirect way by attacking the growth of the national debt, such as through the USA organisation, *Lead or Leave*, which demands that government either take action or leave the scene, and which had nearly a million members in 1994. In other countries, similar protest movements can be predicted and similar generation-related conflicts are to be expected. This will result in an increasing demand for Type 1 and Type 2 DIA.

Not only the issues but also the social settings for DIA are apt to experience change in the near future. Because IA (including DIA) will be increasingly required by law, there will be an increase in DIA at the project, regional, national, and international level. The contribution of DIA to science will also increase. Each year, more DIAs are being verified by time. Increasingly, DIAs which have been undertaken in the past will be audited, particularly with respect to the degree of accuracy they exhibited. As DIA becomes more important and more accurate, it will itself become more important and more valued as a discipline amongst the social sciences. In the next decades, DIA needs more and better social networking. At this moment there is not yet an institutionalised network of the scientists engaged in this kind of applied social research. As soon as DIA becomes compulsory on a legal basis, the demand for trained and experienced scientific personnel in this field will increase drastically.

REFERENCES

Barber, M. and G. Ryder, eds. 1990. *Damming the Three Gorges*. London and Toronto: Earthscan.

Baltes, P.B. and J. Mittelstrass, eds. 1992. *Zukunft des Alterns und gesellschaftliche Entwicklung*. Berlin: De Gruyter.

Beck, U. 1992. *Risk Society, Towards a New Modernity*. London: Sage.

Becker, H.A. 1985. *Dutch Generations Today*. Wassenaar: Netherlands Institute for Advanced Study in Humanities and Social Sciences.

Becker, H.A., ed. 1990. *Life Histories and Generations*. Utrecht: Institute for Social Research, Utrecht University.

Becker, H.A., ed. 1992. *Dynamics of Cohort and Generations Research*. Amsterdam: Thesis Publishers.

Becker, H.A. 1993. "Epilogue," pp. 837–844 in *Solidarity of Generations, Demographic, Economic and Social Change, and Its Consequences*. H.A. Becker and P.L.J. Hermkens, eds. Amsterdam: Thesis Publishers.

Becker, H.A. and P.L.J. Hermkens, eds. 1993. *Solidarity of Generations, Demographic, Economic and Social Change, and Its Consequences*. Amsterdam: Thesis Publishers.

Becker, H.A. 1994a. *Generaties en hun kansen* (4th ed.). Amsterdam: Meulenhoff.

Becker, H.A. 1994b. *De verborgen miljarden van de Baby Boom*. Utrecht: Department of Sociology, Utrecht University.

Beekes, A. 1990. "The development of cohort analysis," pp. 547–562 in *Life Histories and Generations*. H.A. Becker, ed. Utrecht: Institute for Social Research, Utrecht University.

Blossfeld, H.P., A. Hamerle, and K.U. Mayer. 1986. *Ereignisanalyse*. Frankfurt: Campus.

Boudon, R. 1986. *Theories of Social Change*. London: Polity Press.

Bryant, C. and H.A. Becker, eds. 1990. *What Has Sociology Achieved?* London: Macmillan.

Central Planning Bureau (CPB). 1992. *Scanning the Future, A Long-term Scenario Study of the World Economy 1990–2015*. The Hague: Staatsdrukkerij en Uitgeversmaatschappij.

Dewulf, G. 1991. *Limits to Forecasting, Towards a Theory of Forecast Errors*. Amsterdam: Thesis Publishers.

Ester, P., L. Halman, and R.A. de Moor, eds. 1993. *The Individualizing Society, Value Change in Europe and North America*. Tilburg: Tilburg University Press.

Forrester, J. 1971. *World Dynamics*. Cambridge: Wright-Allen Press.

Gillis, J.R., L.A. Tilly, and D. Levine, eds. 1970. *The European Experience of Declining Fertility, A Quiet Revolution 1985–1970*. Cambridge: Blackwell.

Hoehn, Ch. 1991. "Policies relevant to fertility," pp. 247–256 in *Future Demographic Trends in Europe and North America. What Can We Assume Today?* W. Lutz, ed. London: Academic Press.

Inglehart, R. 1977. *The Silent Revolution, Changing Values and Political Styles among Western Publics*. Princeton: Princeton University Press.

Inglehart, R. and P.R. Abramson. 1993. "Affluence and Intergenerational Change: Period Effects and Birth Cohort Effects," pp. 71–114 in *Solidarity of Generations. Demographic, Economic, and Social Change, and Its Consequences*. H.A. Becker and P.L.J. Hermkens, eds. Amsterdam: Thesis Publishers.

Kaa, D.J. van de. 1987. "Europe's second demographic transition." *Population Bulletin* 42(1): 1–57.

Kenis, D. 1995. Improving Group Decisions. Doctoral thesis, Utrecht University.

Kennedy, P. 1993. *Preparing for the Twenty-First Century*. New York: Random House.

Keilman, N.W. 1990. *Uncertainty in National Population Forecasting; Issues, backgrounds, analyses, recommendations*. Amsterdam: Swets and Zeitlinger.

Keyfitz, N. 1982. "Can knowledge improve forecasts?" *Population and Development Review* 8(4): 729–751.

Lutz, W., ed. 1991. *Future Demographic Trends in Europe and North America. What Can We Assume Today?* London: Academic Press.

Malthus, T.R. 1798. *An Essay on the Principle of Population as It Affects the Future Improvement of Society*. London: Dent (1960).

Mannheim, K. 1928–1929. "Das problem der generationen." *Koelner Vierteljahreshefte fur Soziologie* 7: 157–185, 309–330.

Meadows, D., et al. 1972. *Limits to Growth*. New York: Universe Books.

Organization for Economic Cooperation and Development (OECD). 1970. "Simulation Option Model: A simulation model of the education system." Technical report by B. Schwarz. Paris: OECD.

Pendergast, Ch. 1989. "Condorcet's canal study: The beginnings of social impact assessment." *Impact Assessment Bulletin* 7(4): 25–31.

Rawls, J. 1972. *A Theory of Justice*. Oxford: Oxford University Press.

Sanders, K. and H.A. Becker. 1994. "Transitions from education to work and social independence: A comparison between the United States, The Netherlands, West Germany, and the United Kingdom." *European Sociological Review* 10(2): 135–141.

Soerensen, A.B., F.E. Weinert, and L.R. Sherrod, eds. 1986. *Human Development and the Life Course: Multidisciplinary Perspectives*. Hillsdale: Lawrence Erlbaum Associates.

Stoto, M.A. 1983. "The accuracy of population projections." *Journal of the American Statistical Association* 78: 13–20.

Thurow, L. 1993. *Head to Head*. New York: William Morrow.

Wils, A.B. 1991. "Survey of immigration trends and assumptions about future migration," pp. 181–300 in *Future Demographic Trends in Europe and North America. What Can We Assume Today?* W. Lutz, ed. London: Academic Press.

World Commission on Environment and Development (WCED). 1987. *Our Common Future* (the Brundtland report). Oxford: Oxford University Press.

Chapter 7

Health Impact Assessment of Development Projects[1]

MARTIN H. BIRLEY
Liverpool School of Tropical Medicine, UK
GENANDRIALINE L. PERALTA
University of the Philippines at Quezon City, Philippines

INTRODUCTION

The wider objective of development activity is to promote a sustainable, high quality life for all communities and this implicitly includes good health. Development programs and projects derive their immediate objectives from an analysis of the constraints to achieving this wider objective. The constraints include lack of infrastructure, credit, trained personnel, and poor access to markets and appropriate technologies. One important set of constraints is the indirect and unintended impacts that a development in one sector may have on other sectors. Such impacts affect the community, their environment, and their health. Health impacts refer to both positive and negative changes in community health that are attributable to a development project. Although it is important to consider both kinds of impact, we emphasise the negative impacts and seek methods and procedures by which the risks may be assessed and then effectively managed. These methods and procedures are referred to as health impact assessment (HIA). HIA is a multidisciplinary activity that cuts across traditional boundaries between public health, medical services, and environmental and social science. It is a necessary component of project planning in all countries and part of environmental impact assessment (EIA). This paper is limited to the use of HIA in less developed countries and it is primarily concerned with rural rather than urban/industrial projects.

The following account of HIA distinguishes between a health hazard and a health risk. A health hazard has a potential for causing harm. In contrast, a health risk indicates the likelihood of harm occurring to a particular community group at a particular time and place. The identification of a health hazard does not imply that it is going to occur in every development project. The analysis of health risk determines the likelihood of a significant problem arising in the specific project under consideration.

---

[1]*Environmental and Social Impact Assessment* - Edited by F. Vanclay and D.A. Bronstein. Copyright © 1995 by the International Association of Impact Assessment. Published in 1995 by John Wiley & Sons Ltd.

TYPES OF HEALTH IMPACT

Development-related health hazards can be grouped into five major categories. *Communicable diseases* require relatively small doses of exposure to infective agents and then multiply in the human community. There are often many separate sources of infection. *Noncommunicable diseases*, by contrast, may arise from the accumulation of toxic chemicals from a few point sources. *Malnutrition* occurs where communities are deprived of subsistence foods through changes in land use or farming systems, or through uneven economic development. *Injuries* occur through exposure to new technologies, poor working practices, poor dwelling design, improper use of machinery, and poor machine maintenance. *Mental disorder* may be associated with the stress of new ways of living and the disruption of long-established communities. Figure 1 illustrates the linkage between some of these health hazards and development projects.

HISTORICAL PERSPECTIVE

HIA procedures for development projects have evolved independently in several development sectors. The principles used and the problems encountered share many similarities. There are additional similarities to methods and procedures used in EIA and in more general development aid evaluation (OECD 1986). In this section we briefly review the procedures that are used in water supply and sanitation, water resource development, and the chemical sector. We also list some other current programs.

Water Supply and Sanitation

In the water supply and sanitation sector, the World Health Organisation (WHO) has published procedures for analysing nonfunctioning or under-utilised systems and for evaluating the positive health impacts of fully functioning systems (WHO 1983a; WHO 1983b). In this sector, health impact generally refers to the intended health improvements that are assumed to derive from safe water supply and sanitation. These consist of reductions in the prevalence of certain communicable diseases. It is relatively easy to install the 'hardware' of well-engineered systems. It is harder to ensure the community participation required for appropriate use and maintenance. It is harder still to demonstrate a positive health impact.

Figure 1.   **Examples of the linkage between health hazards and development**

- Transport systems provide a conduit for the distribution of communicable disease such as occurred with HIV infection in truck drivers and women working in bars in Uganda in 1986 (Carswell 1987). People who migrate in search of work are especially at risk from sexually transmitted diseases and may carry the infections to new areas.
- During the 1970s half the malaria cases in Amazonia were linked to the narrow area of influence of the Transamazon Highway (Coimbra 1988). The malaria mosquito became more abundant because forest clearance provided new breeding sites. The parasite was transported to the region by migrants.
- Transport-related injuries are more common in developing countries where driving tests may be circumvented, roads may be of poor quality, and vehicles may be poorly maintained. In Papua New Guinea during the 1970s, traffic injuries were estimated to cost 1% of GNP (Havard 1978). In developing countries generally, the annual rate of disabling injuries to workers is 21–34%, compared with 3% in the UK (El-Batawi 1981).
- Miners frequently work in confined spaces that are heavily laden with rock dust. Long-term exposure to this dust may induce permanent lung damage. Damaged lungs are more susceptible to infection with tuberculosis. Infection rates have been reported in South African mines of 800–1,000 per 100,000 (Packard 1989). Miners are exposed to injury from rock falls and fast-moving machinery. In Bolivia during the 1970s, the population of 24,000 mineworkers in large mines had 5,430 injuries (El-Batawi 1981).
- Household cooking on an open fire may be the largest single occupational health problem of women. Burning biomass fuels produces large quantities of pollutants that may lead to respiratory or eye disease.
- In Sri Lanka, a rice development project created breeding sites for mosquitoes that transmit Japanese encephalitis. Pigs near the rice fields provided the virus. The mosquitoes acquired the virus by feeding on the pigs. The virus multiplied in the mosquitoes and was subsequently transmitted to people who were bitten by the mosquitoes. Consequently there was an epidemic of this often fatal disease (IRRI 1987). Again in Sri Lanka: until the 1970s, child labour was common on some tea estates, education facilities were minimal, and water supplies were inadequate. Chronic malnutrition and infant mortality rates were twice the average rate in other rural communities (Laing 1986).
- Urban water supplies are often provided before drainage facilities. Consequently, large areas of polluted water form that are ideal breeding sites for the mosquito that transmits lymphatic filariasis (Macdonald 1991). Manifestations of this disease are common in many Asian coastal cities.
- In Cubato, Brazil, during the 1980s, there were 23 major industrial plants and many small operations. A high rate of respiratory disorders was associated with high levels of water and air pollution. Neonatal mortality and birth deformities increased (Findley 1988; Pimenta 1987; Thomas 1981). Other examples include lung disease due to lack of face masks in fibre factories in Albania, birth defects associated with lead and zinc plants in Bulgaria, and chronic ill health due to air pollution in Hungary (Carter and Turnock 1993).

Water Resource Development

Although water resource developments such as dams and irrigation have a long history of benefiting human health through increased production, they can also have a deleterious effect through chemical pollution, loss of subsistence crops, poor drinking water quality, and vector-borne diseases. Vector-borne diseases, such as malaria and schistosomiasis, are particularly important and have received considerable attention (for reviews see Hunter et al. 1993; Oomen et al. 1988; Service 1989). Such developments change the distribution and flow of surface waters, creating a favourable habitat for the breeding of vectors, such as mosquitoes and snails. Human exposure to biting insects or contaminated waters provides the conditions necessary for an increased health risk. Expensive mitigation measures take the form of vector control through chemical application or environmental modification. In recognition of such increasing health risks, a joint WHO/FAO/UNEP/UNCHS Panel of Experts on Environmental Management (PEEM) was formed in 1981.

The panel members were aware that an important component of environmental management occurred at the design stage. Decisions about infrastructure, location and resettlement could help reduce vector populations or prevent exposure. This, in turn, would require an assessment procedure. The procedure was published as *Guidelines for Forecasting the Vector-Borne Disease Implications of Water Resources Development*, known as PEEM2 (Birley 1991). It covered the subsectors of irrigated agriculture and multipurpose reservoirs and assisted the user to identify (1) the specific vector-borne disease hazards that occur regionally and the important environmental factors; (2) the vulnerable communities; and (3) the capabilities of the health service to monitor, safeguard, and mitigate.

These three components of the assessment were then combined into a statement of health risk. In retrospect, PEEM2 omitted to distinguish between a health hazard and a health risk, reducing the logical flow of the procedure. This chapter focuses on methodologies. The requirement for training programs is being met by the development and testing of task-based multisectoral courses such as *Health Opportunities in Water Resource Development*, jointly developed and offered by PEEM, Danish Bilharziasis Laboratory, and the Liverpool Health Impact Programme. A recent report advocates a simpler procedure than PEEM2, for use by nonhealth specialists during the project screening phase of the development cycle. A set of vector-borne disease hazards was identified and a simple questionnaire was devised to determine whether the health risks appeared to warrant the project manager seeking specialist advice (Bolton et al. 1990).

Chemical Sector

In the chemical sector there are two main sources of health hazards: *noncommunicable diseases* associated with poisoning by routine or accidental exposure to toxic chemicals; and *injury* from fire, explosion, radiation, and corrosive action.

Over 60,000 chemicals are in common use and adequate information about toxicity and reactivity is not available for all of them. A meeting in 1986 established principles and objectives for health and safety assessments (WHO 1987). HIA was viewed as a component of EIA, already a well-established procedure. A program of work was outlined to transfer knowledge and encourage debate to meet the perceived needs for HIA. The meeting also reviewed the experiences of thirteen assessments of chemical plant spanning the period 1973–1983. A number of methodological issues were discussed that can be adapted to HIAs in other sectors. The three main tasks of HIA were listed as: identification of hazard, interpretation of health risk, and risk management.

A procedure referred to as environmental risk assessment (ERA) has been developed to analyse the health hazards arising from chemicals used or produced in industrial projects (ADB 1990, see chapter by Carpenter). ERA focuses on discrete and relatively rare events, such as the unplanned release of toxic and reactive substances. The risks are effectively considered to be zero before construction commences. Impact frequency is tabulated against impact severity. Risk is defined as the probability that an identified hazard will cause harm of a specific severity. For example, there could be 10^{-7} events per year in which the number of fatalities exceeds 1000. Quantification of this kind is not practical in most HIAs. ERA is described as a component of EIA. EIA identifies the hazards and the uncertainty; the more detailed process of ERA then replaces uncertainty by risk estimates. The normal outputs of EIA are viewed as risks with a high probability and an obvious need for mitigation.

Other Initiatives

In 1990, the Health Impact Programme was established at the Liverpool School of Tropical Medicine with support from the British Overseas Development Administration (ODA). The objective was to improve the analysis and management of the health impacts of development projects by developing a procedure for HIA. The program initially focused on generating a detailed review of health hazards, grouped by development sector. It also has undertaken training or advocacy programs and rapid HIAs in nine developing countries. In 1992, the Asian Development Bank commissioned us to prepare for their internal evaluation of projects, *Guidelines for the Health Impact Assessment of Development Projects* (ADB 1992). Other relevant references include the *Australian National Framework for Health Impact Assessment* (Ewan et al. 1992) and *Environmental and Health Impact Assessment of Development Projects, A Handbook for Practitioners* (Turnbull 1992). The World Health Organisation have published various reviews of health and development and these are referred to below.

PROBLEM ANALYSIS

The previous sections illustrate that while there is widespread concern about the need for HIA, there is no universal procedure. In order to illustrate what is required we have constructed the problem tree in figure 2. The central statement in the figure poses the problem that health risks can increase as a result of development. The effects are listed above and the main causes are listed below. The postulated effects include reduced productivity, quality of life and educational achievement. These, in turn, will reduce project sustainability and transfer hidden costs to the health sector. The main cause of the problem is identified as inadequate attention to health safeguards and mitigation measures during project design and operation. This, in turn, is postulated to be the result of no HIA, no budget, and no technical skills for the design of health safeguards and mitigation measures.

Figure 2 is deliberately framed as a set of negative statements to highlight the problem components. The problem analysis proceeds by converting all the negative statements into equivalent positive statements, referred to as an objectives tree. For example, as the result of a HIA we expect health risks to decrease, because there are adequate health safeguards. This, in turn, depends on HIA being routinely applied and an adequate budget being available, where necessary. To ensure this outcome, we require HIA methodologies, widespread appreciation of the health hazards of development, and training programs.

Environmental Impact Assessment

In order to avoid duplication of procedures, HIA should be integrated with the existing procedures for EIA. In theory, EIA already includes a limited concern with health issues. In practice, health is often neglected. The reasons for this are complex. Policymakers are not yet as sensitive to health issues as they are to environmental issues. Or, if they are, they consider it a specialist subject that belongs to the health profession. On the other hand, the health profession is often only trained and interested in matters of development that lie directly within the health sector. In this section, we review the experience and procedure of EIA and incorporate HIA.

Considerable experience of EIA has become available to donor agencies and governments during the last two, or more, decades. This has helped to clarify the nature of the impact assessment process. As well as a formal planning procedure, impact assessment is a process of communication between participants with different perspectives, objectives, and requirements (Spellerberg and Minshull 1992). There are three principal participants.

- *Project proponents* are primarily concerned with gaining consent in the most cost-effective manner. They are frequently line ministries responsible for building and operating infrastructures or relocating and employing labour.
- *Consultants* are employed by the project proponents to produce the impact statement under severe time and financial constraints.
- *Planning authorities* are concerned with effective application of planning policy. In many developing countries such authorities are called Environmental Protection Agencies. They require accurate statements of suitable detail to enable proposals to be assessed for permitting or regulatory purposes. Where they are of recent origin, they may not have functional procedures for either EIA or HIA.

Figure 2.   **Problem tree for the health impact of development projects**

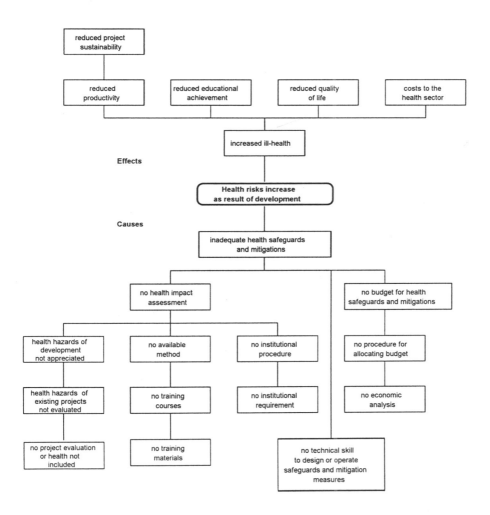

Environmental Protection Agencies

Governments and aid agencies, such as the Asian Development Bank and the Overseas Development Administration, have been instrumental in establishing and strengthening national environmental protection agencies (EPAs). These agencies have a role in safeguarding, mitigating, and monitoring the environmental impacts of development projects. This role includes safeguarding health. A critical appraisal of the capability of the agencies involved is essential to determine whether health safeguards, mitigation, and monitoring measures are likely to be effective and where action is required to strengthen existing capabilities.

One of the largest practical obstacles to HIA appears to be the planning delays which prevent health impact studies from starting at the same time as other studies. Studies that are left until the last minute, or that require long baseline data collection periods, are unlikely to receive proper consideration during the final planning negotiations. From the viewpoint of the line ministries responsible for design and implementation of a development project, intersectoral concerns add costs and delays. The objective of impact assessment is to assist agencies responsible for protecting the environment, or the health of the community, to participate in negotiations with the line ministries, the community, and the donor agencies in a timely and practical manner (Tiffen 1991). The objective is not to block the development but to seek mutually acceptable solutions.

The following experience is typical. An Asian country recently created an Environmental Protection Agency. Legislation required all new projects to receive an EIA. As the agency had limited resources, there was a large backlog of projects to review. Fears were expressed that unless a process could be found to separate projects with major and minor impacts, the whole development process could become paralysed. The primary concerns of the agency were environmental issues, including pollution. The procedure was to establish a commission within each line ministry to review project plans. In principle, representatives from other ministries could be invited to attend and be included in the review process. In practice, as there was no compulsion to do so, the health sector was often not represented. Similarly, there was no requirement to consider occupational health and safety.

HEALTH IMPACT ASSESSMENT PROCESS

A HIA is an early warning. Its purpose is to alert decision-makers to possible health risks that may be reduced, but not totally eliminated, by incorporating safeguards and mitigation measures. It consists of three main steps: (1) health hazards identification; (2) interpretation of the health hazards as health risks attributable to the project; and (3) health risk management.

The first two of these steps are referred to as project screening. Screening is the process of determining whether a development project should be subject to an impact assessment (WHO 1987). It depends on the perception of hazard by the decision maker, the politician, and the general public. Screening is most likely to be conducted internally by administrative staff who need to decide quickly whether each project has a potential for undesirable side-effects. The screening must be kept simple as the decision maker is unlikely to be a specialist in either EIA or HIA. The principal decision is whether to call in specialist support. The additional information to be gained from such support must be offset against the cost and the delay. For example, the British Overseas Development Administration (ODA) uses a screening process to register 'danger signals' and to avoid unnecessary investigation where impacts are likely to be minimal (ODA 1989).

Health Hazard Identification

In most cases the screening process will be based on an identification of the health hazards that are associated with the category of project and its location. Decision makers require a catalogue indicating the kinds of health hazards that have been experienced on similar projects. They also require information about health-sensitive locations—obtained from maps, health records, and knowledge of disease foci. For example, irrigation is associated with schistosomiasis in parts of Africa but not in many parts of Asia because the snail host is absent. The catalogue of health hazards is based on a review of known health impacts and should be carefully classified. Classification systems are important because they have predictive value. Similar projects are likely to share similar health hazards.

Compiling such information provides a useful method of sensitizing the development community to the health issue. It is not, in itself, sufficient. The important next step is to make use of the compiled material within a procedure for HIA. WHO have published several reviews of the impact of development and environmental change on health (Cooper Weil et al. 1990; WHO 1992a; WHO 1992b). Such reviews are inevitably open-ended and relevant new material can always be added. We have compiled our own catalogue in a form readily accessible to project planners, by cross-referencing health hazards to development sectors (Birley in press). Figure 1 illustrated the material available.

There is a range of health hazards between severe and trivial. For example, the odours from factories and piggeries, the drone from construction equipment, the taste of drinking water may pose inconvenience but do not constitute a significant health risk. Therefore, the next step is to identify the health risk associated with the health hazard.

Interpretation of Health Risks

Interpretation of health risks forms the core of the HIA. In order to expedite this procedure it is necessary to divide it into two steps: a rapid assessment followed as necessary by a detailed HIA. The rapid assessment is also referred to as an initial health examination. This integrates with the Initial Environmental Examinations (IEE) used in EIAs. It uses secondary data, checklists and, sometimes, a fact-finding mission. The output is a project classification (World Bank 1991). The detailed HIA is similar in content to the rapid assessment but normally employs specialists working to a terms of reference and is based more securely on primary data (ADB

1990; ADB 1992; Tiffen 1991). The project classification usually consists of three classes:

  a.   Significant impacts, mitigation difficult or requires special budget, requires a detailed HIA
  b.   Significant impact, mitigation practical without special budget component, may require a detailed HIA
  c.   No significant impact

In rapid assessment, it is necessary to determine which geographical areas, project phases and communities to include in the analysis. It is also necessary to consider the environmental factors that determine exposure to the health hazard and the capabilities of the agencies responsible for protecting human health. Adequate consultation should be undertaken among potential stakeholders, including the affected community. Finally, the conclusions reached in each component of the analysis must be combined and presented to the decision makers in a format that will enable them to use the information effectively.

Geographical areas. One important consideration is the geographic boundary, especially downstream and downwind. In the case of communicable disease it is necessary to consider whether migrants will import new diseases from distant localities. If the disease is transmitted by flying insects, it is necessary to consider the flight range of those insects. In the case of contaminated water, the downstream impacts may be important. For example, a large project was completed to supply a city with a new piped sewerage system. The engineering focus was on removing raw sewage from the city streets. The main sewerage pipes led to a primary treatment plant with settling tanks. The septic fluid was then discharged into open drains that were beyond the bounds of the engineering project. There were plans for secondary treatment but these were not scheduled for early implementation. The unfenced drains flowed through densely populated suburbs and the community had direct contact with the septic fluid, even extracting it for domestic use. In this example there was a clear case for the boundary of the HIA to be larger than the boundary of the engineering problem.

Project phases. The most important phases in the project cycle are: construction, early operation, and late operation (after 10 years). Some health risks can increase rapidly when the environment is changed. Examples include malaria, traumatic injury, and acute malnutrition. Other health risks increase slowly, as a result of prolonged exposure of the community to the environmental factor. Such hazards may be regarded as 'time-bombs', ticking away at the heart of the project (Stigliani et al. 1989). A chain of events may result in the delayed and sudden occurrence of harmful effects due to the mobilization of chemicals stored in soils and sediments. Examples of these are conversion of farmlands to forest requiring lime application that releases pesticides, fertilizers, and heavy metals (Munn 1992). Another example is dredging of coastal areas for reclamation freeing sulphides and bound heavy metals. In both cases, the harmful chemicals that are released may accumulate in food plants and animals. People who eat the contaminated food may then receive a harmful dose. Other examples of delayed effects include schistosomiasis, dust-induced lung disease, and chronic malnutrition. Some forms of ill-health, such as genetic defects and carcinomas, may be difficult to attribute to specific projects, because of long time delays and low prevalence rates (Eisenbud 1990).

Communities. There will usually be many different human communities affected by the project and they may have differing vulnerability to the health hazard. There are many procedures for classifying the community vulnerability. For example, in a chemical plant development it may be appropriate to distinguish exposure through occupation, accident, consumption, and residence (WHO 1987). On a resettlement scheme, by contrast, it may be appropriate to distinguish settlers chosen by govern-ment, self-selected settlers, displaced communities, and temporary workers (Roundy 1989). Large construction and mining projects sometimes require a workforce of single adult males. They may live away from their families for years at a time, in remote rural environments. Such communities attract large numbers of 'camp-followers'. These communities may live in temporary settlements without adequate water supply or sanitation and sell goods and services to the newly wealthy work-force. Sexually transmitted diseases may flourish under such conditions.

Environmental factors. Environmental factors determine whether a health hazard could become a health risk through processes such as amplification, concentration, deprivation, and exposure. The hazard may be restricted to point sources or widely distributed. Many industrial developments require the use and disposal of toxic and reactive substances. Exposure occurs through skin contact, inhalation, or ingestion and can be classed by severity, extent, and distribution (ADB 1990; WHO 1987). Concentration of toxins occurs by bioaccumulation, as they pass through the food-chain. Exposure depends on discharge dilution rates, land zonation policies, consumption rates, and occupational safety procedures.

Agricultural projects are associated with vector-borne disease hazards (Birley 1991). Exposure occurs through skin/water contact and arthropod bite. For example, the intermediate hosts of schistosomiasis are aquatic, or semi-aquatic, snails that may proliferate in flooded fields, irrigation and drainage canals, reservoir margins, or swamp. The amplification of exposure sites depends on factors such as rainfall, water velocity, temperature, chemical composition, and fluctuation. Exposure depends on working, bathing, hunting, or playing in infected waters. Construction of field latrines may contribute to the interruption of the transmission cycle. The mosquito-borne diseases associated with agricultural development include malaria, dengue, filariasis, and viral encephalitis. The extension and modification of surface waters amplify the mosquito population by providing breeding sites. Exposure to mosquito bite depends on settlement location, house design, and occupation. For example, in Thailand the rubber plantations provide shaded stream pools that are the habitat of the larval malaria mosquito. Rubber-tapping takes place at night, provid-ing occupational exposure to the biting, adult mosquitoes. Determinants of childhood malnutrition include absolute and seasonal shortages and differential entitlements to

food within the household. A switch from subsistence to cash cropping may para-
doxically increase malnutrition as food purchases compete with other household
priorities.

These examples demonstrate that the environmental factors which influence
community vulnerabilities to health hazards are many and complex. They range
from the most direct consequences of a development project, such as construction
injuries, to the most indirect, such as variation in insect abundance or food security.
The pathways may be described using causal modelling. However, conclusive proof
linking cause and effect may not be obtainable.

Capabilities of health protection agencies. The capability of national health services
to contribute to health safeguards varies greatly between countries, as does the
percentage of government income spent on health. National health statistics may
contain little reliable information and there may be no functional health surveillance
system. Health service capability may be restricted entirely to providing curative and
supportive medical care. Preventative services may not have the authority, staff or
other resources to act as watchdogs to developments in the nonhealth sector. In the
case of diseases such as malaria, there may be a national control program that is
already responsible for combating the disease. The disease may already be prevalent
before the development commences but it is necessary to ensure that the development does not increase the health risk. There is a tendency to assume that such
services have the existing capacity to manage any changes in the health risk that
may occur. In reality, the health service may not be informed, or consulted, and
valuable opportunities to safeguard health may be lost at the planning stage.

In cases of industrial development, a new health hazard is introduced by the
development. However, plans and procedures to contain the release of chemicals
from production plants may ascribe no role at all to the health service in relation to
risk assessment and management (ADB 1990; WHO 1987). At the least, an
adequately equipped and staffed casualty unit will be required if exposure cannot be
averted.

Health protection is not only the responsibility of health ministries. The capacity
of other agencies should also be considered. For example, occupational health and
safety is often the responsibility of labour ministries while water supply and
sanitation may be the responsibility of ministries of works. Many nongovernmental
organisations (NGOs) also have a role in health protection, especially those working
with local communities.

The health risk summary. The final task of the assessment is to reach a conclusion
regarding the change in health risk from examination of the components. In many
cases there will be a health risk even if there is no project and only the risk
attributable to the project should be included. A simple, nominal classification may
suffice such as 'risk increase', 'risk decrease' or 'no change'. The summary should
be accompanied by explanatory text and supported by detailed appendixes, as necessary. The process can be repeated for each vulnerable community group, for each
project stage and for each health hazard. Because of the lack of data or the indirect

nature of the pathways, the assessment is inevitably based on assumptions as well as sound scientific judgement. The process remains valid only as long as the assumptions are fully stated, so that the reader is free to disagree.

HEALTH RISK MANAGEMENT

The health risk assessment is presented to an audience of decision makers. They must evaluate the relative importance of the impacts that have been identified in a wider context. They must decide whether safeguards and mitigation measures are required, negotiate resources, and assign monitoring tasks. Health risk management consists of incorporating safeguards and mitigation measures into project design, construction, and operation. Safeguarding entails modifications to project plans and operations. For example, settlements should be sited so as to separate people from health hazards. Mitigation may entail vigilant monitoring followed by appropriate and timely response to increasing health risks. For example, if the prevalence rate of malaria increases then action may be required to control mosquitoes, detect new cases, and provide additional drug supplies. In order to undertake such mitigation, it is necessary to have an adequate health service capability. However, as health services are often very weak in developing countries, it cannot be assumed that they have sufficient staff, a responsive management structure, supplies of appropriate drugs, or a functioning system of drug distribution.

During the construction and operation start-up phase an environment and safety overseer may be hired to supervise the implementation of safeguards. Additional tasks could consist of coordinating the collection and reporting of health and environment information generated within the project. The cost of onsite supervision must be balanced against the risk of an adverse outcome.

It is always best to adopt a precautionary approach in the context of health risk management—a 'no-regrets' policy. In other words, sufficient action should be taken to ensure that there should be no later stage at which regret is expressed that action was not taken. Certain actions or policies could be taken as insurance measures. Examples of these measures are:

- Basic medical examinations for construction workers, as part of health surveillance
- Environmental management measures for vector control
- Creation of a formal linkage between environmental and health and safety plant personnel for regular consultation and joint monitoring
- Training of community and health workers

Monitoring and Surveillance

Surveillance is often defined as systematic measurement of variables and processes to establish a time trend. Monitoring, by contrast, is the process of collecting data for analysis and action. Monitoring depends on an adequate health information system. Routine health surveillance data is unlikely to be sufficiently accurate, sensitive or to have the coverage needed to indicate changes in health risk associated with a specific development project. It may be necessary to commission special surveys. However, the linkage between the project and community health may be so indirect that health change cannot be conclusively attributed to the project.

Direct monitoring of human health may be expensive, impractical, unreliable, or unethical and there may be a need for proxy health indicators. For example, an urban drainage project might monitor the biting density of the common mosquito, *Culex quinquefasciatus*, as an indicator of drainage obstruction, as well as a factor in filariasis transmission. The state of the ecosystem may provide another proxy indicator. For example, fish consumption is the major route of human exposure to hydrophobic toxic chemicals such as pesticides, which are assimilated in fish lipids and fats instead of being dissolved in water or dissipated in air. If birds and other wildlife are unaffected by pollution, then it may be safe to say that humans are spared of any imminent danger (Mackay 1992).

Monitoring systems are always perceived as additional costs with no direct link to productivity. Optimization of monitoring systems can be achieved through linkage with established environmental networks and health surveillance teams, if any are located within the project's focus. The public, especially in rural areas, could also provide an inexpensive and reliable option. Chemical accidents or spillage are best reported by those on the spot rather than by setting up an elaborate official surveillance network (Whyte and Burton 1980). River or coastal pollution can be monitored by fishermen using indicators such as dead fish, algal bloom, smells, foam, or coloured discharges. Public complaints to an upsurge in vermin or mosquito bites could also be used. However, these monitoring systems rely on public education about the indicators of high risk and require an effective communication system between the public and responsible government officials. To the environmentally conscious public, environment is too important to be left to experts who are no longer seen as omniscient or incorruptible (Lowenthal 1990). Therefore, public participation in matters of health concern could easily gain wide acceptance.

Monitoring is usually done by the project proponent while surveillance is by the regulatory agency. Where both parties agree on the objectives of monitoring and aim to share information regularly, they can be forewarned by certain trends or upsets in the system. In this way, preventive measures can be planned whenever possible. The municipal health officers and environment protection officials are both key persons to alert to possible disturbances. Since monitoring has been proven to be a cost-effective measure to avoid adverse outcomes, it is important to invest time and expertise in developing and accomplishing the monitoring plan. Most often, monitoring becomes an afterthought and relegated to junior employees.

Environmental Management

Many health safeguards consist of physical modifications to the environment. These measures have been considered particularly important for limiting the breeding sites

of disease vectors (WHO 1982). However, similar principles apply to other health hazards. Three forms of environmental management are usually distinguished.
- Permanent modification to the environment, such as canal lining and land zonation
- Repetitive actions, such as weed removal and routine maintenance
- Changes in human behaviour and habitation that are designed to reduce the hazardous conditions or limit exposure.

None of these measures is likely to be incorporated in project design or operation unless previously identified through HIA.

FUTURE DIRECTIONS

At present HIA is a blunt tool with the rudiments of an accepted methodology. It requires progressive refinement through detailed substudies. Refinements are needed on two separate levels: managerial and technical. In developing countries the primary problem is not one of improving the technical capabilities to deliver accurate assessments, although this is a difficult enough task. Rather, we must convince decision makers that health impact should be undertaken and that the results can be used to modify project design and operations. For this latter task we need three kinds of studies. First, we need policy research to explain how health impact decisions are made. Second, we need economic research to ensure that the decisions are made in a forum where the economic costs and benefits of health are assessed alongside all other costs and benefits. Third, we need case studies that can be used as models that are specific to the needs of each country. In the longer term, our objective is to provide more quantitative assessments of more carefully defined accuracy.

CONCLUSION

HIA is applied unevenly in several development sectors. It is not yet a requirement for all development projects. However, recent World Health Assembly declarations and statements in Agenda 21 of UNCED have placed renewed emphasis on impact assessment of all kinds.

If HIA is to be applied widely and successfully it must have a structured approach. Experience of EIA and ERA suggests that long and costly procedures do not appeal to users. Donor agencies are already overburdened by requirements to consider issues such as gender, environment, and poverty. They are unwilling or unable to absorb complex health assessment procedures. HIA must a have a simple entry point based on rapid assessment and project screening. In many cases a rapid assessment will suffice to strengthen the negotiations of the health sector within the multisector planning forum. Health risk management is consistent with sound project management. Human health and the health of the environment are mutually interdependent. If the health of people is safeguarded, then, ultimately, the eco-system itself must be safe.

ACKNOWLEDGEMENTS

Preparation of this paper was partially supported by the ODA through the Liverpool Health Impact Programme. The continuing support of the Joint WHO/FAO/UNEP/UNCHS Panel of Experts in Environmental Management is also acknowledged. Opportunities to develop these ideas were provided during a consultancy to the Asian Development Bank, Office of the Environment, to whom we would like to express our gratitude. Responsibility for the ideas expressed remains solely with the authors.

REFERENCES

ADB. 1990. *Environmental Risk Assessment: Dealing with Uncertainty in Environmental Impact Assessment*. Asian Development Bank. Environmental Paper, Environmental Paper No. 7.

ADB. 1992. *Guidelines for the Health Impact Assessment of Development Projects*. Asian Development Bank. Environmental Paper, Environmental Paper No. 11.

Birley, M.H. 1991. *Guidelines for Forecasting the Vector-borne Disease Implications of Water Resource Development*. World Health Organisation. WHO/CWS/91.3, second edition.

Birley, M.H. (in press) *The Health Impact Assessment of Development Projects*. Norwich: HMSO.

Bolton, D., A.M.A. Imevbore, and P. Fraval. 1990. A Rapid Assessment Procedure for Identifying Environmental and Health hazards in Irrigation Schemes. Hydraulics Research Wallingford. OD 120.

Carswell, J.W. 1987. "HIV infections in healthy persons in Uganda." *AIDS* 1: 223–227.

Carter, F.W. and D. Turnock. 1993. *Environmental Problems in Eastern Europe*. London: Routledge.

Coimbra, C.E.A. 1988. "Human factors in the epidemiology of malaria in the Brazilian Amazon." *Human Organisations* 47(3): 254–260.

Cooper Weil, D.E. et al. 1990. *The Impact of Development Policies on Health: a Review of the Literature*. Geneva: World Health Organisation.

Eisenbud, M. 1990. "The ionizing radiations." In *The Earth as Transformed by Human Action: Global and regional changes in the biosphere over the past 300 years*. B.L. Turner, W.C. Clark, R.W. Kates, J.F. Richards, J.T. Matthews, and W.B. Meyer., eds. Cambridge: Cambridge University Press.

El-Batawi, M.A. 1981. "Special problems of occupational health in the developing countries." In *Occupational Health Practice*. R.S.F. Schilling, ed. London: Butter-worths, 27–46.

Ewan, C., A. Young, E. Bryant, and D. Calvert. 1992. *Australian National Framework for Health Impact Assessment in Environmental Impact Assessment*. University of Wollongong.

Findley, R.W. 1988. "Pollution control in Brazil." *Ecological Law Quarterly* 15(1): 1–68.

Havard, J.D.J. 1978. World Health Organization Assignment Report. World Health Organisation Regional office for the Western Pacific. ICP/HSD/015.

Hunter, J.M. et al. 1993. Parasitic Diseases in Water Resources Development. Geneva: World Health Organisation.

IRRI. 1987. Vector-borne Disease Control in Humans through Rice Agroecosystem Management. International Rice Research Institute, Los Banos, Philippines: International Rice Research Institute and WHO/FAO/UNEP Panel of Experts on Environmental Management for Vector Control.

Laing, R. 1986. *Health and Health Services for Plantation Workers; Four Case Studies*. G. Walt, ed. London: Evaluation and Planning Centre for Health Care, London School of Hygiene and Tropical Medicine.

Lowenthal, D. 1990. "Awareness of human impacts: Changing attitudes and emphases." In *The Earth as Transformed by Human Action: Global and regional changes in the biosphere over the past 300 years*. B.L. Turner, W.C. Clark, R.W. Kates, J.F. Richards, J.T. Matthews, and W.B. Meyer, eds. Cambridge: Cambridge University Press.

Macdonald, W.W. 1991. "Control of *Culex quinquefasciatus* in Myanmar (Burma) and India: 1900–1960." *Annals of Tropical Medicine* 85(1): 165–172.

Mackay, D. 1992. *Multimedia Environmental Models: The Fugacity Approach*. Boca Raton FL: Lewis Publishers.

Munn, R.E. 1992. "Towards sustainable development." *Atmospheric Environment (26A)* 15: 2725–2731.

ODA. 1989. *Manual of Environmental Appraisal*. Overseas Development Administration.

OECD. 1986. *Methods and Procedures in Aid Evaluation*. Paris: Organisation for Economic Cooperation and Development.

Oomen, J.M.V., J. de Wolf, and W.R. Jobin. 1988. *Health and Irrigation: Incorporation of disease-control measures in irrigation, a multi-faceted task in design, construction, operation*. Wageningen: International Institute for Land Reclamation and Improvement.

Packard, R.M. 1989. "Industrial production, health and disease in sub-Saharan Africa." *Social Science and Medicine* 28(5): 475–496.

Pimenta, J.C.P. 1987. "Multinational corporations and industrial pollution in Sao Paulo, Brazil." In *Multinational Corporations, Environment and the Third World*. C.S. Pearson, ed. Durham: Duke University Press, 198–220.

Roundy, R.W. 1989. "Problems of resettlement and vector-borne diseases associated with dams and other development schemes." In *Demography and Vector-Borne Diseases*. M.W. Service, ed. Boca Raton FL: CRC Press, 193–205.

Service, M.W., ed. 1989. *Demography and Vector-Borne Diseases*. Boca Raton: CRC Press.

Spellerberg, I.F. and A. Minshull. 1992. "An investigation into the nature and use of ecology in environmental impact assessments." *British Ecological Society Bulletin* 23(1): 38–45.

Stigliani, W.M. et al. 1989. "Future environments for Europe: Sme implications of alternative development paths." *Science and Total Environment* 80: 1–102.

Thomas, V. 1981. Pollution Control in Sao Paulo, Brazil: Costs, Benefits and Effects on Industrial Location. The World Bank. Staff Working Paper No. 501.

Tiffen, M. 1991. Guidelines for the Incorporation of Health Safeguards into Irrigation Projects through Intersectoral Cooperation. World Health Organisation. WHO/CWS/91.2.

Turnbull, R.G.H. 1992. *Environmental and Health Impact Assessment of Development Projects, A Handbook for Practitioners*. London and New York: Elsevier.

WHO. 1982. Manual on Environmental Management for Mosquito Control, with Special Emphasis on Malaria Vectors. Edited by WHO. Offset Publications. Geneva: World Health Organisation.

WHO. 1983a. Maximising Benefits to Health. An appraisal methodology for water supply and sanitation projects. World Health Organisation. ETS/83.7.

WHO. 1983b. Minimum Evaluation Procedure (MEP) for water supply and sanitation projects. World Health Organisation. Mimeograph, ETS/83.1,CDD/OPR/83.1.

WHO. 1987. Health and Safety Component of Environmental Impact Assessment. World Health Organisation. Environmental Health, Environmental Health Report No. 15.

WHO. 1992a. Health Dimensions of Economic Reform. Geneva: World Health Organisation.

WHO. 1992b. Our Planet, Our Health: Report of the WHO Commission on Health and Environment. Geneva: World Health Organization.

Whyte, A.V. and I. Burton. 1980. *Environmental Risk Assessment*. New York: John Wiley and Sons.

World Bank. 1991. Operational Directive 4.01: Environmental Assessment. World Bank.

Chapter 8
Ecological Impact Assessment[1]

JOANNA TREWEEK
Institute of Terrestrial Ecology at Monks Wood, UK

INTRODUCTION

The ecological consequences of human activities are only partially understood. The extent to which 'natural' ecosystems are buffered against anthropogenic perturbation is unclear, and it is difficult to estimate the risk of irreversible damage to ecosystem components and functions which may be essential for human wellbeing. This uncertainty has prompted much recent debate on the need to preserve biodiversity, and to promote principles of sustainable development based on 'wise-use' of finite natural resources (IUCN 1980; WCED 1987). It has also lent impetus to formalised mechanisms for environmental assessment and management that are designed to regulate the environmental impacts of human activity.

By providing analytical procedures for studying the interrelationships between organisms and their environment, ecological science should have a pivotal role in impact assessment by predicting the consequences of individual development projects on valued ecosystem components and evaluating the environmental consequences of policies, plans, and programmes. Increasingly, ecological impact assessment is also required in industrial regulatory and consent procedures.

In many countries, there is a considerable gulf between the techniques actually used for impact assessment and those researched and promoted in the scientific literature. This paper sets out to establish the range of techniques used for ecological impact assessment and their theoretical basis; to evaluate their effectiveness in identifying and quantifying the ecological consequences of human actions; and to identify areas of ecological science that must be developed further if ecological impact assessment techniques are to be improved.

THE HISTORY OF ECOLOGICAL IMPACT ASSESSMENT

As a formal discipline, ecological impact assessment had its origins in the National Environmental Policy Act (NEPA) which became law in the United States in 1969 and established a legislative requirement for proponents of an action to assess potential environmental impacts. The implementation of the act set a precedent for

[1]*Environmental and Social Impact Assessment* - Edited by F. Vanclay and D.A. Bronstein. Copyright © 1995 by the International Association of Impact Assessment. Published in 1995 by John Wiley & Sons Ltd. A version of this chapter will appear in *Impact Assessment,*the quarterly journal of IAIA.

adoption of similar forms of legislation on environmental impact assessment (EIA) in other countries. Mechanisms for EIA are now generally accepted to be prerequisites for effective environmental planning and management, even in countries that previously possessed well-developed planning systems, like the UK (Burdge 1991). Introduction of legislation for EIA throughout the world has greatly increased the demand for techniques which can be used to predict and evaluate the ecological consequences of human action, whether at the local, regional, national, or global level.

Ecological impact assessment has been used primarily to predict the consequences of development activities for organisms other than people. Many of the techniques used were originally developed for purposes of nature reserve selection or wildlife management, and required some modification before they could be adapted for purposes of impact assessment. The application of ecological principles within the boundaries imposed by EIA legislative frameworks has been fraught with difficulties, many of which were identified very early in the evolution of EIA procedures and have still not been resolved 25 years later. Key problems identified have been the lack of availability of relevant data, inadequate understanding of complex ecological processes, and procedural difficulties largely due to the lack of temporal flexibility in EIA procedures (Beanlands and Duinker 1983; Spellerberg 1993). The following sections outline the main components of ecological impact assessment and summarise some of the problems that have been encountered in developing effective techniques.

ECOLOGICAL IMPACT ASSESSMENT: DEFINITION AND DESCRIPTION

In its simplest definition, ecological impact assessment is a formal process of defining, quantifying and evaluating the potential impacts of defined actions on ecosystems. It therefore requires identification of environmental components and understanding of the factors or processes which determine their interactions with each other, so that impacts of specific activities can be superimposed on baseline conditions and their potential effects forecasted. Where it is possible to predict these outcomes in quantifiable terms, their relative importance can then be evaluated for decision-making purposes.

As a formal discipline, ecological impact assessment has been developed primarily in support of EIA. For administrative and research purposes, it has therefore been customary to consider ecological impact assessment as a stepwise series of procedural stages commensurate with those in EIA and applied predominantly on a site- or project-specific basis. Ecological impact assessment has potential as an environmental management tool with far broader applications, however, and is best considered as an iterative process with a transboundary perspective which incorporates monitoring and feedback at all stages to characterise the 'state of the environment'.

The basic components of ecological impact assessment are baseline studies (which may or may not incorporate ecological scoping and screening procedures), impact assessment, impact evaluation, mitigation and monitoring. The process of ecological impact assessment relies in the first instance on standard techniques of survey, taxonomic classification, monitoring and predictive modelling. These techniques are fundamental to the academic discipline of ecological science which is dispassionate and objective and seeks simply to quantify ecosystem components and the processes that link them. Thus surveys might be carried out to estimate populations of species linked by processes such as nutrient cycling, energy flow between trophic levels or population dynamics which can be measured as rates and used as a basis for modelling.

In addition, ecological impact assessment generates a requirement for techniques of ecological evaluation which can be used to estimate the significance or importance of ecological impacts and to convert measurements of ecosystem state or function into a form which can be used for decision making. Evaluation of ecological impact significance or importance depends on the definition of objectives and evaluation criteria which can be used to interpret ecological measurements in terms of policy-aims which may be strongly influenced by socio-economic or political factors. This 'fluid boundary between science and politics' (Pritchard 1993) has been the source of much inconsistency in the application of ecological impact assessment procedures and their effectiveness in influencing planning decisions.

It is common, but not axiomatic, for 'ecological impacts' to be interpreted in terms of 'nature conservation value'. Scales of value determined on the basis of other social or economic considerations could also be used (Brown, Moran 1993), but there is an ongoing debate about the ability of economists to measure the value of 'natural' or ecological resources in monetary terms (Krutilla and Fisher 1975; Bergstrom 1980).

BASELINE CONDITIONS: SURVEY AND INVENTORY

Nearly all EIA legislative guidelines refer to the need for a description of the 'existing environment', some characterisation of baseline conditions being essential before the potential ecological impacts of an action can be predicted. Ideally, baseline studies are based on repetitive sampling to statistically define the spatial and temporal variability of measured variables (Beanlands, Duinker 1983). However, the scope of baseline studies carried out for EIA is invariably constrained by pressure to complete studies within a reasonable time-frame and a restricted budget. In the European context, it is still common for environmental impact statements (EISs) to include descriptions of the 'existing environment' based on 'one-off' surveys, often carried out at a relatively superficial level to identify habitats or communities which merit more intensive survey and/or monitoring. By failing to quantify inherent variation (in either spatial or temporal terms) such approaches negate attempts to attribute any subsequent change in ecosystem parameters to specific impacts, resulting in impact assessments which are fundamentally flawed.

While identification of ecosystem components is prerequisite for character-isation of baseline conditions, it is generally accepted that it will be impossible to establish comprehensive, long-term monitoring programmes for all ecosystem components. As a result, much research has focused on methods by which 'valued ecosystem components' (VECs) and key biological processes can be identified at the beginning of the impact assessment process, with subsequent studies being designed to investigate potential changes in these. Straightforward inventory can therefore be complemented by a process of 'focusing' to rationalise the subsequent impact assessment (Beanlands and Duinker 1983; Kennedy and Ross 1992). However, in practical terms, no general consensus has been reached about appropriate methods for impact focusing.

IDENTIFICATION OF VALUED ECOSYSTEM COMPONENTS

'Key' species have been selected on account of their economic importance, their protected status, their rarity, their sensitivity to specific impacts, or their representa-tiveness of other species which use a common environmental resource (guilds). There are attendant difficulties with all these selection criteria. Those species most sensitive to specific impacts, for example, may be relatively rare and difficult to survey (Eberhardt 1976). While survey efficiency can be increased by application of guild theory (Johnson 1981; Severinghaus 1981), the ability to assign species to guilds and select appropriate indicator species for each guild is complicated by the fact that no two species share precisely the same habitat requirements, life-strategies or responses to impacts.

Application of guild theory is most effective when acceptable levels of homogeneity are maintained within individual guilds whilst also allowing maximum generalisation about species similarities. Johnson (1981) therefore used key growth-form, physiological, and dispersal characteristics to assign plant species to guilds and avoided use of attributes likely to engender variable responses to impacts. Detailed knowledge of the full range of species present and their autecology is essential for such selective approaches to be valid. Surveys can therefore be rationalised by use of guild theory only if existing information is comprehensive and reliable, which is often not the case. The practical application of guild theory to ecological impact assessment has been negligible, despite its theoretical potential in providing efficient analytical frameworks for impact prediction.

Expediency may necessitate the concentration of survey effort on easily recorded taxonomic groups and on those for which effective evaluation mechanisms are available. In the UK, for example, higher plants are sampled far more frequently than other groups (Treweek et al. 1993). It is also common for Lepidoptera to be sampled/surveyed in preference to other invertebrate groups because there are standard survey methods available in the form of the 'Butterfly Monitoring Scheme' (Pollard and Yates 1993) and because the widespread use of this scheme has generated a lot of comparative data. Nevertheless, attempts to rationalise subsequent studies by focusing only on preselected ecological components do create potential bias due to premature assessment of 'importance' or 'value' and the tendency to exclude 'unfashionable' and under-recorded species. The validity of impact focusing techniques therefore depend on the availability of comprehensive and up-to-date data. It is possible that more use should also be made of historical data on distributions of habitats or species to improve understanding of baseline conditions and to project future trends in the light of current, or new combinations of threats and impacts.

Environmental components meriting subsequent survey should obviously also include those most likely to be affected by specific development actions. These may be identified using 'impact scoping' procedures (Kennedy and Ross 1992) to further rationalise the range of components that will be covered by the impact assessment.

IMPACT ASSESSMENT

Impact Identification/Focusing

The first requirement for impact prediction, is to identify those activities likely to interact with ecosystem components or processes and to define their expected range in spatial and temporal terms.

In addition to straightforward checklists, matrices constructed with project activities on one axis and potentially affected environmental components on the other (Leopold et al. 1971; Kennedy and Ross 1992) are widely used for impact identification. They are particularly useful in cases where large amounts of data need to be summarised and can be used to screen out impacts likely to be insignificant or to rationalise the immediate exclusion of some environmental impacts from further evaluation (Kennedy and Ross 1992).

However, two-dimensional matrices suffer from problems of identifying the secondary or higher order impacts which characterise most ecosystems. These have been tackled by use of a component interaction technique (Shopley and Fugle 1990) in which the environment is modelled as a list of components, ranked in order of their ability to initiate secondary impacts on the basis of chains of dependence between them. Computerised matrix powering procedures are employed to structure the data and facilitate an investigation of the secondary impact potential of the system. In this method, dependencies are scored as simply present or absent in order to measure the extent to which ecosystem components are linked. The degree of disruption caused by removal of a component can be traced through its associated linkages (using the assumption that indirect dependencies only operate through the shortest path) in order to identify critical components which might therefore merit further study. However, such approaches depend on a thorough understanding of how key components of ecosystems interact with one another. For many ecosystems, this understanding is very limited.

Increasingly, geographical information systems (GIS) are emerging as tools for the identification and quantification of potential ecological impacts. Linked to comprehensive databases on the distributions of abiotic and biotic variables, they offer powerful techniques for addressing 'where' and 'what if' questions about the likely location and magnitude of interactions between ecosystem components and defined activities. Simple overlay techniques, for example, can be used to identify areas where the risk of ecological impact is likely to be greatest, or to plan for avoidance of key areas. Appelman (1991) assigned values for importance, sensitivity and vulnerability to ecosystem components in relation to defined impacts in a GIS so that areas of high ecological value and high sensitivity to impact could be screened out in the early stages of route selection for proposed road developments. In other studies, plume-modelling techniques have been superimposed on land cover maps derived from satellite imagery to identify areas where sensitive vegetation communities are most likely to be subjected to high levels of pollutant deposition from industrial installations (Radford et al. 1994).

Impact Measurement

There is a widespread belief that current understanding of the relationship between increasing anthropogenic stress and declining ecosystem function is inadequate for predicting subtle or incremental changes (Cairns and Niederlehner 1993). Nevertheless, the fact that impact assessment requires study of cause–effect relationships between ecosystem components and defined actions cannot be ignored.

Traditional approaches to quantifying impacts use data collected in control and impact zones before and after an impact occurs (Green 1979), which lend themselves to straightforward techniques of analysis of variance to measure impact significance. However, classical inferential methods cannot be applied to pre- and post-operational data on only one impacted site (Eberhardt 1976) with no formal experimental control. Many impact assessments have been criticised on account of lack of replication and randomisation of impacts to sites (Hurlbert 1984).

In order to get round this problem, the use of 'control' sites situated as close as possible to the test site has been proposed, with 'baseline' studies being carried out to establish, *inter alia*, the ratio of population density in the test site to that on control site(s) before and after the test site is developed. However, in countries where habitats have been greatly modified by human activity or management and are already highly fragmented, the identification of suitable 'control' sites can be very difficult. The statistical problems associated with ecological impact assessment studies are considered in more depth by Smith, Orvos, and Cairns (1993). As a result of the difficulties inherent in treating impact assessments as controlled field experiments which measure functional attributes before and after a defined impact, statistical analyses are rarely even attempted in the vast majority of ecological impact assessment reports.

Methods for measuring ecosystem function are also much less well developed than those for defining ecosystem structure (Cairns and Niederlehner 1993). This has reinforced the tendency for ecological impact assessment to evolve as a descriptive, rather than an analytical process. However, it is sometimes possible to interpret ecosystem structure in functional terms. Noss (1990), for example, recommended the use of quantifiable indicator variables for inventorying, monitoring, and assessing terrestrial biodiversity according to hierarchical levels of organisation. Indicator variables were chosen to reflect the composition and structure of ecosystems (for example, species frequency, richness, and diversity) and also function (for example, biomass productivity, herbivory, parasitism, predation, rates of colonisation, and extinction), with methods for measurements being chosen according to the scale, or level of organisation of study.

GIS-based techniques also offer considerable potential for combining impact identification and quantification with predictive modelling for impact assessment. In theory, they can be used to combine data on the spatial distributions of environmental components with inferential rules derived from an understanding of ecosystem processes in order to model, or predict the spatial consequences of different impact scenarios.

As well as facilitating measurements of impacts such as direct habitat loss, GIS can be used to calculate indices based on spatial attributes which may be related to functional aspects of landscape ecology in predictable ways. In countries with predominantly anthropogenic landscapes, remaining wildlife habitat is often restricted to isolated 'islands' in a highly modified 'matrix'. The viability of wildlife habitats and species in such fragmented landscapes is jeopardised by increasing distance between fragments (isolation), increased edge:interior ratios and decreasing patch size. Where the average distance between fragments exceeds dispersal distances, genetic exchange between populations may be reduced; while increased edge:interior ratios are often associated with increased pollution effects, or ingress of noncharacteristic species.

Geographical information systems (GIS) have made it practically feasible to quantify incremental, or cumulative impacts on the regional or national availability of habitat for associated species. Treweek and Veitch (in press) found that the regional, cumulative impacts of new road development on woodland and lowland heathland habitats could be effectively estimated using a GIS based on a land cover map generated from satellite imagery (Fuller, Groom, and Jones, in press; Fuller and Parsell 1990). The challenge now is to improve understanding of the relationships between spatial descriptors of habitat distribution and the functional behaviour of species and populations, for example, through studies of metapopulation dynamics.

Predictive Ability

Characterisation of the existing environment, definition of project activities, and understanding of ecosystem function lay the foundations for the prediction of future conditions relative to baseline conditions. Impact assessment is essentially an exercise in prediction, but predictive ability is generally recognised as weak, largely due to the complexity of ecosystems and a shortage of long-term datasets.

Predictions made during ecological impact assessment should be regarded as hypotheses that can be tested using monitoring data (Buckley 1991). For this to be possible, predictions must be defined in quantitative terms, preferably presented in the form of a time-series covering the predicted duration of project activities and accompanied by estimates of probability of predicted impacts occurring. Reviews of impact statements, however, have revealed a widespread lack of quantified predictions and an emphasis on verbal forecasts and 'vague or contradictory statements' about impact significance (Culhane et al. 1978). Buckley (1991) reviewed 181 Australian EISs and found that predictions were less than 50 percent accurate on average, and occasionally over two orders of magnitude out. Culhane et al. (1978) found biological impacts to be the least frequently discussed, the least often quantified, and the least accurately predicted of impact categories. If made at all, ecological predictions in UK EISs are generally so ambiguous as to be virtually nontestable (Spellerberg 1991; Treweek et al. 1993). Because of the uncertainty that surrounds ecological predictions, predictive statements should always be accompanied by a discussion of the limitations of the analyses used and a statement of the confidence which can be attached to the predictions made (Culhane et al. 1987; Glasson 1994)—but this is rarely the case.

It is universally acknowledged that the ability to predict ecological impacts depends on a thorough knowledge and understanding of ecosystem function. For some species, sufficient autecological information exists for it to be possible to identify generic responses to different categories of stress. Some species possess identifiable sets of attributes which determine their likely response to broad categories of impact such as disturbance, eutrophication or dereliction, for example. The effects of such impact-types can therefore be predicted, whether for individual species or for communities. Hodgson (1991) used Grime's C-S-R theory of life-history strategies of plant species (Grime 1974; 1979) to predict their responses to different impact scenarios and to identify those species likely to be promoted or inhibited by each.

Cairns and Niederlehner (1993) found community species composition to be a consistent and sensitive indicator of a range of stress types. A number of environmental quality models have been structured on the premise that the presence of a particular suite of species can be linked to certain habitat conditions and that changes in environmental quality, for example as a result of pollution, may be reflected in predictable community changes. Without followup monitoring of the effects of defined impacts, however, such knowledge and understanding has been slow to develop.

MONITORING

The ability of ecologists to predict the ecological consequences of proposed actions will remain limited unless the use of operational and post-operational phase monitoring becomes more prevalent. The need for comprehensive baseline surveys and structured monitoring programmes has been repeatedly emphasised by ecologists (Eberhardt 1976; Beanlands and Duinker 1983; Bernstein et al. 1993). As a general rule, however, ecological assessment procedures end with the submission of the impact statement and remarkably few studies have been carried out to test the accuracy of impact predictions (Culhane et al. 1978; Buckley 1991) or to monitor the longer term consequences of development actions. As a result, EIA has failed to provide a mechanism for the development of ecological knowledge and understanding which could be used to enhance predictive capabilities for impact assessment. As long ago as 1983, Beanlands and Duinker recommended that 'environmental impact assessment agencies undertake whatever procedural changes are necessary to have monitoring formally recognised as an integral component of the assessment process'. However, formal requirements for monitoring remain the exception rather than the rule.

EVALUATION

Ecological factors represent only a portion of the total range of factors involved in environmental impact assessment and cannot be evaluated effectively in isolation from prevailing social, economic, aesthetic and cultural conditions. It is often the tradeoff between 'biophysical and socio-economic impacts that is crucial in decision making' (Glasson 1994).

These other factors influence both the importance attached to ecological concerns and therefore the funds likely to be available to mitigate adverse impacts or protect important ecosystem components. Ecological evaluation is therefore necessary both to assess the consequences of development for ecological components per se (which might be termed 'ecological impact significance') and to provide a basis for comparison with other categories of impact (which can be referred to as 'ecological impact importance').

The need to base impact assessments on attributes which can be measured effectively and also translated into management goals (Cairns and Niederlehner 1993) has already been referred to. Disparate impacts can only be optimised during evaluation if they have been quantified in some way, and can be converted to common scales of comparison (Culhane et al. 1987). Bias in evaluation is minimised if the use of subjective evaluation criteria is delayed until the last stage in the decision-making or impact assessment process, after any quantification of impacts has been carried out. As a result, estimates of impact significance should always precede assessment of impact importance.

Evaluation of impact significance might be based on comparison between a measured state (following impact) and a reference standard, which might be an 'optimal state' or a quality objective. The significance of habitat loss for a species might therefore depend on the relationship between habitat availability and species population-dynamics and it might be possible to establish thresholds or standards reflecting the extent of habitat loss which can be accepted before the viability of a species will be jeopardised. On the other hand, the importance of habitat loss depends also on the perceived value of the species in question. Loss of habitat is likely to be regarded as more important for species which are rare, threatened, or protected by law. However, it is important that such criteria should be applied in a rigorous and transparent way, if policy-related priorities are to be translated into effective regulatory action.

Habitat evaluation procedures as developed in the US (Atkinson 1985; Crance 1987) use autecological information on the habitat requirements of species to define their minimum habitat requirements and assess the quality and amount of available habitat. Potential losses of habitat (in terms of both area and quality) can then be compensated or mitigated by the provision of equivalent habitat in order to ensure that 'target' or 'valued' species are conserved. The procedures focus on the calculation of habitat suitability indices (or HSIs) which express the ratio between the habitat conditions provided by a study area for an 'evaluation species' (which may be a single species, group of species, individual life-stage, or 'life requisite') and its optimum habitat conditions. The HSI is multiplied by the area of available habitat in order to calculate habitat units (HUs), which can later be adjusted on the basis of ranking criteria which include scarcity, vulnerability, and replaceability.

By deriving biological expectations for 'healthy systems' or 'optimum system states', it is possible to evaluate 'before and after conditions' against a common standard, or index. Indices of biotic integrity, or IBIs, for example, have been used effectively for freshwater systems in the US. Such indices can put into a regional context by assigning 'regional reference habitats' (Karr 1991).

In many circumstances, the application of such procedures is limited by lack of information on the habitat requirements of species. Minimum viable habitat sizes are known for only a very few plants or invertebrates, for example. Furthermore, in highly anthropogenic and fragmented landscapes it can be very difficult to determine the 'optimum' habitat requirements of species. In such landscapes, where 'wilderness areas' are highly restricted, there has tended to be much more emphasis on the identification, selection and preservation of remaining, important sites for nature conservation purposes. In many countries, losses of wildlife habitat have already been so extensive, that the ability to locate 'replacement' habitat is extremely limited.

EVALUATION CRITERIA/INDICATORS

The use of evaluation criteria has been researched in depth, but in terms of practical application, no overall consensus has been reached. In the UK, considerable use has been made of Ratcliffe's (1977) criteria for the selection of sites of nature conservation importance. These include size, rarity, diversity, naturalness, fragility, and typicalness as well as 'recorded history, position within ecological unit, potential value and intrinsic appeal'.

Others have included 'resilience' as a criterion relating to the ability of an ecosystem to return to some approximation of its predisturbance condition (Cairns and Niederlehner 1993). For ecosystems with low resilience, the probability of 'irreversible or irretrievable commitment of resources' (CEQ 1986) is higher.

Combined terms have also been used. Species 'security', for example, combines measures of rarity and degree of threat, such that species which are rare and threatened throughout their range are regarded as very insecure. The ability to restore or reinstate ecosystems or habitats ('restorability') is also increasingly being used as an evaluation criterion in impact assessment.

The use of species as indicators of, for example, habitat quality and/or age, naturalness, pollution status offers considerable potential for ecological impact assessment, but is relatively under-researched. In practical terms, species-based analyses are prone to bias towards more popular, or 'emblematic' species at the expense of those which might give a better indication of habitat quality. In the UK, the lemon slug (*Limax tenellus*), for example, is among the most stenotopic ancient woodland indicators (Eversham, pers. comm.). It seems to be incapable of colonising recent (even 200-300-year-old) plantations, possibly because of their poor soil microflora relative to truly old woodlands. It is therefore a good indicator, but may have less appeal as a candidate for survey than conventionally more attractive invertebrate species (usually the butterflies).

MITIGATION

Ecological mitigation measures fall under three main categories: impact avoidance, impact amelioration and habitat replacement/restoration. It is important that mitigation measures should be tied closely to an understanding of ecosystem function and to defined 'performance standards', so that it is possible to evaluate their effectiveness. Thus, an impact which causes a reduction in overall environmental quality, or a measured decline in the status of specific ecological components should trigger appropriate mitigative measure to restore standards back above thresholds of acceptability.

In some ecological impact assessments, limits of acceptable change (LACs) have been defined for specific environmental components, with agreement on subsequent mitigative action if they are exceeded (Bayfield 1990, pers. comm.). For example, LACs were defined for percentage vegetation cover, the average size of bare patches, and the width of footpaths in order to assess and regulate erosion in a Scottish skiing area. Such iterative approaches to impact mitigation have been rare, however, mainly because of the lack of formal requirements for followup monitoring.

Undertakings to transfer habitats and communities to new sites, to reinstate habitats which are damaged or lost, or to construct replacement habitat are frequently made in impact statements. However, successful examples are hard to find. Notable exceptions include the reinstatement of species-rich grasslands on UK road verges (Department of Transport 1993) and the restoration of lowland heathland communities following pipeline construction (British Gas 1988). There would not appear to be any sound evidence that it is possible to reconstruct most target communities or habitats in their entirety. Habitat restoration is fraught with problems, mainly because of limited understanding of successional processes.

Some habitats and communities have proved extremely difficult to restore and others, such as ancient woodland, cannot by definition be restored within human timescales. Recommendations for mitigation should therefore be accompanied by some indication of their likely success.

RELATIONSHIP TO OTHER FORMS OF IMPACT ASSESSMENT

To be successful, ecological impact assessment depends on a fully integrated approach involving elements of risk assessment (what is the likelihood of an adverse impact occurring?) and economic impact assessment (what are the costs of mitigating an adverse impact, or of allowing it to proceed without mitigation?) as well as techniques for ecological survey, monitoring, prediction, and evaluation.

Ecological impact assessment should be structured so that cross-comparison with other categories of impact is possible. It is often necessary, for example, to evaluate ecological impacts using socio-economic criteria and there are also circumstances in which different categories of impact may be closely interrelated. However, little research effort has been invested in methods for analysing such interrelationships.

The consequences of undertaking a development action, for example can have a 'precedent-setting' effect, altering the rate of development of new activities in an area (Contant and Wiggins 1991), or leading to shifts in regional development structures which can have considerable 'knock-on' effects for ecological resources. In order to develop ecological impact assessment as a mechanism for environmental regulation and management which takes account of cumulative change, it will be necessary to monitor development actions by type and location over time and also to take account of future development by collecting data on those socio-economic system parameters which influence the nature and rate of development activity.

There is a growing need to coordinate scientific knowledge, public concerns, and management information in establishing clear objectives for environmental monitoring and assessment. This is likely to require a shift away from impact assessment procedures that take a compartmentalised approach and deal with sectional interests separately.

THEORETICAL/PRACTICAL PROBLEMS

The need for ecological impact assessment techniques arises from a political or socioeconomic motivation and there has been a tendency for scientists to doubt whether 'it is an acceptable forum in which to rigorously apply the scientific method' (Beanlands and Duinker 1983). As a consequence, the potential of ecological impact assessment to provide much-needed empirical data on impact-responses has not been realised.

The distinction between theoretical and practical problems is not always straightforward. The lack of clear ecological rules to guide impact prediction, for example, has direct consequences for practical decision making in relation to ecological issues. However, it is possible to identify a number of practical or institutional constraints which have made the effective application of rigorous ecological assessment methods almost impossible, and which have contributed towards a number of commonly identified shortcomings.

1. Independent Review

In many countries, ecological impact assessments are commissioned or carried out by the proponents of projects or policies without any independent review. This tends to result in ecological impact assessments with restricted scope (in both spatial and temporal terms), which lack objectivity and rigour, and are often misleading. Studies should be timed to coincide with key life-cycle periods (for example, main breeding or feeding periods) and to take account of seasonal variation. Where possible, repeated samples should be carried out throughout the year, with one year being the basic, minimum planning unit (Eberhardt 1976). However, in practical terms, impact assessments structured in this way are the exception rather than the rule. Common shortcomings identified include inappropriate survey times, lack of replication, and neglect of key components. Without an independent review body to enforce compliance with best-practice techniques, such shortcomings are likely to persist.

2. Guidance

Lack of independent review may be compounded by lack of official guidelines on appropriate methodologies. This has implications for practitioners, proponents and decision-making authorities as well as the overall quality of the process. Lack of official guidance has contributed towards the marked inconsistency in approach that characterises ecological impact assessment and impact assessment in general in most countries, and has hampered the ability to learn from experience and define best-practice methodologies.

3. Monitoring

The lack of any formal requirement for followup monitoring has greatly hampered the development of predictive methods and effectively renders ecological impact assessment useless as a mechanism for environmental regulation (Hollick 1981). The need for monitoring has already been discussed. As an integral component of the impact assessment process, its formal inclusion is both urgent and vital.

4. Information Handling/Availability

Proactive environmental assessment and conservation must be based on the effective management of relevant data to provide usable information in support of land-use decision making. Information-systems approaches to the management of geographically referenced data allow powerful analytical techniques to be applied (Pellew 1989) and have also made it possible to utilise sources of data, such as remotely sensed data, which were not previously accessible to most EIA practitioners. Satellite imagery, for example, has proved to be of great value in providing comprehensive, up-to-date, objective, and readily quantifiable data on vegetational characteristics or land cover (Veitch, Treweek, and Fuller, in press). The ability to access ecological information rapidly without the need for expensive and time-consuming site-surveys has removed one of the main barriers preventing early consideration of potential ecological risks in the EIA process. Satellite-derived information on land cover has also made it possible to quantify regional and cumulative impacts of development actions and to model the implications of different development scenarios in broad spatial terms. Subsequent impact assessments can then focus more detailed field studies on those areas identified as being most at risk, or where ecological value is considered greatest.

The use of new techniques for the storage, manipulation and analysis of spatially referenced ecological data has varied considerably between countries, however. GIS technology is widely and routinely used in the Netherlands, Australia, the United States, and Canada, but only sporadically in other countries.

There are major gaps in knowledge concerning the distribution and status of biota and ecosystems in most countries. Assessment of ecological impacts at anything other than local scales is impossible without the appropriate reference data, but comprehensive, national data-sets on species- and habitat- distributions are rare.

5. Resourcing Constraints

The pressures of time and restricted budgets also create difficulties, making it difficult to initiate the types of baseline survey and monitoring programmes which would provide sound ecological data. Ecological studies are frequently under-resourced and carried out on shoe-string budgets.

6. Sampling/Statistical Problems

Ecological impact assessments have often been criticised for relying too heavily on subjective evaluation and approaches based on quantitative methods have frequently been advocated by ecologists (Eberhardt 1976; Spellerberg 1991). The problem of taking account of inherent variation in baseline studies has already been alluded to. One consequence of inadequate baseline data is that it becomes difficult to determine whether statistically significant changes are directly attributable to defined impacts (Eberhardt 1976). Classical experimental designs used to detect and measure effects are often inappropriate for ecological impact assessment because of the lack of scope for true randomisation and replication (Eberhardt 1976). It is important that more use should be made of statistical advice in designing impact assessment studies. Otherwise, considerable time and effort may continue to be invested in the collection of data which cannot be effectively analysed.

7. Decision making

The need for effective incorporation of ecological issues into decision making is increasing as threats to natural resources multiply. Incomplete understanding of certain conceptual and practical aspects of ecosystem management and biodiversity conservation can serve as a barrier to greater incorporation of these issues into decision making (CEQ 1993). Effective communication is essential, particularly where decisions are taken by non-ecologists.

8. Strategic Assessment

In a number of countries (including European countries), the lack of legislation requiring assessment of plans, policies and programmes makes effective ecological assessment very difficult. Ecology is a discipline which requires a longer view in order to place short term, or immediate effects in the context of long-term trends or cycles. The need to take a more strategic and synoptic approach to ecological impact assessment is fundamental to its maturation as an effective tool for environmental management. It is therefore considered in more depth in the following section.

CONTRIBUTION TO POLICY MAKING AND ENVIRONMENTAL MANAGEMENT

Ecologists are united in their belief that ecological impacts should be assessed at all relevant scales. However, the majority of ecological impact assessments have been carried out for individual projects and have been strongly site-specific. There has been an almost universal failure to assess cumulative or long-term effects, despite the fact that these are often a highly significant cause of environmental degradation. Individual human activities produce many impacts that are insignificantly small when considered in isolation, but which have significant incremental effect when considered in conjunction with other past, present and future activity (Eberhardt 1976). It is becoming increasingly common for planning authorities to incorporate estimation of ecological resources into structural and development plans at local and regional levels, but insufficient attention is given to the ecological implications of changes in policy at higher tiers.

The process of predicting and minimising the consequences of a single action has not adequately considered the accumulative nature of some effects, the nonlinear responses of some natural systems, nor the linkages between a single action and other related activities (Clark 1986; Roots 1986; Hirsch 1988; Contant and Wiggins 1991). Along with providing accurate predictions of the direct consequences of an action on an environmental parameter, cumulative impact analyses require a proposed action and its impacts to be placed in the context of other existing or expected actions and existing or expected environmental conditions (Cowart 1986). Broadening the scope of ecological impact assessment to take account of cumulative effects would thus require combined modelling of development patterns and natural systems' responses (Sonntag et al. 1987; Contant and Wiggins 1991).

In most cases, an individual project proposal is part of a larger programme of activities. In the US, environmental assessment regulations (US CEQ 1978) require that connected, similar and cumulative actions should be grouped together in one analysis, but in most countries, procedures for impact assessment of connected actions have been neglected. There has also been a tendency for developers to segment large projects into smaller units in order to downplay potentially major impacts of the whole project and focus on the relatively minor, separate impacts of sub-units. In a practical sense, it is important that cumulative effects should be considered early in the planning process and within the context of larger programmes of related activities.

Natural systems rarely react in a direct, or straightforward way to external pressures. A great many ecological impacts are, in effect, cumulative and interactive. There are many examples where incremental accumulation (Baskerville 1986; Clark 1986) or delayed responses (Baskerville 1986) have led to discontinuous impacts, exceedance of thresholds, or the crossing of stability boundaries, which cannot be attributed to any one, single action (Preston and Bedford 1988; Sonntag et al. 1991). There are also many examples of nonlinear functional relationships (Bedford and Preston 1988) and synergistic or interactive effects (Vlachos 1985; Beanlands et al. 1986). Beanlands et al. (1986) identified 'time crowding' effects, which relate to the inability of systems to recover from earlier perturbations before repeated perturbations occur and 'space crowding' effects resulting from the incremental insult of repeated actions on an area over time.

Incorporating assessment of cumulative and wider ecological effects into project-level decisions is likely to require significant improvements in existing administrative and managerial systems. A major limitation in most cumulative impact analyses is the lack of detailed monitoring information on previous development projects and several key environmental parameters.

Cumulative impacts expressed at a regional scale can only be controlled through planning processes directing development at that scale (Sonntag et al. 1987). In many cases, effective control of cumulative impacts therefore requires regional planning and cooperation. For this reason, separate legislation for strategic environmental assessment (SEA) has been recommended for the UK (Lee and Walsh 1992; Therivel et al. 1992; Cuff and Ruddy 1994). Some provision for impact assessment of procedures, policies, and plans already exists in a number of countries, including the US (Webb and Sigal 1992), Australia and New Zealand (New Zealand Government 1992; Wood 1992), and the Netherlands (Verheem 1992). However, there are few examples of strategic use of ecological impact assessment. These would have to be based on a translation of principles such as sustainable development, or conservation of biodiversity into measurable ecological attributes such as carrying capacity, minimum viable population sizes, species/area relationships, and so on.

In particular, a more strategic approach to ecological impact assessment is considered important to take account of the different scales of impact generated by linear developments such as roads (Sheate 1992; Treweek et al. 1993) and pipelines and also transboundary pollution effects.

LIKELY FUTURE

Progress in data handling/interpretation should improve ecological impact assessment techniques by facilitating the compilation of relevant databases and enhancing the ability to model potential impacts and communicate complex findings to nonexperts. Comprehensive and up-to-date information on the state of the environment is essential if a more strategic approach is to be taken to environmental assessment and management. Ecologists and nature conservationists continue to advocate the adoption of more integrated approaches to environmental management which would allow development to be planned in order to avoid ecologically sensitive locations or to maintain damage or loss within acceptable limits which allow natural resources to be maintained (Pritchard 1993). However, the development of national datasets that will allow the 'state of the environment' to be audited will require considerable commitment and investment by governments which has so far been lacking in the majority of cases.

CONCLUSION

It is possible to identify a basic set of requirements that must be met if ecological impact assessment techniques are to be applied effectively for purposes of environmental assessment and management. Current legislative frameworks for EIA have not been structured so as to permit the effective application of ecological principles. In many countries, investment in the collation of ecological data for EIA is effectively wasted, because there is no scope for its analysis or evaluation.

In particular, there has been a tendency for environmental assessment procedures to be applied in a relatively restricted sense to individual projects. The challenges of devising monitoring programmes which will detect widespread, subtle and cumulative impacts on a regional basis require a shift away from prevalent space- and time-limited approaches (Bernstein et al. 1993).

The need for more integrated and strategic approaches to ecological impact assessment continues to grow. Widespread declines in biodiversity, for example, have increased dependence on regulatory mechanisms such as EIA to complement traditional approaches to nature conservation which relied primarily on protective land designations to secure key areas for wildlife preservation (Pritchard 1993). In the United States, promotion of general biological diversity is now officially recognised as a requirement under the National Environmental Policy Act (NEPA) in any circumstances where it is 'possible to both anticipate and evaluate' the effects of federal actions on biodiversity (CEQ 1993).

Ecological impact assessment has developed largely in tandem with EIA. As currently administered in most countries, however, EIA has failed to provide an effective mechanism for the regulation of ecological impacts. This would require the adoption of ecological quality standards or objectives, with protective mechanisms designed to prevent reduction in ecosystem quality below these standards, and mitigation measures which would come into play if this did occur.

However, practical constraints have resulted in ecological impact assessments that are both temporally and spatially restricted and which often fail to provide meaningful data; while lack of consensus as to appropriate theoretical approaches has resulted in marked inconsistency of method and a failure to capitalise on experience.

REFERENCES

Appelman, K., ed. 1991. Landschapsecologische Kartering voor milieu-effect rapportage bij aanleg van rijkswegen. Eindrapport methode-ontwikkeling. Rijkswaterstraat Meetkundige Dienst, Dienst Weg-en Waterbouwkunde: Delft.

Bailey, J. and V. Hobbs. 1990. "A proposed framework and database for EIA auditing." *Journal of Environmental Management* 31: 163-172.

Beanlands, G.E. and P.N. Duinker. 1984. "An ecological framework for environmental impact assessment." *Journal of Environmental Management* 18: 267-277.

Bedford, B.L. and E.M. Preston. 1988. "Developing the scientific basis for assessing cumulative effects of wetland loss and degradation on landscape functions: status, perspectives and prospects." *Environmental Management* 12(5): 751-771.

Bergstrom, J.C. 1980. "Concepts and measures of the economic value of the economic value of environmental quality: a review." *Journal of Environmental Management* 31(3): 215-229.

Bernstein, B.B., B.E. Thompson, and R.W. Smith. 1993. "A combined science and management framework for developing regional monitoring objectives." *Coastal Management* 21: 185-195.

Environmental Advisory Unit 1988. *Heathland restoration: a handbook of techniques.* Southhampton: British Gas.

Brown, K. and D. Moran. 1993. Valuing biodiversity: the scope and limitations of economic analysis. Centre for Social and Economic Research on the Global Environment Global Environmental Change Working Paper 93-09.

Buckley, R.C. 1991. "How accurate are environmental impact predictions?" *Ambio* 20(3-4): 161-162.

Burdge, R.J. 1991. "A brief history and major trends in the field of impact assessment." *Impact Assessment Bulletin* 9(4): 93-105.

Cairns, J. Jr. and B.R. Niederlehner. 1993. "Ecological function and resilience: neglected criteria for environmental impact assessment and ecological risk analysis." *The Environmental Professional* 15: 116-124.

Cocklin, C., S. Parker, and J. Hay. 1992. "Notes on cumulative environmental change I: concepts and issues." *Journal of Environmental Management* 35: 31-49.

Cocklin, C., S. Parker, and J. Hay. 1992. "Notes on cumulative environmental change II: a contribution to methodology." *Journal of Environmental Management* 35: 51-67.

Commission of the European Communities. 1985. Council directive of 27 June 1985 on the assessment of the effects of certain public and private projects on the environment. *Official Journal of the European Communities* L175: 40-48

Contant, C.K. and L.L. Wiggins. 1991. "Defining and analysing cumulative environmental impacts." *Environmental Impact Assessment Review* 11: 297-309.

Council on Environmental Quality. 1978. National Environmental Policy Act-Final Regulations. Federal Register 43(230): 55978-56007.

Council on Environmental Quality. 1986. Regulations for implementing the procedural provisions of the National Environmental Policy Act. 40CFR 1500-1508.

Council on Environmental Quality. 1993. Incorporating biodiversity considerations into environmental impact analysis under the National Environmental Policy Act. CEQ, Washington DC.

Cowart, R.H. 1986. "Vermont's Act 250 after 15 years: Can the permit system address cumulative impacts?" *Environmental Impact Assessment Review* 6: 135-144.

Cuff, J. and G. Ruddy. 1994. "SEA—evaluating the policies EIA cannot reach." *Town and Country Planning* 63(2): 45-48.

Culhane, P.J. 1987. "The precision and accuracy of US environmental impact statements." *Environmental Monitoring and Assessment* 8: 217-238.

Culhane, P.J., H.P. Friesema, and J.A. Beecher. 1987. *Forecasts and Environmental Decision-making: The content and predictive accuracy of environmental impact statements.* Social impact assessment series 14. Boulder CO: Westview Press.

Department of the Environment. 1989. *Environmental Assessment: a guide to the procedures.* London: HMSO.

Department of Transport. 1993. *The wildflower handbook.* Advice note HA67/93 in section 4, volume 10 of the Design Manual for Roads and Bridges. London: Department of Transport.

Eberhardt, L.L. 1976. "Quantitative ecology and impact assessment." *Journal of Environmental Management* 4: 27-70.

Fuller, R.M., G.B. Groom, and A.R. Jones. (In press.) Photogrammetric Engineering and Remote Sensing.

Fuller, R.M. and R.J. Parsell. 1990. "Classification of TM imagery in the study of land use in lowland Britain: Practical considerations for operational use." *International Journal of Remote Sensing* 11: 1901-1917.

Gilliland, M.W. and P.G. Risser. 1977. "The use of systems diagrams for environmental impact assessment: Procedures and an application." *Ecological Modelling* 3: 183-209.

Glasson, J. 1994. "EIA—only the tip of the iceberg?" *Town and Country Planning* 63(2): 42-45.

Gray, J.S. and K. Jensen. 1993. "Feedback monitoring: A new way of protecting the environment." *TREE* 8(8): 267-268.

Green, R.H. 1979. *Sampling Design and Statistical Methods for Environmental Biologists.* New York: John Wiley and Sons.

Grime, J.P. 1974. "Vegetation classification by reference to strategies." *Nature* 250: 26-31.

Grime, J.P. 1979. *Plant Strategies and Vegetation Processes.* Chichester: John Wiley.

Hirsch, A. 1988. "Regulatory context for cumulative impact research." *Environmental Management* 12(5): 715-723.

Hodgson, J.G. 1991. "The use of ecological theory and autecological datasets in studies of endangered plant and animal species and communities." *Pirineos* 138: 3-28.

Hollick, M. 1981. "The role of quantitative decision-making methods in environmental impact assessment." *Journal of Environmental Management* 12: 65-78.

Holling, C.S. 1978. *Adaptive Environmental Assessment and Management.* New York: John Wiley and Sons.

International Union for Conservation of Nature and Natural Resources (IUCN). 1980. World conservation strategy: Living resource conservation for sustainable development. Gland Switzerland: IUCN.

Johnson, R.A. 1981. "Application of the guild concept to environmental impact analysis of terrestrial vegetation." *Journal of Environmental Management* 13: 205-222.

Julien, B., S.J. Fenves, and M.J. Small. 1992. "An environmental impact identification system." *Journal of Environmental Management* 36: 167-184.

Karr, J.R. 1991. "Biological integrity: A long-neglected aspect of water resource management." *Ecological Applications* 1: 66-84.

Kennedy, A.J. and W.A. Ross. 1992. "An approach to integrate impact scoping with environmental impact assessment." *Environmental Management* 16(4): 475-484.

Kozlowski, J.M. 1990. "Sustainable development in professional planning: A potential contribution of the EIA and UET concepts." *Landscape and Urban Planning* 19: 307-332.

Krutilla, J.V. and A.C. Fisher. 1975. *The Economics of Natural Environments: Studies in the valuation of commodity and amenity resources*. Baltimore: John Hopkins University Press for Resources for the Future.

Lee, N. and F. Walsh. 1992. "Strategic environmental assessment: An overview." *Project Appraisal* 7(3): 126-137.

Leopold, L.B., F.E. Clarke, and B.B. Henshaw. 1971. A procedure for evaluating environmental impact. Geological Survey Circular, Washington DC.

Morgan, R.K. and A. Memon. 1993. Assessing the environmental effects of major projects: A practical guide. Environmental Policy and Management Research Centre Publication no 4. Dunedin: University of Otago.

New Zealand Government. 1992. Resource Management Act.

Noss, R.F. 1990. "Indicators for monitoring biodiversity: A hierarchical approach." *Conservation Biology* 4(4): 355-364.

Odum, H.T. 1972. Use of energy diagrams for environmental impact statements. *In*: Tools for Coastal Management. Proceedings of the 1972 Conference, Marine Technology Society, Washington DC, 197-213.

Pellew, R.A. 1989. "Data management for conservation," pp. 505-522 in *The Scientific Management of Temperate Communities for Conservation*. I.F.Spellerberg, F.B. Goldsmith, and M.G. Morris, eds. Oxford: Blackwell Scientific Publications.

Pollard, E and T.J. Yates. 1993. *Monitoring Butterflies for Ecology and Conservation*. London: Chapman and Hall.

Pritchard, D. 1993. "Towards sustainability in the planning process: The role of EIA." *Ecos* 14 (3/4): 10-15.

Radford, G.L. et al. 1994. Site assessment of four exemplar compressor stations. Institute of Terrestrial Ecology Report to British Gas.

Severinghaus, W.D. 1981. "Guild theory as a mechanism for assessing environmental impact." *Environmental Management* 5(3): 187-190.

Sheate, W.R. 1992. "Strategic environmental assessment in the transport sector." *Project Appraisal* 7(3): 170-175.

Shopley, J., M. Sowman, and R. Fuggle. 1990. "Extending the capability of the component interaction matrix as a technique for addressing secondary impacts in environmental assessment." *Journal of Environmental Management* 31: 197-213.

Smith, E.P., D.R. Orvos, and J. Cairns Jr. 1993. "Impact assessment using the before-after-control-impact (BACI) model: Concerns and comments." *Can. J. Fish. Aquat. Sci.* 50:627-637.

Sonntag, N.C. et al. 1987. *Cumulative Effects Assessment: A context for further research and development*. Canadian Environmental Assessment Research Council, Canada: Minister of Supply and Services.

Spellerberg, I.F. 1991. *Monitoring Ecological Change*. Cambridge: Cambridge University Press.

Treweek, J.R., S. Thompson, N. Veitch, and C. Japp. 1993. "Ecological assessment of proposed road developments: A review of environmental statements." *Journal of Environmental Planning and Management* 36(3): 295-307.

Treweek, J.R. and N. Veitch. (In press.) "Use of GIS and remotely sensed data for ecological assessment of proposed new road schemes." *Journal of Biogeography*.

Veitch, N., J.R. Treweek, and R.M. Fuller. (In press.) "The land cover map of Great Britain: A new data source for environmental planning and management." In *Advances in Environmental Remote Sensing*. F.M. Danson and S.E. Plummer, eds. New York: Belhaven Press.

Verheem, R. 1992. "Environmental assessment at the strategic level in the Netherlands." *Project Appraisal* 7(3): 150-157.

Vlachos, E. 1985. "Assessing long-range cumulative impacts." In *Environmental Impact Assessment and Risk Analysis*. V.T. Covello, ed. Berlin: Springer-Verlag.

Webb, J.W. and L.L. Sigal. 1992. "Strategic environmental assessment in the United States." *Project Appraisal* 7(3): 137-143.

Wood, C. 1992. "Strategic environmental assessment in Australia and New Zealand." *Project Appraisal* 7(3): 143-150.

Wood, C. and C. Jones. 1991. *Monitoring Environmental Assessment and Planning*. London: HMSO.

World Commission on Environment and Development (WCED). 1987. *Our Common Future*. Oxford: Oxford University Press.

Chapter 9
Risk Assessment[1]

RICHARD A. CARPENTER
Consultant, Virginia USA

INTRODUCTION

Uncertainty is present in all environmental problems, but it is not always dealt with explicitly. Most environmental impact assessments (EIA) use a single number to represent the range of values that a measured parameter actually can have. The decision may be to use the average (mean) or expected value or, alternatively, to use a worst-case value. The implied choice may be conservative or optimistic and is usually internally consistent. When uncertainties are large and important to the outcome of the problem analysis (e.g., the chance of an accidental spill of a toxic material), the assessment is not completely informative, and it can be potentially quite misleading to express biophysical measurements or modeling results with single numbers. The correct and appropriate way to characterise data is to describe the statistical distribution of a range of values and the confidence with which that range is held to be true. Environmental risk assessment (ERA) makes it practical to carry throughout the problem analysis, following the rules of probability theory, an expression of the likelihood of all possible values of each parameter. ERA is also known as probabilistic risk assessment (PRA) and probabilistic quantitative risk assessment (PQRA).

EIA and ERA advise managers and decision makers about the *frequency* and *severity* of adverse consequences to the environment from their activities or planned interventions. If these officials are not comfortable with these predictions, then changes can be made to mitigate or eliminate the impact and/or to reduce the risk; e.g., to use a different site or alternative technology, to implement risk management or emergency response capability. Since risk assessment adds to the costs of EIA, close coordination between analyst and manager is necessary to decide when and how much to do. There should be only one environmental assessment report; EIA should include ERA when risk is important. This chapter presents the elements of performing ERA and illustrates how it can be effective in improving environmental protection and management.

All societies become aware that there are more requests for government action and expenditure in the area of public health and safety, than revenue to pay for

[1]*Environmental and Social Impact Assessment* - Edited by F. Vanclay and D.A. Bronstein. Copyright © 1995 by the International Association of Impact Assessment. Published in 1995 by John Wiley & Sons Ltd. A version of this chapter will appear in *Impact Assessment*, the quarterly journal of IAIA.

them. Some method of allocating available resources, not only money but personnel and management attention, is essential. Since health and safety are sensitive political issues, choices and priority decisions are often made in response to alarming events and perceived risks, rather than actual risks. Decisions in these circumstances can result in wasted resources, unjustified fears, and social disruption (Ahearne 1993). Risk assessment is now seen as a tool for more rational and effective environmental management (Glickman and Gough 1990). A risk-based strategy can show which actions will result in the most risk reduction per unit expenditure and which uncertainties are most important for additional scientific study (US EPA 1993). ERA is objective advisory information that can enhance the participative political process where values and preferences are properly integrated into the final administrative program for protecting lives and ecosystems. This is the opportunity in public policy for environmental risk assessment.

HISTORY

Risk has been a vital part of management information in insurance and investment for centuries. Technological risks began to be specially analysed during World War II in military operations research, and thereafter in the nuclear energy and space exploration fields. The concern was mainly with infrequent but catastrophic events. Since then, the number of severe industrial accidents that have captured headlines has increased. At the same time, environmental concerns have become a central theme in public policy discussions. Factory explosions, oil tanker spills, chemical tank car derailments, and petroleum product fires have generated a public demand for prevention and a profound concern for victims and damage to the natural environment. Official responses have included:

1980 The Scientific Committee on Problems of the Environment (SCOPE) of the International Congress of Scientific Unions (ICSU) published the landmark report *Environmental Risk Assessment* (Whyte and Burton 1980).
1982 The European Economic Community issued the *Seveso Directive* on potential industrial hazards, following a serious dioxin release incident in Seveso, Italy (EEC 1982).
1984 The World Bank, after the Bhopal (India) methyl isocyanate disaster, issued guidelines and a manual to help control major hazard accidents (World Bank 1985a, 1985b).
1987 The Organisation for Economic Cooperation and Development compiled a report (Hubert 1987) on risk assessment in the OECD countries with sections on the nuclear industry, chemicals, petroleum processing, transportation of hazardous materials, and dam-reservoir projects.
1987 The much-referenced Brundtland Report of the World Commission on Environment and Development (WCED 1987) called for the further development of technology assessment and risk assessment methodologies in pursuing sustainable development.

1992 More than 50 commercial banks signed a statement that, as part of their credit risk procedures for both domestic and international lending, they would recommend EIA and ERA (UNEP 1992).

The most frequently cited risk assessment framework was developed by the US National Research Council (1983) for use by the US Environmental Protection Agency (EPA). About the same time, the Hazard and Operability (HazOps) study method evolved, among others, in the chemical process industry as an outgrowth of quantitative probabilistic risk assessment that had matured in the nuclear industry. The US EPA made a decision to base its programmatic priorities on comparative risk and asked its scientific advisory board to review the state of the art of risk assessment. The subsequent reports set the stage for risk assessment to enter into public policy in a major way (US EPA 1987, 1990, 1992). As of May 1994, more than 25 US states and regions have undertaken projects in comparative risk assessment for risk-based strategic planning of government programs in environmental protection (*The Comparative Risk Bulletin* 1994).

Many US laws call for environmental regulations that are based on risk and their implementation has motivated much of the methods development in the field. The clean-up of hazardous waste sites under the so-called Superfund program has been accompanied by a substantial research program on all phases of risks to human health and ecosystems (US EPA 1988a).

Because cancer is such a dreaded disease, it attracts considerable political interest and as a result, the US risk assessment history is dominated (and distorted) by carcinogenic chemical concerns. The increased incidence of cancer, with death as the expected outcome, is the focus of attention, while risks from other diseases have been given less research. Ecological degradations deal with complex, self-organising communities of plants and animals, and are relatively less quantifiable.

DEFINITIONS AND SCOPE

Hazard is a danger, peril, or source of harm. *Risk* is an expression of chance, combining both frequency and severity of damage from hazards. *Uncertainty* is caused by natural variation and the lack of knowledge or understanding about cause–effect relationships in an existing or future condition. *Assessment* is an evaluation in order to decide.

Environmental Risk Assessment addresses four questions:
1. What can go wrong to cause *adverse consequences*?
2. What is the probability of *frequency* of occurrence of adverse consequences?
3. What is the range and distribution of the *severity* of adverse consequences?
4. What can be done, at what cost, to *manage and reduce unacceptable risks* and damage?

Environmental impact assessment should answer the first question and give at least a qualitative expression of the magnitude of the impacts. Thus, the major additional consideration in ERA is the frequency of adverse events. Risk management is integrated into ERA because it is the attitudes and concerns of decision makers that set the scope and depth of the study. ERA attempts to quantify the risks to human health, economic welfare, and ecosystems from those human activities and natural phenomena that perturb the natural environment.

The 5-step sequence in performing ERA is:

1. Hazard identification
2. Hazard accounting
3. Scenarios of exposure
4. Risk characterisation
5. Risk management

This sequence shows that ERA, like EIA, is a management advisory process with iteration and continuity, rather than a single analytical report (Carpenter et al. 1990). It systematises the approach to hazard, determines what is more or less risky, and optimises risks as compared with benefits (Rimington 1993).

Step 1. Hazard Identification

Hazard identification is akin to the qualitative prediction of impacts in EIA and begins to answer the question of what can go wrong. It lists the possible sources of harm, usually identified by experience elsewhere with similar technologies, materials, or conditions. This is, in fact, a preliminary risk assessment and is immediately useful to managers in appraising the project or activity upon which they are embarking.

Typical hazards associated with economic development projects are:

- Toxic chemicals
- Failure of mechanical equipment
- Flammable or explosive materials
- Failure of critical controls
- Highly corrosive or reactive substances
- Natural disasters
- Extreme conditions of temperature or pressure
- Ecosystem damage (eutrophication, soil erosion)
- Collisions in transportation

When to include risk assessment. The need to extend an EIA to include ERA depends on whether identified uncertainties are large and important. Of course, if the uncertainties can be resolved by readily acquiring more information, then the assessor should do so. Examples of questions or uncertainties about the above hazards that might trigger an ERA are:

- Potential release of hazardous materials (rate and amount)
- Accidental fires and explosions
- Dilution and dispersion mechanisms and rates
- Transport and fate of pollutants in the environment
- Failure rates of equipment and structures
- Dose/response relationships based on animal studies

- Human behavior (reactions, errors)
- Natural hazard occurrence and frequency
- Alterations in the landscape due to changes in landuse patterns

Uncertainties arise from:

- Lack of theory, explanatory paradigms, and basic understanding
- Inadequate monitoring of parameters of environmental conditions
- Sampling and analytical errors
- Lack of baseline environmental data at a project site
- Models that do not completely correspond to reality because they cannot consider all variables and are therefore (over)simplified
- Novelty of technology, materials, or siting
- Inherent variation and stochastic events in complex natural systems
- Control and replication problems in ecological research

Hazardous chemicals are a major topic for ERA and elaborate screening procedures have been devised to judge when a chemical merits full investigation (Carpenter et al. 1990). The US EPA (1988b) and the World Bank (1985a) issue threshold guidelines based on frequently revised lists of highly toxic chemicals, and the amounts of each if present at any one location, that trigger risk assessment and emergency planning. Similar quantity-related guidelines are issued for highly reactive and flammable materials.

Step 2. Hazard Accounting

Hazard accounting considers the total system of which the particular problem is a part, and begins to answer the questions of how frequent and how severe are the likely adverse impacts. It also sets the practical boundaries for the assessment. The scoping of an EIA will cover much of hazard accounting. For example, a hazardous chemical may pose a risk in any stage of its life cycle, i.e., from mining and refining or synthesis through manufacturing, processing and compounding, storage and transportation, use and misuse, and finally, to post-use waste disposal or recycling (Smith et al. 1988).

Risk managers must state their concerns and indicate possible linkages of operations to mitigation measures. Some of the scoping choices to be made are:

- Geographic boundaries
- Time-scale of impacts
- Stages of the causal chain of events
- Phase or phases of the technological activity
- Routine releases or accidents
- Workforce, neighbouring community, or wider population
- Definitive end points for health or ecosystem effects
- Cumulative effects and interactive risks that result from interaction with other projects

The scope should include the social and natural systems around a project and not just a single pollutant path. For example, it would be wrong to assess the risk posed by small concentrations of halomethanes produced incidentally to the chlorination of drinking water without comparison of the risks to the same public from *not* killing the pathogenic organisms.

The time covered should include all phases of an activity where risk is important and not just the operational period. Construction, maintenance, and dismantling may pose special hazards; for example, it is well known that the Chernobyl nuclear reactor was being tested and normal safety systems were disabled at the time of the disaster (Park 1989). Toxic effluents such as heavy metals may circulate for a long time and nuclear wastes may have half-lives of thousands of years. It is common practice to look at least one lifetime into the future, about 75 years. The important point is that the time horizon should be consciously chosen and recorded as one of the assumptions of the ERA.

A causal chain for a risk may stretch from an original decision to satisfy some wants and needs, through the choice of technologies, to adverse events, to exposure conditions, and finally to health impacts. In a sense, the Bhopal accident originated with India's desire to be self-sufficient and to invite the local manufacture (incidentally by a multinational concern) of pesticides necessary for the protection of food crops. Such a comprehensive analysis of all related human activities is difficult and infrequently attempted.

Step 3. Scenarios of Exposure

Scenarios of exposure are experiential or imaginative constructions of how the hazard might be encountered. For the environmental pathway, the bodily dose/ response calculation is only one step. Knowledge of earlier parts of the exposure sequence can reveal chances to reduce risk. For example, while a toxic chemical may ultimately poison people by inhalation, ERA seeks information on a wide range of related variables, such as:

- Inventory of the type, amount, location, and storage conditions
- Releases to the environment, both deliberate or accidental
- How people are exposed and for how long
- Ambient concentrations
- The actual bodily dose
- The physical condition of victims that might affect how they respond

Reasonable sequences of events and environmental pathways are devised through which the source of harm could impact health and welfare, including the condition of ecosystems. For example, a toxic chemical might move from any point in its life cycle through air, water, plants and animals, or soil to cause an exposure by skin contact, inhalation or ingestion.

Frequency and severity

Methodically observing or estimating the likelihood of occurrence and the severity of impacts for each scenario can produce curves plotting the probability of *frequency* of adverse events of a given *severity* versus the severity per event, e.g., the number of fatalities (see figure 1). Known as *F/N* curves, they present the 'how often' and 'how bad' aspects of risk. However, the integral, or area under the curve, is not the whole story. In figure 1, the hypothetical project (policy, technology, or site) A has a lower mean risk than does B, but A also has a larger probability of a catastrophic accident. In an example of the siting of a chemical plant, Site B in a rural area will have risks associated with a spill or fire from a tank truck in transit, while Site A, an urban site will have risks associated with the explosive dispersion of a toxic material in a highly populated area. There is no objective way to combine these two criteria and different societies or individuals will make different choices between the two. However, the explicit depiction of risk is valuable information.

Risk may also be indicated by the breadth and shape of the distributions or probability densities of the severity values. If the standard deviation is small and the distribution approximately log-normal (bell-shaped), then the mean can adequately represent the impact. But if the standard deviation is large and there is a pronounced positive skew (tail) with low frequency but high severity outcomes, then an expression of this risk and a more thorough investigation are warranted.

Even a qualitative presentation of risk is useful. It is obvious that whenever frequent occurrence is combined with catastrophic or critical severity, the risk must be reduced if the project is to proceed. Occasional or infrequent adverse events that have only negligible or marginal consequences may be acceptable because of the benefits of the project or activity. (See the section on ecological risk for an application using expert judgment to express relative risk when quantification is not possible.)

Accident scenarios and their likelihood are analysed with methods that evolved from the nuclear energy industry. Hazard and operability studies (HazOps) investigate deviations from the intent of engineering design. A multidisciplinary team identifies all credible accident scenarios using detailed design information, operating characteristics, and actual operating experience with engineering components and systems.

The Fault Tree procedure begins with an accident and determines with 'reverse analysis', the equipment failures or events that could lead to it. The Event Tree procedure begins with a component failure and follows a 'forward analysis' to determine if a major accident could result. Maintenance of publicly available databases on industrial accidents is carried out by the European Economic Community, the US EPA, and the American Institute of Chemical Engineers. In the USA, it is estimated that 140,000 plant sites will perform some sort of process hazard analysis by 1997 under requirements of the US Occupational Safety and Health Administration (Illman 1993).

Figure 1.

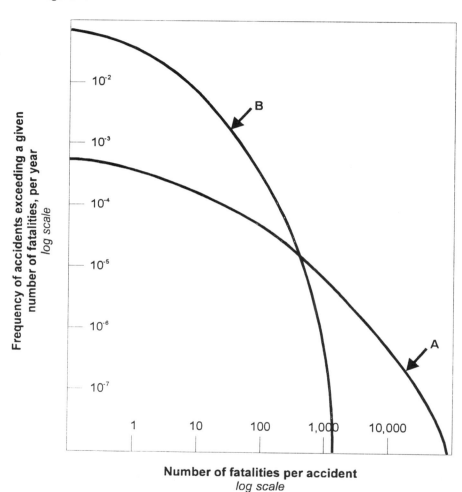

Number of fatalities per accident
log scale

This figure shows the risk distribution for two hypothetical alternative industrial facilities. Plant A has a lower mean risk (expected value of damage) than Plant B. Plant A, at the same time, has a considerably larger probability (although still small) of causing a large accident that kills many people. There is no objective way to combine these two criteria (expected value and distribution), and different groups of people will make different choices.

Step 4. Risk Characterisation

Risk characterisation facilitates the judgment of the acceptability of the risk. Risks to health are typically characterised in terms of:
- Exposure period
- Potency of a toxic material
- Number of persons involved
- Quality of models
- Quality of data, assumptions, and alternatives
- The uncertainties and confidence in the assessment
- Appropriate comparisons with other risks

Useful risk characterisation expressions include:
- Probability of the frequency of events causing some specified number, or more, of prompt fatalities
- Annual additional risk of death for an individual in a specified population
- Number of excess (additional) deaths per million persons from a lifetime exposure
- Annual number of excess deaths in a population
- Reduction in life expectancy

Risk acceptability by the parties at interest is a fundamental management goal to which ERA contributes. Acceptability depends on a complex set of psychological factors that are beyond the scope of this chapter (see Slovic 1987). Whyte and Burton (1980) suggest the following comparisons to interpret the findings of ERA:
- Elevation of the risk above the natural background level
- Risk of alternative actions to achieve the same goal
- Other familiar risks
- Benefits of continuing the project and taking the risk

Comparisons must be carefully chosen because unacceptability is increased (Covello et al. 1988) when:
- Risks are involuntary or controlled by others
- The consequences are dreaded and delayed
- The benefits and risks are inequitably distributed
- The project is unfamiliar or involves complex technology
- Children are threatened
- Basic human needs (clean air, drinking water, food) are at risk

The communication of ERA results should take the form of decision analysis; i.e., what options are available, and for each option what are the risks, costs, and benefits, and how are these distributed within society. Proper comparison and communication can actually change laypersons' misperceptions of risks so that participative decision making may proceed on a more rational, less emotional basis.

Step 5. Risk Management

Risk management is the use of the ERA results to mitigate or eliminate unacceptable risks. It is the search for alternative risk reduction actions and their implementation that appears to be most cost-effective. Most human activities are undertaken for obvious and direct benefits and risks are intuitively compared with these benefits. Avoiding one risk may create another (risk transference) and so net risk is a consideration facilitated by ERA.

There is a strong reiteration and feedback between risk management and hazard accounting because changes in the scope of the ERA may be necessary to fully answer the questions of management, and because relatively simple changes in the project may alter the hazard and reduce risk (e.g., a different site).

HUMAN HEALTH RISK ASSESSMENT METHODS

Cancer

About one out of every three persons in the USA can be expected to contract cancer during their lifetime. Use of tobacco and dietary excesses (e.g., fats) account for about 90 percent of the cancer burden in the western world (Weisburger 1993). Cancers traced to direct, involuntary exposure to environmental pollution are estimated to constitute about 2 percent of total cancer risks (Doll and Peto 1981). The other major causes of cancer are lifestyle choices and genetic mutations.

The US EPA has sponsored the development of methods for estimating the *additional* risk of an individual contracting cancer from exposure to some human-introduced carcinogen in the environment. Several hundred chemical compounds have been tested using animals. The procedure used is to expose a group of test animals to high (maximum tolerated) dosage levels of the suspected carcinogen (by inhalation, ingestion, or skin absorption) for a period of time (usually the animal's normal lifetime). Postmortem inspection of each test animal is made for tumours or malignant neoplasms (i.e., cancers).

The percentage (or probability) of animals developing cancer at each dosage level is recorded. The slope of this curve is then extrapolated to the low doses expected to be encountered by human beings who may be exposed to the chemical. Various assumptions are made as to the shape of the extrapolated dose/response curve as it approaches zero dose and zero response (e.g., whether a threshold exists or not). The slope becomes a *unit cancer risk factor*, or potency, expressed in terms of $(mg/kg/day)^{-1}$. There is considerable controversy within the scientific community as to the validity of this method because of the high doses and extrapolation assumptions used (Ames and Gold 1990), let alone the ethical concerns about the use of animals for scientific experimentation and vivisection. Another assumption is that human beings react in the same way as do the animal test species. Nevertheless, currently there is no alternative available for estimating the dose/response relationship.

When the unit cancer risk factor is multiplied by a dose to an individual, expressed as mg/kg/day, the units cancel and the resulting number is the additional risk to the individual (chance or probability) of contracting cancer during a lifetime of exposure at that dose level to the toxic agent in question. This represents an excess risk over the sum of all other risks of contracting cancer (0.3333). Most cancer risk (other than genetic mutations) is caused by lifestyle choices; e.g., smoking, drinking, diet, and sunbathing. If, for example, a lifetime exposure of a person to a carcinogenic material in the environment is calculated to yield an added risk of 1×10^{-4}, then the new total cancer risk for that person is 0.3334. This is a small additional risk but, but it is important to determine. Individuals are exposed to many natural and artificial carcinogenic substances, some of which are highly potent. Many naturally occurring substances in food are more carcinogenic than are the contaminants from industrial chemicals (Ames et al. 1987). It is assumed that risks are additive unless there is strong evidence for synergism or antagonism. Some exposures of the general public are involuntary (e.g., polluted air) and not avoidable by individual choice. Such risks, no matter how small, are generally not acceptable, because they are perceived to be unaccompanied by any benefit, and people feel helpless.

The expressed risk of contracting cancer is not the same as the risk of death. All cancers are not ultimately fatal. Toxic substances cause tumours at various sites in the body and different cancers have different mortality rates. About 50 percent of cancer patients survive at least five years. One out of four deaths in the USA is caused by cancer, and so the overall risk of death by cancer is 0.25 (American Cancer Society 1992).

Another useful way of expressing risk is the annual cancer incidence due to exposure to some specific carcinogen; i.e., the additional number of new cases of cancer in a population each year. This carcinogen-specific incidence depends on the number of people exposed to varying concentrations of that carcinogen. The US EPA (1990) has made estimates for the national population for a number of cancer-causing agents.

Non-Cancer Diseases

Risks of contracting diseases other than cancer from exposure to toxic agents in the environment are estimated (according to US EPA guidelines) by using a potency factor called a reference dose (RfD) or reference concentration (RfC); the maximum daily exposure *unlikely* to cause deleterious health effects. The RfD is also derived from animal test data as described earlier. Because of the variety of effects short of death that are associated with non-cancer diseases, morbidity categories also are established for comparing health effects:

Observable effects—detectable but may not show a disability (e.g., a change in enzymes or blood biochemistry, or low weight gain in infants)

Serious effects—development or behavior abnormalities and/or dysfunction of an organ

Catastrophic effects—prompt death or shortened life, severe disability

An average reasonable exposure scenario (not a worst-case scenario) for the general population is usually chosen for the risk characterisation. In some instances, a specially identified subpopulation may be exposed differently and its risk assessed separately. An exposure from a maximum plausible accident sometimes may be included to test the sensitivity of a scenario.

Criteria for Comparative Ranking of Health Risks

The large uncertainties in risk assessment preclude reliance on the absolute quantitative risk numbers that are generated. For example, a scenario following US EPA procedures may produce an estimated exposure dose to a population, which, when multiplied by a unit cancer risk factor, infers that 100 excess cancer cases may be expected due to a specific carcinogen. This is based on the so-called '95 percent confidence limit, upper bound' estimate, which contains various conservative assumptions, so that it is certain not to be exceeded more than 1 in 20 incidences. Actually carrying through all uncertainties of exposure, potency, and individual response might produce a wide range of possible excess cancer cases, and that range might not be normally distributed. There might well be a positive skew, or tail, of low frequency/high cancer incidence outcomes. The upper bound would then be in this area of remotely possible, very bad news. Thus, the mean and the standard deviation should also be given in order to provide the appropriate information to the decision maker.

If consistent assumptions and models are carefully used for different scenarios, carcinogenic substances, and populations, then a comparison among the similarly calculated risks is a valid way of ranking risks. Uncertainties tend to cancel out. There can be confidence in estimating the relative future risk of a project scheme that is revised in order to reduce risk. A difference of at least an order of magnitude (e.g., 100 vs. 10 calculated cases) is significant management information. All else being equal, more risk-reduction attention should be devoted to the scenario that yields a calculated 100 cases than to the scenario that yields 10. If the difference is less than a power of ten, then other factors in the environmental problem analysis may dominate the choice of action to be taken.

Despite these limitations, it is necessary to loosely anchor comparative (higher or lower) risk rankings to an absolute scale. This can avoid absurdities such as ranking a risk of 10^{-8} excess cancer cases as higher than a risk of 10^{-9} when both are *de minimus* (trivial); or arguing whether to reduce a 10^{-3} risk or a 10^{-2} risk when both need urgent attention.

Throughout the world, different communities have adopted generally similar absolute values for acceptable risks to human life from involuntary, technology-caused, anthropogenic hazards. Lifetime excess risks greater than 10^{-4} (1 in 10,000) for the individual are considered high and unacceptable (Travis et al. 1987). Excess risks less than 10^{-6} (1 in a million) are low and acceptable—*de minimus* in legal terms—'the law does not deal with trivialities'. As a reference, the risk of being struck by lightning is about 1 in a million in the USA. For excess risks of between 10^{-6} to 10^{-4}, decisions as to whether to take reduction action are based on additional considerations such as direct project benefits or costs of avoidance.

For non-cancer disease risks, the ratio of the exposure dose to the reference dose determines acceptability, with the latter being defined (i.e., $<10^{-6}$) to prevent the risk of unacceptable health damage. If the ratio is less than 1.0, the risk is acceptable. The likelihood of exposure and the type of endpoint (morbidity or death) considered by the RfD are also factors in acceptability.

Risk Comparison

The effectiveness, and efficiency, of risk management depends on deploying limited resources where they are most needed. Comparing risks and the costs for their reduction is a valuable decision tool (US EPA 1993). For example, hazardous waste sites are perceived by many citizens as posing a high health risk and large expenditures are made to clean them up. Yet, when quantitative probabilistic risk assessment is conducted in relation to these sites, they usually turn out to be relatively lower threats. This is because, in most cases, the chance of exposure is slight, due to their isolation from drinking water supplies and prevention of public access to these sites. In contrast, the risk from indoor air pollutants is found to be relatively high and worthy of greater reduction efforts than the public might demand. People spend most of their time indoors, often in poorly ventilated areas, exposed to vapours of household products that are often hazardous, sometimes exposed to tobacco smoke (both actively and passively), and, in some locations, to radon.

The US Department of Energy is undertaking an enormous environmental remediation effort, treating and disposing of hazardous and radioactive wastes at hundreds of sites all over the world, and at a cost of hundreds of billions of dollars. Risk assessment will be used to set priorities for the decades of work to be done. The public will be involved in the risk assessment from the outset in order to improve acceptability of the plans (National Research Council 1994).

An example of misallocation of risk management effort is reported on the London Underground (Rimington 1993). Fire as a hazard continues to receive the largest preventative expenditures, but fire actually constitutes only 1 percent of the residual risk, given existing precautions. This illustrates the essential importance of updating ERA as continuing management information is acquired, and separating the residual risks from the total risk of any one hazard. Finding a small residual risk does not mean that the management activities that have brought the risk down should be decreased, although they should be reviewed for cost effectiveness. It is the further expense of reducing the small residual risk that is subject to question. For example, in the case of public water supply in most western countries, the low risk is testimony to good sanitation and water treatment practices. However, often proposed drastic and expensive measures necessary to remove trace amounts of pesticides that may pose only a small residual risk should be judged against other opportunities for protecting public health. Cost effectiveness may be illustrated by calculation of the cost per life saved. For example, the banning by law of unvented kerosene-fuelled spaceheaters costs $100,000 per premature death averted, whereas regulations keeping petroleum refinery sludge out of land fills cost $27.6 million per life saved (Ahearne 1993).

Guidelines for Comparison

The following guidelines can help bring consistency to the ranking exercise. A periodic comprehensive intercomparison of all major hazards in a governmental jurisdiction can lead to improved budget allocation and administration of risk reduction efforts. Where quantitative, probabilistic risk assessment is possible, comparison is straightforward. Where only qualitative information is available, group consensus among technical professional experts is attempted.

An ordinal ranking would be inappropriate considering the uncertainties of ERA, and is unnecessary. Three levels of risks are sufficient to suggest priorities for government attention. Environmental health problems (sources of hazard) are assigned to comparatively higher, medium, or lower risk levels, but are not ranked within each level. The demarcations are not distinct and scientific uncertainties enter into the judgment of comparative risk. In general, where available data indicate that two risks are about the same calculated magnitude, the more certain of the two would be given the higher priority for attention. This guideline recognises that health protection standards can vary in absolute risk by as much as four orders of magnitude. Also, some regulations include consideration of the cost and feasibility of risk management. The laws mandating these standards usually specify that human health is to be protected with an adequate margin of safety. Therefore, whenever government regulations, environmental quality standards, or exposure limits are based on risk and deemed to be complied with, the risk must, *ipso facto*, be placed in the lower category for comparison purposes. If there is evidence of non-compliance, then the risk should be based on a much higher exposure, of course.

The factors considered in health risk ranking are:

Cancer
a. The excess risk of contracting cancer for an individual in the general public (or in a specially exposed population) from lifetime exposure to the estimated reasonably expected concentration of a carcinogen in the environment;
b. The excess annual incidence of cancer in the general population; and
c. The degree of epidemiological association of an environmental contaminant with excess incidence of cancer.

Non-cancer diseases
a. The ratio of the estimated reasonable average exposure dose to the reference dose or reference concentration;
b. The morbidity category for the toxic agent; and
c. The proportion of the population exposed.

For cancer to warrant a higher priority ranking:
a. The risk to an individual should be greater than 1×10^{-5} (1 in 100,000);
b. The predicted excess annual incidence in the USA population should be greater than 2,500 cases; or
c. The epidemiological association should be strong.

A medium ranking would result if:
a. The annual incidence was increased by more than 250 cases;
b. The individual excess risk in the general population was between 1×10^{-6} and 1×10^{-5}; or
c. The individual risk in a special population was 1×10^{-4} or greater.

A lower ranking would be given where:
a. Health protection regulations are met; or
b. The individual lifetime excess risk was less than 1×10^{-6}.

For non-cancer diseases to warrant a higher priority ranking:
a. The ratio of dose to reference dose should be greater than 1;
b. The morbidity category should be catastrophic; and also
c. The general population consistently exposed.

A medium ranking would result if:
a. The dose/RfD was greater than 1, but
b. The morbidity category was serious but not life-threatening, or only specially exposed populations were at risk, or the exposure was sporadic.

The risk would be ranked at the lower level if:
a. The dose/RfD is less than 1 (i.e., regulations are met); or
b. The morbidity category is only 'observable'.

Issues in Environmental Health Risk Assessment

1. Animal testing of potential carcinogens. The transferability of animal response data to humans continues to be a matter of scientific controversy. More importantly, perhaps, the use of the 'maximum tolerated dose' (MTD) concept has raised the question as to whether an observed tumour is the consequence of carcinogenicity, or the rapid cell division that occurs as the animal attempts to repair general toxicity damage of the test. A two stage model of carcinogenicity separating pre-neoplasmic changes from cancer induction is gaining favour. Although MTD may continue to be one level of testing, a lower level may be necessary too. But additional lower levels might require unacceptably greater costs in using more animals and time to obtain statistically significant findings. Thus, the fundamental basis for estimating the potency of chemical hazards is now being debated (Ames et al. 1993).

2. Overly conservative exposure assumptions. The use of the 95 percent upper bound of statistical distributions and 'worst-case' (often inconsistent) scenarios is challenged as being unwarranted and philosophically pessimistic. Even more, these assumptions are seen to be potentially counter productive because they may shift limited risk reduction efforts from some significant risks to small, but exaggerated risks. Risk management is, in this view, a zero sum game and transfers among risks are inevitable. Being overly conservative inhibits optimisation of risk management. The mean values should be carried and presented throughout an ERA to show more about the magnitude of the uncertainties, just as the low probability–high conse-quence information is necessary when there is evidence of non-normal distribution.

3. Risk communication, perception, and acceptability. The public appears to expect and demand a zero risk along with the benefits of a highly technological civilisation and burgeoning economy. Risk professionals must patiently do more and better explanation of the probabilistic nature of science and the realities of risk in our complex civilisation. The unwillingness to consider taking even infinitesimal and vague levels of involuntary risk (Kaplan and McTernan 1993) must somehow be overcome in a participatory democratic way. The correct choice of familiar risks with which to compare new risks is all important. Risk cost-benefit analysis should be further developed as a presentational technique for the results of ERA.

4. Biomarkers. Sophisticated microbiological techniques are being developed to ascertain effects of toxic substances on fundamental processes such as enzyme production and cell level behaviour in humans and animals. These indicators may offer short cuts to predicting health impairment, but the skills and equipment required for these analyses are formidable.

ECOLOGICAL RISK ASSESSMENT

The US National Research Council in its 1993 report, *Issues in Risk Assessment*, notes:

> Ecological risk assessments have no equivalent of the lifetime cancer-risk estimate used in health risk assessment. The ecological risks of interest differ qualitatively between different stresses, ecosystem types, and locations. The value of avoiding these risks is not nearly so obvious to the general public as is the value of avoiding exposure to carcinogens. . . the function of risk assessment is to link science to decision making, and that basic function is essentially the same whether risks to humans or risks to the environment are being considered.

The objectives of comparative environmental risk assessment as applied to ecosystems are to rank a comprehensive set of environmental problems and ecosystem sites relative to one another into broad groupings of risk, and to target response actions toward those geographical areas or ecosystem sites that are at greatest risk (Barnthouse et al. 1986). The US EPA and the Oak Ridge National Laboratory (ORNL), among others, have sponsored conferences, reports, and research to foster a consistent approach for the developing field of ecological risk assessment. There is as yet no widely applicable, established procedure. Several approaches have been used.

Reductionist ecological methods: These attempt to compartmentalise ecological processes and effects into a myriad of understandable units and linkages. Generally these lack, and do not allow, evaluation of synergism. An example is tracking the flow of nutrients from their sources through the ecosystem. Quantification of the flow is possible, but drawing implications about effects at the regional level is complex. Reductionist approaches are useful in defining how individual stressors affect individual species and (sometimes) biological communities, and how they can be detected and monitored. With two or more stressors operating on the same system, the analysis and interpretation become increasingly more difficult.

Bottom-up methods: These rely on the use of models and laboratory data to quantify biological and ecological processes and impacts, primarily at the species and community levels. This can be useful at site specific locations, but extrapolating the results to ecosystem and regional levels is more difficult, especially if two or more ecosystems and stressors are involved. A standard water column model comprising many biogeophysical parameters is used at ORNL "to extrapolate the results of laboratory toxicity data into meaningful predictions of ecological effects in natural aquatic ecosystems" (Bartell et al. 1992).

Top-down methods: These evaluate structural and functional changes at the ecosystem and regional levels and are most easily applied where there is large-scale homogeneity in both the ecosystem and the stressor that affects it. Conversely, these methods break down when a region is a mosaic of many stressors and ecosystems. Normally there is a lack of sufficient data from a broad region to allow quantification.

Practical methods: A recent review has recognised the need to design and accomplish practical comparative ecological risk assessments useful to decision makers, politicians, and nonscientists. To meet this goal, comparative ERA need not be quantitative, and it may actually be preferable to keep it qualitative. A combination of the best judgment of ecologists and professional land and water managers with on-site experience, and the systematic evaluation of risks from available information, should be pursued. Effective communication to decision makers is accomplished through use of maps, simplified scoring systems, clearly defined evaluative criteria, and a manageable set of ecological stressors. Defining the specific problem areas and classifying the ecosystems of the study region are important early steps in this approach to comparative ecological ERA.

Health risk assessments (with heavy emphasis on public health) differ from ecological risk assessments in several significant ways. For ecosystems, the ERA must consider effects beyond just individual organisms or a single species. No single set of ecological values and tolerances applies to all of the various types of ecosystems. Stressors are not only chemicals or hazardous substances, they also include physical changes and biological perturbations. For public health purposes, all humans are treated equally, but with ecosystems some sites and types are more valuable and vulnerable than others. Accommodating these factors complicates comparative ecological risk assessments and renders them more subjective.

Qualitative Methodology

Risks to ecosystems are based on the values (intrinsic and anthropocentric) of actual individual sites and the probability that stressors from human activities will significantly degrade these values in the near future. Uncertainties about value, frequency of adverse impacts, and severity of response to stress are identified and evaluated as a part of the ERA. The ability of the ecosystem occurrence (site) to recover is also considered (Carpenter et al. 1992; Maragos 1992). Just as the individual human being is the focus of health risk assessment, the individual ecosystem site is evaluated in ecological risk assessment. Ecosystems are bounded biotic communities in interaction with their physical surroundings of energy, air, water, minerals, soil, and other ecosystems. Usually, an ecological risk assessment treats only natural, more or less intact, ecosystems that are lightly managed or essentially undeveloped. Urban and agricultural areas that are substantially modified, intensively managed, and where economic value dominates all others are considered dedicated.

Procedure for Ranking Risks

The following steps create relative rankings that can be used to inform public debate, set budget allocations, focus administrative attention, and establish site specific priorities for remediation, restoration, or protection. Since uncertainties are explicitly recognised and preserved in the assessment, areas where further research and monitoring would be worthwhile to decision makers are also illuminated. A multiple site comparative ERA can also reveal which stressors are the most common and damaging, which activities generate the most stress, and which types of ecosystems are most vulnerable.

1. *Establish an ecosystem classification:* Define and select a manageable number of ecosystem types that are identifiable (mappable) through currently available databases and reports, and categorised by biophysical properties (climate, rainfall, topography and elevation, vegetation, geology and geomorphology, hydrology, soils, etc.). Examples of marine and terrestrial ecosystem types are: coral reef, freshwater stream, wetland, lowland dry scrub, montane wet forest, and subalpine dry grassland.

2. *Inventory and map the ecosystem occurrences (sites):* Gather data concerning the location, extent, and status of resources, degree of disturbance, and level of protection. Previous research and monitoring and personal interviews may suffice for the inventory, but new field studies are often necessary. In the USA, the state 'heritage programs' and surveys performed by The Nature Conservancy are a good source.

3. *Develop criteria of value for each ecosystem type:* Determine individual criteria for the components of value for each of the different ecosystem types. Value components include: economic productivity, recreation, biodiversity, cultural and aesthetic significance. Criteria would include, for example, the wetlands classifications of the US Fish and Wildlife Service, the presence of endangered species, rarity, ratio of native to exotic (introduced) species, and tourism visitor counts. Seek out previous valuation studies, measurable attributes, and changes to those attributes which degrade the resources.

4. *Estimate the value of each ecosystem occurrence:* Assign numerical scores to each value component at each site on the basis of a simple scalar using quantitative measurements of the criteria where available and professional judgements. The certainty of the score of each value component is recorded. Sum the component scores for an overall value score.

5. *Develop a list of stressors:* Determine which consequences of, and perturbations from, human activities may plausibly cause unwanted negative impacts on the natural ecosystems. Examples of stressors are: exotic species, toxic chemicals, excessive nutrients, erosion and sedimentation, water diversion, fire, and human crowding.

6. *Gather data on stressors and estimate the risk from each:* Collect information on past, present, and near-future human activities that impact the specific ecosystem sites chosen for this study. Environmental experts and site managers estimate the frequency (F) of occurrence and severity (S) of damage from stressors to each site with which they are familiar. The uncertainty of the estimates is also recorded. Scalars are used to roughly quantify these judgments and the product $(F \times S = R)$ becomes a risk score for that stressor at that site.

7. *Map all of the information:* Manually create map overlays or use computerised geographic information systems to display all data relevant to risk at each site. Examples of data to be displayed include: location and site boundaries, values, stressors, risk scores, and geographic attributes such as present land use, native forest distribution, rare or endangered species habitat, historic/cultural sites, exotic species distribution, public recreation areas, and concentrated fisheries.

8. *Rank sites comparatively according to risk:* Compare a site's overall priority-for-attention score, which is the product of a site's value score and total risk score. Scores should differ by at least 20 percent of their absolute value to be regarded as different in priority.

9. *Rank stressors and ecosystem types:* Compare the stressors as to importance in a region, and compare different ecosystem types as to degree of risk. Remedial actions can be guided by learning which stressors are widely felt and which ecosystem types are most susceptible to damage.

Example of Ecological Risk Assessment

In a Hawaii study (Carpenter et al. 1992), several hundred selected sites were evaluated throughout the islands but a detailed assessment was made only of Molokai. A well-known area, Hanauma Bay on Oahu, is briefly presented here to illustrate the method. Value components were rated on a 3-level scalar: 1 = low, 2 = medium, and 3 = high. Certainty about all ratings was indicated by a 2-level scalar: 1 = poor and 2 = satisfactory.

The scores for Hanauma Bay were:

Component	Value	Certainty
Biodiversity	3	2
Recreation	3	2
Economic productivity	3	2
Cultural/esthetic	3	2

The total value score = 12, with uniformly satisfactory certainty.

This is the highest possible value score and is to be expected for this bay and its fringing reef, a prime tourist destination that is designated an underwater park with no fishing allowed. Stressors or sources of harm were identified as toxic chemicals, excess nutrients, erosion/sedimentation, and human crowding. A panel of environmental scientists familiar with the bay was convened to rate the likelihood and severity of these stressors causing adverse impacts to the values of the bay.

A 6-level scalar for frequency (F) of occurrence was: 1=remotely possible, 2=plausible, 3=likely in the near future, 4=occasionally, 5=ongoing, and 6= progressively increasing.

A 6-level scalar for the severity (S) of damage was: 1=minor loss with rapid recovery, 2=partial loss with recovery, 3=partial loss with long-time recovery, 4= major loss with long recovery, 5=total loss with eventually some recovery, and 6=irreversible loss of the unique resource.

Each expert independently rated the stressors on the site and the results were then discussed by the group. Consensus was reached on the following ratings:

Stressor	Frequency	Severity	Certainty	F x S
Toxic chemicals	2	3	Satisfactory	6
Excess nutrients	5	4	Poor	20
Erosion/ sedimentation	4	2	Satisfactory	8
Human crowding	6	4	Poor	24

The F x S scores (only the top three stressors for any site) were totalled to yield a risk score of 52. This was multiplied by the value score of 12 to yield a priority-for-attention score of 624. The average of all priority scores in the study was 300 so that Hanauma Bay clearly warranted prompt action. The agencies with juris-diction decided to close the park one morning each week, pump all sewage out to a collector sewer and close all septic tanks, ban feeding of fish, begin a good-behavior education program for visitors (e.g., do not walk on the reef), and increase monitoring and research at the site in order to reduce uncertainties about the impacts and their synergism.

In many of the site rating meetings, the experts were divided into two teams and their results compared. The method appears to have a replicability of about $\pm$ 10 percent of the priority score, and so it was decided to ignore differences in risks of < 20 percent in assigning overall action priorities. Despite the qualitative nature of this risk assessment, it has been well received and found useful by government agencies, opinion leaders, environmental groups, and in public forums on future planning for Hawaii.

Issues in Ecological Risk Assessment

1. Endpoints. In dealing with ecosystems there is no equivalent to the premature death of an individual human being that so dominates health risk assessment. Species extinction is analogous but whole communities of many species and their surround-ings are of interest. Most risks have to do with the probability of observing an unwanted effect as a result of exposure to a toxic chemical (Bartell et al. 1992). But an endpoint must specify a measurable effect that is relevant to society's concerns. The measurement problem is often complicated by uncertainty. For example, it may be impractical, because of the large number of observations required for statistical significance, to determine the decrease in biomass of a species population more closely than 25–50 percent whereas a decrease of 10–15 percent might be serious from an ecological viewpoint. The important endpoint might be the disruption of an ecosystem function that was controlled by a diverse group of organisms.

2. Single specie, single chemical data. The laboratory experiments to determine exposure–response relationships are usually restricted to one or a few species and one or a few suspected toxic materials, whereas in the real world, pollutants are more often a mixture of chemicals and many species, even entire ecosystems, are at risk. Further, the mixture usually varies over time and space so that estimating actual exposure levels is even more difficult. Research is proceeding on bioassays using actual polluted water, and on the fundamental chemistry and toxicology of the most toxic constituents of pollutant mixtures (Bartell et al. 1992).

3. Recoverability. It is important to risk managers to know the extent to which a damaged ecosystem may recover, with or without assistance. Different species show different behavior. Recovery is also related to the endpoint chosen. There may be no recovery because ecosystems can have more than one stable state. Where native communities are displaced by exotic species, recovery is unlikely.

4. *Scale*. The spatial and temporal scales that can be practically covered in an ERA are much smaller than those of interest to managers. Extrapolation from observed areas and experimental time periods to regional scale and the long term future requires models which are not yet reliable, partly due to the inadequacy of ecological theory.

5. *Uncertainty*. In addition to measurement uncertainties, ecosystems appear to be self-organising and nondeterministic. This 'true uncertainty', where even the probability of an event is uncertain, cannot be reduced by more effort. Ecosystems are extremely complex and evolve in unpredictable, near chaotic fashion.

6. *Validation of predictive tools*. The lack of long term monitoring and research has hampered the improvement of models. More retrospective examination of environmental management actions is needed to establish confidence and to reject misleading relationships.

Risks to Economic Welfare

Ecosystems have economic value because people derive utility from their use or their existence. A healthy ecosystem generates market values by providing goods and services, and also provides nonmarket values such as clean air, soil and water quality, flood protection, biological diversity, recreational and educational opportunities, aesthetics, and quality of life. Although some of these economic values are nonmonetary, they are important. Pollution, resource extraction, overuse, and exotic species invasion can degrade an ecosystem and reduce its economic value. International concern for the health of the global environment has been growing and attention has focused on tropical forests and fragile island and marine ecosystems. Market values for timber and nontimber forest products, particularly foods and medicines, are used to quantify and monetise the values and damages to existing ecosystems. Analyses of the services provided by the environment, such as soil stabilisation, watershed protection, climate regulation, and flood protection, are far less developed.

Recent work in risk to economic welfare because of environmental degradation has indicated the importance of conserving biodiversity by demonstrating the contribution that biological resources make to social and economic development (and losses to society when they are damaged). The dollar value of biological resources may be estimated using actual, option, and existence values. Dixon and Hufschmidt (1986) illustrate the use of benefit-cost analysis and other techniques in a number of case studies, including assessment of water resources and watershed protection, and lake and marine bay fisheries projects. Aesthetic vistas, clean air view distances, and presence of wilderness or specific flora and fauna are far more difficult to value and are often discussed in qualitative terms or estimated with existence valuation methods.

Economic studies on environmental quality specifically in tourist destination areas address three major issues: (1) the impacts of pollution on tourism-related economic values; (2) environmental degradation caused by tourism; and (3) ecotourism. Evaluation has been more qualitative than quantitative because of difficulties in monetising the large number of services provided by the natural environment. The impacts of pollution on tourist areas are well documented in a US Department of Commerce (1983) study of the Amoco Cadiz oil spill in Brittany, France. Due to oiled beaches, major losses in tourist industry revenues and consumer welfare were recorded. Tourists today are far more environmentally aware (and concerned) than their predecessors and are demanding higher standards of environmental quality. Tour operators now call for boycotts of degraded sites in favour of other destinations. Ecotourism (environmentally conscious visitors to unique or outstanding natural areas) is a small but growing sector of the tourism market that is increasing the value of biological diversity, endangered species, and naturally functioning ecosystems. Successful ecotourism depends on a high-quality environment; however, tourism development itself may degrade those qualities. The problem of maintaining a balance between tourism and long-term environmental health (the issue of carrying capacity) was documented by the OECD (1980), where environmental deterioration in Majorca, Spain, caused a shift of tourists to other destinations.

Valuation Methods for Economic Damage to Ecosystems

The literature on economic valuation techniques for environmental degradation is considerable, and applied theory is continually evolving. For comprehensive overviews of techniques, the reader is referred to reviews on contingent valuation and comprehensive summaries of techniques by OECD (1989) and Hufschmidt et al. (1983). Economic damages represent the monetary value of environmental impacts from residual pollution problems. Direct ecosystem values include production (e.g., fishery, forestry, agriculture), commercial services, and unpriced amenities. Other values are indirect or involve potential use (option value) or nonuse (existence value) of ecosystems. Where ecosystem damages cannot be valued in monetary terms, damages should be discussed qualitatively.

The primary method for valuing productivity losses from environmental degradation is *change in productivity*. The method calculates the difference in production, valued at market prices, from a natural system with and without degradation. The resource restoration cost method calculates actual or predicted expenditures to restore the damaged resource to its former condition.

The *loss in income* method can be used to estimate welfare damages to commercial firms affected by environmental degradation. This method calculates the difference in the net income of commercial enterprises with and without resource degradation.

The most generally applicable method of valuing such amenity losses is *contingent valuation*. This approach involves direct questioning of consumers to ascertain the willingness of individuals to pay for environmental improvements, or, alternatively, their demand for compensation for environmental losses.

Travel cost and *property value* are two other methods of estimating amenity values. The travel cost approach utilises information on differences in travel costs and visitation rates from different communities to estimate a demand curve for a recreation area. The property value method uses multiple regression analysis to estimate how proximity to amenities such as good beaches or urban parks influences surrounding property values.

Indirect ecosystem values often benefit society at large rather than individuals or businesses. For example, the indirect ecosystem values of watershed and wetlands include regulation of freshwater supplies, nutrient cycling, protection of soils, maintenance of atmospheric quality, and climate control. *Option value* measures the willingness of individuals to pay in order to retain the option of having future access to a species or resource. *Existence value* is the value people attach to the existence of species or habitat that they may have no intention of ever using or visiting, but get satisfaction in knowing that they exist.

Contingent valuation is a common method economists use to estimate indirect, option, and existence values. Contingent ranking is a related approach but provides an ordinal ranking rather than cardinal values.

Many of the valuation methods mentioned have theoretical and practical limitations and require careful interpretation. However, uncertain scientific knowledge about what services ecosystems provide and how ecosystem services are affected by stressors is probably a more serious problem in actually performing economic welfare analysis. Uncertainties in cause–effect relationships and quantification of impacts preclude useful economic valuation in many instances of ecosystem degradation from human activities.

Oil Pollution: An Example of Environmental Risks to Economic Welfare

Recent studies (Lee 1992; Carpenter et al. 1992) in Hawaii illustrate the policy relevance of ERA when economic valuation is added to the biophysical assessment. The Hawaiian islands depend on imported oil and also concentrate considerable additional shipping traffic. Thus, Hawaii's environment is particularly vulnerable to offshore oil spills. Between 1987 and 1991, collisions, groundings, and accidental leaks in Hawaiian coastal waters caused 250 oil spills ranging from 1 to 120,000 gallons. The risk to economic welfare is a function of the costs of a large spill and its probability of occurrence. Costs include at-sea response, clean-up, waste oil disposal, vessel damage, lost visitor revenues, and environmental damage remediation. No Hawaii-specific studies of visitor revenue losses have been done but extrapolation from other coastal spill episodes indicates that most of the 10 billion dollar per year tourist industry would be drastically affected for many months. The probability of a 10,000- to 20,000-gallon spill in Hawaiian waters is estimated, from shipping experience, to be about once every two years; a 40,000- to 50,000-gallon spill every 4.5 years; and a devastating 10 million-gallon spill (equivalent to an Exxon Valdez spill) every 135 years. The current oil response capability of the US Coast Guard in Hawaii is 42,000 gallons and the state has no spill prevention plan. The recommendations of the ERA were for a prompt review of regulations and response capability, and to undertake surveys of local businesses and visitors to refine the estimates of tourism losses.

Issues in Risks to Economic Welfare

Nonmarket and even nonuse values are important to this element of ERA but the social science research techniques to establish the monetary worth of such amenities are still being developed. The dilemma of increasing tourism revenues, rising visitor counts, and environmental stability that allows continued economic benefits presents one of the major issues for future research in risks to economic welfare. Contingent valuation surveys and interviews are dependent on how questionnaires are phrased and administered. Also, the willingness to pay is affected by the ability to pay so that socioeconomic factors must be taken into account. The subjective nature of this field is an obstacle to its utility.

CONCLUSION

Environmental risk assessment is developing rapidly and there is no one clearly superior approach to its performance. Adherence to probability theory is the one essential in adding this explicit presentation of uncertainty to the management information package begun about 20 years ago as Environmental Impact Assessment. Frequency and severity of adverse impacts should be investigated and interpreted for decision makers and the interested public if policies and action programs are to be effective and efficient.

Health risks and ecosystem risks differ substantially in the endpoints chosen for risk characterisation (i.e., the individual person compared with the biological community), and in the uncertainties accompanying experimental data. Human health risk assessment is far more advanced in methodology. However, both are dominated by concern with toxic chemicals. Both require close communication between environmental scientists and risk managers. A common but flexible framework for hazard identification, exposure pathway analysis, and hazard accounting is useful because the underlying treatment of uncertainty and the decision process is the same for both.

A major application of ERA is in allocating limited budgets and setting priorities for risk reduction programs. Therefore, the economic valuation of damages from adverse events and the cost of reducing or avoiding the risks is essential. Public acceptability of risks is assisted by ERA if proper comparisons are made with the benefits of technological activities, the risks of alternative economic development strategies, and familiar risks in other parts of modern lifestyle. ERA can help correct misperceptions of risk and avoid unnecessary public anxiety. Environmental risk assessment is maturing as a practical and valuable addition to the set of management and policy tools needed in a complex world.

ACKNOWLEDGEMENTS

The sections on health and ecosystem risk assessment are based on the report (Carpenter et al. 1992) of the Hawaii Environmental Risk Ranking study, which I directed, and which is copyrighted by the East–West Center. Excerpts are used with their permission. The introduction, scope, and definition sections are drawn, in part, from the book for the Asian Development Bank (Carpenter et al. 1990), of which I was the principal author.

REFERENCES

Ahearne, J. 1993. "Risk analysis and public policy." *Environment* 35(2): 16.

American Cancer Society. 1992. Cancer Facts and Figures.

Ames, B., R. Magaw, and L. Gold. 1987. "Ranking possible carcinogenic hazards." *Science* 17(Apr.): 271.

Ames, B. and L. Gold. 1990. "Too many rodent carcinogens." *Science* 249(31 Aug.): 970–971.

Ames, B., M. Shigenaka, and L. Gold. 1993. "DNA lesions, inducible DNA repair, and cell division: Three key factors in mutagenisis and carcinogenesis." *Environmental Health Perspectives* 101(5): 35–44.

Barnthouse, L. et al. 1986. *User's Manual for Ecological Risk Assessment*. Oak Ridge TN: Oak Ridge National Laboratory. Pub. No. 2679 ORNL-6251.

Bartell, S., R. Gardner, and R. O'Neill. 1992. *Ecological Risk Estimation*. Ann Arbor MI: Lewis Publishers.

Carpenter, R. et al. 1990. Environmental Risk Assessment: Dealing with Uncertainty in EIA. ADB Environmental Paper No. 7. Asian Development Bank, Manila.

Carpenter, R. et al. 1992. Report of the Hawaii Environmental Risk Ranking Study. Department of Health, State of Hawaii, Honolulu.

The Comparative Risk Bulletin. 1994. "Project News." *Comparative Risk Bulletin* 4(5&6).

Covello, V., P. Sandman, and P. Slovic. 1988. *Risk Communication, Risk Statistics, and Risk Comparisons: A Manual for Plant Managers*. Washington DC: Chemical Manufacturers Assn.

Dixon, J. and M. Hufschmidt, eds. 1986. *Economic Valuation Techniques for the Environment: A Case Study Workbook*. Baltimore: The Johns Hopkins University Press.

Doll, R. and R. Peto. 1981. *The Causes of Cancer*. Oxford University Press, and *J. National Cancer Institute* 66: 1193–1308.

European Economic Community (EEC). 1982. Council Directive of 24 June 1982 on the major accident hazards of certain industrial activities (82/501/EEC), *Official Journal of the European Communities* L230, 5 August 1982.

Glickman, T. and M. Gough, eds. 1990. *Readings in Risk*. Washington DC: Resources for the Future.

Hubert, P. 1987. *Risk Assessment and Risk Management for Accidents Connected with Industrial Activities*. Paris: Organisation for Economic Cooperation and Development.

Hufschmidt, M. et al. 1983. *Environment, Natural Systems, and Development: An economic valuation guide*. Baltimore MD: Johns Hopkins University Press.

Illman, D. 1993. "New initiatives take aim at safety performance of chemical industry." *Chemical and Engineering News* 29 Nov: 12.

Kaplan, E. and W. McTernan. 1993. "Overview of the risk assessment process in relation to groundwater contamination." *The Environmental Professional* 15: 334–340.

Lee, D. et al. 1992. *Impacts of a Catastrophic Oil Spill on Tourism in Hawaii.* Honolulu: Univ. of Hawaii Sea Grant College Program.

Maragos, J. 1992. In *Environmental Risks to Hawaii's Public Health and Ecosystems,* Carpenter et al. Honolulu: Dept. of Health, State of Hawaii.

National Research Council. 1983. *Risk Assessment in the Federal Government: Managing the Process.* Washington DC: National Academy Press.

National Research Council. 1993. *Issues in Risk Assessment.* Washington DC: National Academy Press.

National Research Council. 1994. *Building Consensus.* Washington DC: National Academy Press.

Organisation for Economic Cooperation and Development. 1980. *The Impact of Tourism on the Environment.* Paris: OECD Publications.

Organisation for Economic Cooperation and Development. 1989. *Environmental Policy Benefits: Monetary Valuation.* Paris: OECD Publications.

Park, C.C. 1989. *Chernobyl: The Long Shadow.* London: Routledge.

Rimington, J. 1993. "Overview of risk assessment," *Process Safety and Environmental Protection* 71(May): B2.

Slovic, P. 1987. "Perception of risk." *Science* 231(17 Apr.): 280–285.

Smith, K., R. Carpenter, and M. Faulstich. 1988. Risk Assessment of Hazardous Chemical Systems in Developing Countries. Occasional Paper No. 5, East–West Center, Honolulu.

Travis, C. et al. 1987. "Cancer risk management." *Environmental Science and Technology* 21: 415.

United Nations Environment Programme. 1992. Statement by Banks on Environment and Sustainable Development. Geneva: UNEP.

US Department of Commerce, National Oceanic and Atmospheric Administration. 1983. Assessing the Social Costs of Oil Spills: The Amoco Cadiz Case Study. Washington DC.

US Environmental Protection Agency (US EPA). 1987. Unfinished Business: A Comparative Assessment of Environmental Problems. Washington DC.

US EPA. 1988a. Superfund Exposure Assessment Manual. EPA-540/1-88/001. Washington DC.

US EPA. 1988b. Emergency Planning and Notification. US Code of Federal Regulations, Part 40, Sect.355 (plus latest updates). Washington DC.

US EPA. 1990. Reducing Risk: Setting Priorities and Strategies for Environmental Protection. EPA SAB-EC-90-021. Washington DC.

US EPA. 1992. Framework for Ecological Risk Assessment. EPA 630/R-92-001. Washington DC.

US EPA. 1993. A Guidebook to Comparing Risks and Setting Environmental Priorities. EPA 230-B-93-003. Washington DC.

Weisburger, J. 1993. Letter in *The Scientist,* Aug. 23, p. 12.

Whyte, A. and I. Burton. 1980. *Environmental Risk Assessment.* (SCOPE). New York: John Wiley and Sons.

World Bank. 1985a. *Guidelines for Identifying, Analyzing, and Controlling Major Hazard Installations in Developing Countries.* Washington DC: The World Bank.

World Bank. 1985b. *Manual of Industrial Hazard Assessment Techniques.* Washington DC: The World Bank.

World Commission on Environment and Development (WCED). 1987. *Our Common Future.* Oxford: Oxford University Press.

Chapter 10
Public Involvement:
From consultation to participation[1]

RICHARD ROBERTS
Praxis, Inc., Calgary, Alberta Canada

THE RISE OF PUBLIC INVOLVEMENT

Public involvement in decision making is not new, or necessarily North American. Nor is it a fad. Although its western origins are linked to the rise of democracy and citizens' relationships to the Athenian city state, its antecedents are much older and more varied, forming the basis for decision making in many ancient and contemporary indigenous societies. The current practice of public involvement is, in many ways, the byproduct of a cross-fertilisation of populist ideas, the information revolution and widespread disenchantment with a society where neither industry nor elected officials appear to act 'in the public interest'. Although it is very difficult to ascertain a definite beginning, some analysts believe that the current form of public participation began at a 'grassroots' level and that community development and participation 'just happened' with the initiative coming from the people. Sometimes, however, the impetus for involvement came from government, through such things as the Farm Radio Forum in Canada and community development programs (Draper 1978).

In his manual on public participation, James Creighton (1984) lists the world wars and the Great Depression, along with the rapid rise of the consumer society, mass education, and the media in the 1950s and 1960s, as some of the factors contributing to the demand for greater participation in decision making. Whereas the events of the first half of the century—the world wars and the Depression—greatly increased government's influence over people's lives, widespread education and greater access to information through a variety of media created a newly critical population. Creighton argues that while the children of the postwar period were taught that their feelings were important, they were being depersonalised by mass education. In a like manner, although the media effectively spread news and information, the vision of the Global Village being created was cold and impersonal.

Not only were individuals changing, but government institutions were evolving as well. Local communities were being supplanted by impersonal urban centres. Government was becoming more centralised and bureaucratic, extending its

[1]*Environmental and Social Impact Assessment* - Edited by F. Vanclay and D.A. Bronstein. Copyright © 1995 by the International Association of Impact Assessment. Published in 1995 by John Wiley & Sons Ltd. A version of this chapter will appear in *Impact Assessment,*the quarterly journal of IAIA.

influence over individuals' lives. Farrell et al. (nd: 3), the authors of a government report, *Involvement: A Saskatchewan Perspective*, noted that "the centralization process has entirely changed public perceptions of involvement in the decision-making process and has dramatically changed patterns of participation."

The effects of centralisation on decision making were not confined to any one geographical area. William Hampton (1977: 142) argues that:

> . . . the rapid growth of bureaucracy in the past twenty years in industrialized countries has resulted in the delegation of increasing amounts of decision-making to appointed officials who have no direct responsibility towards, nor communication with, the public.

The media served to advertise the mistakes of government. Events such as Watergate and the Vietnam War created distrust and disillusionment while the peace and civil rights movements challenged traditional attitudes. The perception of government as acting in the best interests of those it served was steadily eroded. This erosion was further enhanced by the rise of the consumer and environmental movements which held agencies and corporations accountable to the consumer (Creighton 1980) and exposed the pitfalls of technological development driven solely by economic considerations. For the first time in history, when poverty, oppression, and environmental degradation in a western world of apparent plenty provoked anger, frustration, and direct action by a minority, millions of people had access to the images and reacted. The public appeared to possess a conscience that government and industry seemed to lack. All of these factors contributed to the Zeitgeist that demanded a more participatory democracy.

In Canada, public involvement efforts concentrated on alleviating the effects of poverty and oppression, but with less violence and racial tension than in the US (Draper 1978). Often issues had no geographical boundaries, making it difficult to define the stakeholders. People became linked by causes rather than communities.

The legitimisation of public involvement as part of the decision-making process came as a pragmatic defence against the power of the public and as a reaction to increasingly complex issues. Whereas at one time, a project was undertaken if it was economically feasible, now a whole gamut of factors had to be considered, from social impacts to effects on wildlife (Creighton 1980; Connor 1985). Neither scientific nor technological specialisation could guarantee a solution to a problem. In order to avoid conflict, decision makers were often forced to take the public's opinions into account, or suffer the consequences. Particularly where the decision had environmental repercussions or great economic impact, people banded together to oppose both big business and government. Methods such as sit-ins, tree-huggings, product boycotts, and litigation forced decision makers to listen and pay attention to the demands of the public.

The process of public involvement continues to evolve, not in a linear fashion, but, in the words of Francis Bregha (nd: 1), as "a true meander of many rivers, cross-currents and under-currents still increasing in depth and complexity." By the end of the 1960s, governments were showing signs of acceding to the demands for public involvement. While the Town and Country Planning Act of 1968 included public involvement in the planning procedure in England and Wales, the US National Environmental Policy Act of 1969 (NEPA) began the trend of public consultation in environmental decision making. Canada followed the USA in 1973 with the Environmental Assessment and Review Process which required public involvement as an integral part of the assessment process. Most recently, international funding agencies such as the World Bank (1993) have implemented requirements for consultation and participation at various stages of the assessment process. These requirements are becoming the standard for public involvement in countries with developing transitional economies.

PARTICIPATION AND DEMOCRACY: THE ONGOING DILEMMA

> . . . I know of no safe depository of the ultimate powers of this society but the people themselves; and if we think them not enlightened enough to exercise their control with a wholesome discretion, the remedy is not to take it from them, but to inform their discretion. (Thomas Jefferson, Letter to William Charles Jarvis, September 28, 1820)

The debate over how much participation should be encouraged has been going on since the dawn of democracy. Two of the most common positions on the nature of democracy argue for either a participatory democracy, based on equality and individual autonomy, or a representative democracy, based on allowing elected representatives to act on behalf of the citizenry. Whereas the former holds that each person is an adequate judge of their own interests, the latter believes that people are unable to judge for themselves for various reasons, and thus need an informed representative to rule in their best interest. Whereas the former is praised for its egalitarian ideology, the latter is criticised for its elitism. Obviously, direct democracy is impractical, the problem, however, is that as decision making becomes more and more complex, representation of the public interest can become less and less satisfactory. In addition, decisions that affect us are not always being made by elected officials, as in the case of a corporation or bureaucracy. With the increase in public involvement, decision making approaches the direct model of democracy.

Ideally, public involvement bridges the gap between participatory and representative democracy by allowing individuals some opportunity to influence decisions normally decided by higher authorities (Olsen cited by Gougeon 1986). The expression of individual opinions, needs and wants in the public involvement process is indeed an expression of free will, and participation will flourish only in a climate of liberty and compromise (Bregha 1978). In fact, according to the most idealistic observers, participation is a flowering of our full human potential. Philosophers and political theorists from Aristotle and Rousseau to J.S. Mill have taught us to believe that individuals attain their full stature as rational, responsible moral beings through political participation, and that not to have an active interest in politics is not merely to fall short of one's duty as a citizen, but to show oneself defective as a human (Benn 1979).

Public involvement does not guarantee that democracy is served, or that citizens are well represented; however, it does bring people closer to driving the democratic machine than simply casting a vote.

PUBLIC INVOLVEMENT: A DEFINITION

This rapid rise in influence of public involvement has led to some confusion surrounding the definition of public involvement. Part of this confusion is the result of the tendency to use the terms *public involvement, consultation,* and *participation* interchangeably, with no recognition of the subtle and not-so-subtle differences between each term. *Public involvement* is a process for involving the public in the decision-making process of an organisation. This can be brought about through either *consultation* or *participation*, the key difference being the degree to which those involved in the process are able to influence, share, or control the decision making (World Bank 1993). While *consultation* includes education, information sharing, and negotiation, the goal being better decision making by the organisation consulting the public, *participation* actually brings the public into the decision-making process. Typically, public involvement has focused primarily on *consulting* the public, with no options for greater involvement. The process has relied heavily on education and information sharing. However, a growing number of organisations have begun experimenting with public participation in the form of joint planning and delegated authority, where the public actually controls and directs the process and the ultimate results.

In one of its most significant policy changes, the World Bank recently declared that funding for projects will be contingent on the inclusion of public consultation in project development (World Bank 1993). Such a mandate cannot be ignored by corporations, countries, or other lending institutions. The inevitable conclusion is that in the very near future, public involvement will affect an even broader field than it does currently. The implication is that the public, in the broadest sense, has gained a voice in matters that affect it. What was once described as a grassroots movement is rapidly becoming the basis for the way that government and industry conduct their day-to-day business when faced with proposals or situations that might be seen as contentious. The message from the public on every front is clear: we will not be left out of the decision-making process. What is not so clear is which groups make up this newly awakened public, and how to involve them.

THE BENEFITS OF PUBLIC INVOLVEMENT

Philosophical reasons aside, there is a strong case for involving the public. From the point of view of the decision maker, soliciting input from the public may help an organisation avoid countless problems, including conflicts with a public that has not been consulted, or unforeseen situations arising from inaccurate information; the public may hold information or creative solutions not yet considered by those charged with making the decision. In the end, the cost of a badly informed decision could far outweigh the cost of involving the public. From the public's perspective, being consulted generates commitment to an issue while increasing confidence in and lending credibility to an organisation that is open about their plans.

As a general rule, most decision-making processes benefit from some degree of public involvement. Public consultation and participation can help avert confrontation between organisations and affected groups and can achieve a higher level of local support for the decisions reached during all phases of planning, development, and implementation. Public involvement can help improve project planning and decision making by an organisation as well as help resolve problems resulting from ongoing operations in a community. By working with the public, an organisation can identify concerns and resolve them before they escalate into major problems. It can reduce opposition to a project and lead to more efficient and effective planning, thereby reducing the likelihood of costly delays or project cancellation. Specifically, in the context of environmental impact assessment (EIA), public involvement provides a different perspective which can lead to better and more sustainable decisions.

Involvement can inform and educate the public about proposals and their potential consequences and create channels for the type of open, honest two-way communication which has been shown to help avoid worst case confrontations. It can help to obtain information for the organisation's development that only those who live and work in the community or region can effectively provide. This special, local or indigenous knowledge can help in the planning and management of projects. When properly designed and implemented, a process involving the public in decision making can raise the level of community commitment to a project and establish a sense of ownership that can help communities create positive developmental impacts.

It is important to realise that many of the benefits of involvement are long term. The process is an investment with a wide range of risks, costs, and benefits. While the initial public involvement costs may seem high, there are considerable benefits that an effective public involvement program may have in the long term. Public involvement is not, however, a panacea. The unpredictability of human behaviour means that problems may develop despite the best of plans and intentions.

RELATIONSHIP WITH ENVIRONMENTAL AND SOCIAL IMPACT ASSESSMENT

In the evolution of planning and impact assessment, an increasing number of factors has become important in ensuring a successful project. Earlier in this century, as far as legislation was concerned, developers needed only to consider economic feasibility and gain political favour in order to obtain approval for a project. Legislation and policy requiring environmental impact assessments (EIA) appeared in the 1960s in both the US and Canada. Soon after the National Environmental

Policy Act made public involvement required by law (US 1969), legal interpretations of the document made social impact assessment (SIA) a requirement of the
greater EIA (Burdge 1990: 123; see chapter by Burdge and Vanclay). Creighton,
Chalmers, and Branch (1983) assert that the expansion of what "has to be paid
attention to" is a result of the internalisation of impacts that were once ignored as
externalities. More simply put, changing values in society is one cause for the
change in approach to development.

     In the early stages, however, the process of assessment was largely technical
or scientific in nature. This approach concentrated on prediction rather than
planning. Organisations called upon experts—most often living outside of the area
in question—to make judgements on the economic or environmental feasibility of a
new development. Donna Craig points out that a positivist approach to development,
based on empirical judgements and geared towards economic growth, dominated the
policy sciences (1990). As with EIA, initial attempts at SIA relied heavily on the
expertise of social scientists and statisticians to predict the effects of a policy,
program or project on a community, region or country. It did not take long, however, for those involved to realise that despite the broadening of scope from
economic to environmental to social concerns, the approach was too narrowly
focused, too one-dimensional. Experts can gather information, make educated judgements and predict how people might be affected. They cannot, however, determine
how people feel, let alone what they desire or value. Experts in isolation cannot see
the total picture.

     At this point the distinction between SIA and public involvement becomes
somewhat confused. Burdge points out that "SIA began to evolve both separately
and as a component of environmental impact assessments," but that the practice
"has been hindered by the unfortunate equating of SIA with public participation and
public involvement" (1990: 81). Unfortunate or not, SIA and public involvement
developed out of the same values—that placed people ahead of economic progress,
but are distinct in themselves. Demands for public involvement reoriented SIA
methodology to focus on the human, living community, not just data, statistics, and
projections. This became known as the community or participative approach to SIA
(Connor 1985). Public involvement has allowed for a deepening of the assessment
technique. How deep the assessment goes depends on the amount of public
involvement utilised in a study. Simply gathering information from the community
is like scraping the surface, whereas giving the public decision-making authority
might be akin to diamond mining—possibly dangerous but extremely rewarding!

     Public involvement, however, is not just a tool for SIA specialists. It also
plays an integrating function within planning and assessment practices (Creighton,
Chalmers, and Branch 1983). One of the problems arising from impact assessment
is that while there is an expert responsible for each aspect of an impact study, there
is a lack of integration between the various study groups. Full-scale public
involvement necessarily creates interaction between all portions and at all stages of
the study. Initially, the need to educate the public forces the organisation to gather
and present the facts. Next, information is gleaned from the public. People can be
an extremely important resource, providing both value judgements and privileged

information that only a resident would be aware of, such as local political issues of the area and personal histories. Creighton, Chalmers, and Branch assert that "value issues are the heart and soul of public involvement" and that without public input, the 'right' and 'wrong' of a project, and who takes which position might never be discussed or determined. During the whole process, communication among the study groups, and with the public, is essential in order for information to be passed to the appropriate recipient. Finally, inviting the public directly into the planning and impact assessment process ensures that the project gains credibility in the public's opinion, the regulators' purview, and within the organisation itself. A lack of confidence, internally or externally, in the study's findings could mean the downfall of the project.

WHO IS THE PUBLIC?

Many organisations approach the public as if it is a homogeneous and stable population. Instead, what they are faced with is a constantly shifting multiplicity of affiliations and alliances that group and regroup according to the issues and their understanding of the issues, perceptions of risk, and the natural evolution of informal structures. There is no single 'public'; instead there are a number of publics, some of whom may emerge at any time during the process depending on their particular concerns and the issues involved. The major concern from an organisation's perspective—and in turn, the regulators'—is that everyone is entitled to the opportunity to participate and, even for those who do not participate, that their interests are protected. Strategies to encourage involvement must be appropriate to the individual, the community or the region potentially being affected.

The entire public in a community or a region will not become involved in each and every process that comes along, because people do not always have the time, energy, or inclination. They have to prioritise what is most important to them. People participate when they believe they will be affected by a certain decision. This means that an individual or group might move through a spectrum of involvement over the life of a project as the consequences of proposed actions become clear, or in relationship to the proximity of critical 'decision points'. Everyone makes choices about whether and when to participate. Everyone is potentially part of a visible and vocal minority, as well as the 'silent majority'.

In any community or region, there are many different and diverse individuals and groups. It is essential that an organisation has different strategies and techniques to involve many of these publics within the community: no single formula will work. A successful public involvement process must take the characteristics of the potential publics and their changing views of contentious issues into consideration.

Organisations must be aware if their 'public' is—

Experienced in public involvement. Different strategies may be necessary according to whether a particular community or group of stakeholders has had previous experience with public processes.

Informed or uninformed about the issues. If the public is already informed about the issues, it will be much easier to bring people 'up to speed'. If they are not well informed, much more time may be required to educate them about the issues before informed decision making can effectively occur between all parties.

Hostile or apathetic. If previous public processes have been contentious or unsuccessful, or if the volatility of the issue or the organisation's own history has led to a hostile situation, then the public must be allowed to 'vent' as the first step toward building or rebuilding trust. This may take a long time and must be handled with care. If the public appears to be apathetic and not interested in becoming involved in the process, the organisation must still provide people with enough information about the issues so that they can make an informed decision not to be involved. If they are truly uninterested in participating it may be because—

- They do not feel adequately represented;
- They do not believe the project justifies involvement;
- They are unaware that they will be affected by the organisation's actions;
- They do not believe they can influence the decision; or
- They believe the decision has been made and have chosen other ways to make their concerns felt, commonly through direct action or litigation. (This last point is becoming a more common reason for nonparticipation in a growing number of areas of the world. It is one of the most dangerous and challenging situations organisations and regulators can face, as direct action often means action that is outside the law.)

United or divided. If the public is united over an issue, the organisation knows where it stands and can develop appropriate strategies to work toward a resolution of the issues. However, it is more difficult for the organisation to decide what approaches to take if the public is divided. Whatever strategies and decisions are taken will adversely affect one group or another. It is still more difficult to determine the best strategies if it is the process or the project itself which unites or divides the public. This usually occurs when the public believes that the process is seriously flawed, noninclusive, or simply a public relations exercise.

Organisations must also be aware of their public's—

Local, regional, provincial (state), national, or international interests. No project is 'local'. The geopolitical interests the public represents also influences how a process should proceed. Recent experiences of international groups 'acting globally' at a local level are causing more local projects to take on international significance. The possibility of streetside demonstrations occurring thousands of kilometres away from the site of a specific project, or stockholder rebellion over an organisation's activities means that organisations must be willing to adopt a different approach to working with these experienced and often much more sophisticated publics.

Ethnic, cultural, and geographical diversity. Indigenous and immigrant ethnic groups must be involved in different ways than 'dominant culture' publics. Their potentially different views and experiences with public processes mean that each group will require special consideration or the process will likely fail. The same is true when involving both rural and urban populations.

While this may seem like an extensive set of differences between the publics, it is very realistic. An organisation must understand the diversity of the public and develop strategies that reflect this diversity if they are to succeed. It must also understand the history of an issue—how long it has been of concern to the public and how well it was handled in its early stages. It is always easier to work with stakeholders and community groups before their positions have had an opportunity or reason to harden, than it is after a process has evolved to a state of conflict and confrontation.

This further supports the need for a number of strategies rather than a single formulaic approach. 'We held an open house [or public meeting] and no one came' is not acceptable. If an organisation has found itself in this situation, then it is not committed to, or does not understand, public involvement.

WHEN WILL THE PUBLIC BE INVOLVED?

It is a good rule of thumb to expect that the public will be concerned when something new, something large, or something different is proposed in or near their community. Concerns will even be greater if the project or activity is located close to their places of residence. Even relatively common activities which an organisation does not see as a problem itself can cause concern amongst the public when it has not been informed and personal concerns have not been addressed. Given the high profile of environmental issues and the public concern for health, safety and quality of life, there will almost certainly be public concern if developments are located near people, or near environmentally sensitive areas such as parks, bodies of water, or regions which are thought to be environmentally sensitive or ecologically significant.

DEGREES OF PUBLIC INVOLVEMENT

When an organisation begins to think about how it is going to develop a public involvement process, it often starts with a public relations exercise. While public relations may have been acceptable in the past, today it can only be considered as one component in a process. However, there is a continuum of degrees of possible involvement, ranging from persuasion (PR), through education and information sharing (consultation), and two way consultation, to participation. More and more, the public is being involved in joint planning activities with an equal say in the process, and in some cases, interested individuals are asked to undertake the process directly with all decision-making authority delegated to them. The continuum is outlined as follows:[2]

- *Persuasion:* The use of techniques to change public attitudes without raising expectations of involvement.
- *Education:* The distribution of information to the public to create an awareness of an organisation's project and issues.
- *Information feedback:* The distribution of information concerning an organisation's position by that organisation with the intent of receiving and considering the public's comments on that position.
- *Consultation:* Use of two-way communication between an organisation and the public based on established, mutually accepted objectives.
- *Joint planning or shared decision making:* The public is represented in the decision-making process and is given voting and decision-making authority.
- *Delegated authority:* The transfer of responsibilities normally associated with the organisation, to the public.
- *Self-determination:* The undertaking of the process directly by the public with the organisation accepting the outcome.

DEVELOPING A PUBLIC INVOLVEMENT STRATEGY

The need to involve the public in the decision-making process is directly related to the significance of the decision to the public, and the extent to which the issue under consideration is controversial. The problem for any organisation is to determine which issues will be considered controversial or significant by the public. An issue that appears relatively insignificant to management and staff, may be viewed in an entirely different light outside the organisation.

As a rule, any organisation should seriously consider public involvement when—

- Reaching a decision requires choosing between important social values;
- The results of a decision will significantly affect the interests—whether economic, political, environmental, social, or cultural—of some people or groups more than others;
- The public perceives that it has a lot to gain or lose by a decision;
- The issue to be decided is already a source of controversy;
- The organisation needs positive public support to implement a decision; and
- Considerable environmental or social impacts may be expected.

---

[2] This section is based on work originally prepared by G.M. Farrell et al. (nd), *Involvement: A Saskatchewan Perspective.*

In the context of EIA, the rules and regulations of the applicable jurisdiction may shape much of the public involvement process.

The Canadian experience has shown that one of the reasons certain involvement programs have been successful is that the organisers invested sufficient time and energy to develop a strategic plan for the project and associated public involvement process. This strategy has been further supported by detailed action plans describing in detail all of the requirements (such as location, materials, budget) and responsibilities before the public becomes directly involved. These plans can be distributed to members of the organisation and the public for their review and comment, thereby promoting trust, internal coordination, and support.

More than just simply preparing to do a good job, the planning provides a structure for scheduling and a framework for discussing the inevitable 'what ifs'. Developing the strategy can act as a practice run. If it is well designed and flexible in its application, it will provide an ongoing opportunity for evaluating the process, adjusting timing and events, and updating managers and corporate partners on the project's progress. Most importantly, if done well, it can result in the involvement of the organisation's senior management in the process.

Failed involvement processes are often associated with inappropriate methods, strategies and expectations, demonstrated by the all-too-common view: 'Be in the field tomorrow' (to do what?) and 'Hold a stakeholder meeting next week and get a decision' (with which groups and about what?), which leave little time to prepare strategies and action plans or establish realistic resource and timing requirements. Planning the public involvement process should be a fundamental part of any development project.

A COORDINATED FRAMEWORK FOR PUBLIC INVOLVEMENT

Few organisations have a consistent and coordinated approach to work with the public, whether it be the proponents themselves, the regulatory bodies, or consultants undertaking the EIA process. Organisations have been developing their own public involvement programs on an *ad hoc* basis with little internal cooperation or consultation.

What is needed is an overall institutional framework so that a coordinated program can be developed. Such a framework can create—

- An expectation of, an openness toward, and willingness to accept public involvement;
- Internal expertise in public involvement processes;
- Internal processes and procedures that facilitate involvement;
- Information and other resources to assist in involvement;
- A focus for sharing lessons, ethics, tools, and principles;
- Consistent methods to coordinate, communicate, and collaborate on individual initiatives;
- An opportunity to reduce time and financial requirements;

- An ongoing relationship with clients, stakeholders, and the public as a means to an end, not an end in itself.

If an organisation takes the time—especially when the time is available and they are not in a crisis developmental situation—to develop its own coordinated framework, then the framework can provide a solid base from which to launch any public involvement initiative.

STAGES IN PUBLIC INVOLVEMENT

Most organisations assume that public involvement is a single, albeit ongoing, activity. However, in developing a strategy, it becomes apparent very quickly, that there are a number of major stages that comprise the process. A simplified strategy should include the following stages.

Stage I: Early Consultation

In many processes, this is actually a 'scoping' stage. The key people and organisations need to be identified and consulted with informally to help identify major issues and other interested parties, and to provide an estimate of the level of public interest. The objectives at this stage should include:

- Identifying the publics, stakeholders, and groups with an interest or concern in the situation;
- Identifying the publics', stakeholders', governments', as well as the organisation's issues, concerns and values as they relate to the decision-making process;
- Reducing misinformation, rumour, and gossip;
- Gathering relevant economic, environmental, and social information;
- Informing and educating the publics, both internal and external, and stakeholders about intended actions and the expected consequences of projects, legislation, policy, or regulations;
- Developing two-way contacts, communication, discussion and negotiation;
- Developing and maintaining credibility; and
- Improving decision making within the organisation and by the publics.

These objectives should be based on a set of principles for participants which state that:

- All parties involved in the process have important contributions to make;
- All parties must understand and agree on the mandate and terms of reference;
- All parties must be involved in the design of the consultation process;
- Consensus is possible but requires flexibility, compromise, trust, and sufficient time and resources; and
- All parties must have access to a common information base that is updated on a regular basis.

Stage II: Initial Planning

In Stage II, it is essential to determine the process of consultation itself, now that some initial contact has been made, and preliminary information is available. There is no substitute for charting the decision-making process for both the regulator and the organisation, to identify how they will meet the requirements imposed by the development activity. Identifying the different steps, 'decision points' and publics that will likely be involved, helps everyone understand when, and how, they can best be involved. It is also essential to establish the goals for the process and decide what information will need to be exchanged among all parties. Graphically portraying the decision points makes explicit to all those managers and others in the organisation that a consistent approach is agreed to by all internal participants.

There are a number of relatively simple charts that have been devised to assist organisations in the planning process (e.g., Praxis 1988). Taking these charts to the public and asking how and when they want to be involved, quickly opens up the process and can illustrate that the organisation is being honest in involving them in the process.

Stage III: Developing the Public Involvement Action Plan

It is essential to:
- Carefully choose the consultation or participation methods that will be used. There is no single best technique for a given public process. In fact, it is advisable to use several techniques in combination in order to meet the needs of divergent groups at different stages of the process, and to accomplish the organisation's various objectives. 'Triangulation' is the term commonly used to describe the use of several different techniques.
- Identify the means of internal and external communication;
- Establish a practical method for evaluating the process; and,
- Commit the necessary staff and monetary resources.

The selection of an overall approach and techniques is generally a matter of judgement based on experience. The organisation can benefit in the design stages through the use of continuous quality improvement techniques such as 'benchmarking', where successful processes are reviewed and analysed, and one is adopted as the standard for the organisation's project. However, the organisation must be aware that the more complex and expensive the methods employed in involving the public, the greater the public's expectations that the results of the process are going to be used. It is important to remember that involvement processes send signals of one kind or another to the public. An organisation must know what signals are being sent to the public and needs to determine whether they are the appropriate signals.

Following are some suggestions when selecting an approach:
- Ask each public, stakeholder, or interest group how they want to be involved. This will provide some sense of the range of techniques the process will need to incorporate. It might be surprising how easy it is to obtain cooperation.

- Avoid consulting only those groups known to support the initiative. Ensure that all relevant groups are offered an opportunity to participate.
- Avoid underestimating the technical and professional competence of citizens.
- Use the process to increase management expertise and develop team-building opportunities both within the organisation and the public.
- Work to develop public expertise, creativity, and consensus.
- Do not ignore the organisation's 'internal publics' or the professional publics from associated departments or regulatory agencies that may be involved in the process.
- Always strive to maintain credibility and legitimacy.

The issue that often derails the process is not the external publics, but those within the organisation that have not been consulted about the process. These internal publics must be accepted as part of the process, and must support the process when some aspect of the process involves their involvement. It is very easy for internal publics to subvert the process (deliberately or accidentally) if they are not aware of, or involved, from the beginning. Ultimately, it is not the approach that will 'win over' the public, it is the sincerity of the people conducting the program. Their commitment, integrity, and ethical standards will gain the trust and commitment of the public, far faster than any technique.

Stage IV: Implementation

While it may sound simplistic, in Stage IV it is necessary to implement and monitor the process, evaluate the results, and adjust the process as necessary. If the organisation has planned all the steps, agreed on the strategy and approaches, and has informally involved the publics in obtaining their support of the approach, then the likelihood of success is high. It should be remembered that no process is ever linear. Events will happen throughout implementation that are unexpected and may take the organisation by surprise. It is easier to meet these surprises in a proactive manner if flexibility is built into the process. As internal or external events and factors affect implementation, it is necessary to go back to the strategy and action plans, and review, revise and reflect on the implications of the changes.

In addition to reviewing the strategy and action plans as events occur, it is also advisable to have set review points, or times where the consultation team and the public come together to review and reflect on the overall approach, the effectiveness of the strategies, and the resource requirements expended or planned. Public involvement can be time consuming and expensive. As such, it should be evaluated for results in the same way any other process or project components would be evaluated. Who conducts the evaluation and what reports are generated should be decided early and included in the work schedule.

Stage V: Post-Decision Follow-up

Currently, even the best public involvement programs usually extend no further than the final decision. Very few consider what to do when it is all over, and yet this is a point in the development and implementation of a project where—due to a lack of continuity and an ignorance of shared history—things commonly go awry. Plans and people need to be in place to smooth this transitional stage. For example, on projects where a regulatory agency gives its approval to proceed, the organisation is often left scrambling to put a process in place for implementation. Those who planned the project, consulted the public, built the relationships between the organisation and the interest groups, and struggled from issue to issue, are no longer in-place, having been seconded or hired for the sole purpose of involving the public in the decision-making phase. When the new team is appointed, it is expected to pick up where the others left off, often with little or no understanding of what has already occurred, who the players are, or which issues have been left simmering on the back burner waiting for the implementation phase.

A plan for the post-decision phase is equally important when the regulatory body does not approve the project. With the planning team disbanded, the public, some of whom have committed hundreds of hours of unpaid time and energy into reaching a mutually accepted decision, is left with a press release and little or no understanding of what has happened, nor what may happen in the future. Without debriefing, they will be left frustrated and may consider that they have been treated unfairly, and will develop a lack of respect for the process and will be less willing to cooperate with the organisation or other organisations in the future.

KNOWING WHEN TROUBLE IS AROUND THE CORNER

Almost everyone involved in public processes has had the experience of badly under- or overestimating the level of controversy generated by a particular situation. Everyone struggles to recognise (and to respond appropriately) when an issue is 'heating up'.

The following is a list of activities or actions that indicate that the controversy surrounding an issue may be escalating (Praxis 1988).

- Issues begin to proliferate, with new groups adding new issues.
- Issues move from the specific (a project) to the general (development).
- Criticism of a proposed action turns into attacks on the organisation or individuals.
- People considered moderate become vocal about the issue.
- Political leaders begin to use the issue for political gain.
- The leadership of established groups becomes more radical.
- Previously established channels of communication shut down and people begin to talk only with others who agree with them.

If some of these indicators are present, the process is in trouble. In fact, some experts feel that these indicators signify a path of escalation that, if nothing intervenes to change direction, will lead to failure. Only a significant commitment to rebuilding trust is likely to provide that change of direction. People will forgive a lot of mistakes if it is apparent that intent is genuine. However, they are quick to get the message that an organisation is only going through the motions and is not committed to the outcome of a negotiated process.

CURRENT ISSUES IN PUBLIC INVOLVEMENT

This section describes a number of the major issues facing public involvement practitioners. While by no means exhaustive, the issues highlighted here, pose some of the most immediate and critical challenges in designing and implementing successful processes.

Consulting the Internal Publics

In the rush to 'go public', permanent and contracted staff and consultants working for the organisation undertaking the process are often left in the dark with only an inkling of the organisation's consultation activities. This can result in serious problems. Consultants and contractors are often thought of as being representatives of the organisation, and staff are usually the first line of contact for the public. If staff are approached by an individual or group regarding a process that the staff are either unaware of or misunderstand, both parties will have reason for concern. The staff will feel that they have been left out of a process designed to be inclusive, and the public will suspect that they are not being told the truth about the organisation's commitment to the process. Therefore, a major component of any public involvement process should include consulting, educating and sharing information with staff and consultants. In some recent involvement processes, up to 20 percent of the resources and time have been allocated to internal consultation.

Unrealistic Expectations

Managers organising their first involvement process often have the illusion that the process is easy: How difficult can it be to plan and hold an open house, or a public meeting? What is actually necessary to support the process? Yet the complexities are often a surprise. Although some processes are straightforward, even those that appear to be simple can be time and resource intensive, and involve complicated political issues. More than likely, if the process is perceived as being simple, it is not being done, or has not been done, properly. There is no acceptable alternative to advance planning, and the better the organisation's strategy plan, the more likely that the problems will be anticipated and worked out in advance.

Championing

Although it is a well-established management philosophy, 'championing' of public involvement is not common in organisations. In general terms, championing is the process by which an individual within an organisation who has the authority, responsibility, drive, belief, and commitment, promotes and defends an activity to both the public and senior management. That person must be able to explain the activity, explain and defend decisions to a board of directors, build links between departments, and provide leadership to teams. Champions and teams are two parts to a whole. Public participation needs champions within organisations to promote the value of participation. However, these champions do not speak for the public, only for the public's right to speak.

Public Overload

The growing trend toward requiring public involvement in decision making, in both the public and private sector, is overloading the public and its ability to respond. This has led some organisations to help provide the resources necessary for the public to participate. However, in some jurisdictions, there are so many actions requiring input and a commitment of time and resources from the public, that the public is increasingly 'shutting down' and withdrawing. This is especially true in relation to environmental issues where a flood of public processes has led to organisations competing for the public's attention. However, a negative reaction to a request to participate should not always be interpreted as a lack of interest in the public process. Implementing a decision that the public has not had an opportunity to comment on is often the best way for an organisation to reinvigorate the opposition.

Paying the Public to Participate

Whether to pay the public for involvement typically generates considerable discussion. In jurisdictions where public hearings and reviews have become common practice, there has been a demand for so-called 'intervener funding' to assist those deemed to be directly affected by a project, so as to offset the public's costs of presenting and researching their case. In some jurisdictions, 'directly affected landowners' has been narrowly interpreted to mean only those whose property is directly adjacent to the proposed project. In other situations, legislation and regulations allow for a broader definition so that cultural and indigenous groups as well as environmental interest groups are eligible for funding, depending on the 'cause' or situation. In some jurisdictions, the funding is provided from the public purse, while in others, the proponent is required to pay some, if not all, of these costs as determined by the regulating authority. In some cases, regulatory agencies require that all 'directly affected landowners' are compensated for involvement, while in other cases, regulatory agencies may request proposals for funding from groups, irrespective of location, to support specific research or for the actual costs of participation.

In most jurisdictions, regulatory agencies do not require the proponent to cover the costs of other parties such as the public. This, however, is not one of the major issues facing government and regulators. More and more government agencies are establishing advisory committees, stakeholder groups, or multiparty committees, to negotiate issues and agreements, such as air or water quality standards, regional or national environmental standards, or strategies to harmonise the differing environmental regulations across jurisdictions. With this broad stakeholder participation, there is no broad-based policy covering the cost of participation. Some government agencies cover only expenses, while others provide honorariums and expenses. This lack of consistency is causing problems between various government agencies and those stakeholders participating in these processes.

Participation by the public is essential to the effective and efficient operation of a development project. Participation costs time and money of those participating, and it is only appropriate that developers reimburse the out of pocket expenses of those participating, if not paying them for their participation.

Public Scepticism

The public is becoming jaundiced with some processes because people do not see the consulting organisation using results of their input. The question they are asking is, "Why should we bother with participating this time, when you didn't use our input last time?" It is a very good question and needs to be responded to in each and every process.

Staff and Decision Maker Overload

While there have been many staff reductions in both the public and private sectors, the workload of those who remain has only increased. In some cases, this has resulted in staff burnout. If public involvement processes are added on top of too large a workload, there will be a tendency to skim the surface of issues and miss valuable opportunities.

Technical and Scientific 'Fact' Versus Public Perception

'Perception is reality' is being recognised as the truth in almost every public process. The public gathers information from a multitude of sources—local, national and international media, interest group networks, and the local neighbourhood—and may form its opinions based on information that may be biased, incorrect or intentionally misleading. Technical experts at odds with each other over apparently factual evidence, and the tendency of Eurocentric bureaucracies to discount inherited, traditional, or local knowledge, only complicate this issue and work to isolate the participants. Once again, a well-thought-out strategy that provides recognition to the contributions each party can make to the process will go a long way to avoiding the process floundering on the 'fact' issue. Credibility, fairness, and a lack of bias in scientific experts and advisers also can go a long way in achieving a balanced and realistic perception amongst the public.

Data Overload

Many organisers vastly underestimate the public's response to an invitation to participate. Very few organisers can cope with, let alone manage, the volume of written, verbal, electronic, and other forms of input from the public. Most organisations are familiar with managing quantitative or statistical information; however, they are not used to managing qualitative or open-ended information in large volumes. How does an organisation process, manage, analyse, and use the information gathered from thousands of written submissions, many days of public hearings, survey results, dial-in telephone services, fax, e-mail, or other responses from the public? Time must be spent up front designing methods, staffing and allocating resources to handle the qualitative information which may be received in very large volumes.

This is important because of the ease with which important material or suggestions crucial to the organisation can disappear when it is immersed in such a large quantity of information. Recent futurist publications indicate that the coming revolution is the 'textual revolution'—a revolution that will see databases built entirely of text (Gudbranson 1994). Public processes are on the cutting edge of this revolution and are already using such databases in a number of areas (Praxis 1990). Management systems, including the computerisation of input, search and analysis, and output capabilities are now developed or being developed. How they can aid the process should be part of the discussion of the organisation's strategic plan.

Indigenous and Ethnic Group Consultation

To date, in most developed societies, there has been a strong focus of consultation programs involving Caucasian middle class publics. With changing ethnicity in many jurisdictions coupled with the internationalisation of consultation and participation, there is a much greater emphasis on the development of approaches and techniques appropriate to different cultures, communities and individuals. The rise of influence and power in First Nations, indigenous or Aboriginal communities around the world, and greater concern about issues that affect them, is requiring a rethinking of the way consultation and participation is undertaken both within and between communities, as well as between indigenous and ethnic organisations and the various levels of the public and the private sector. The World Bank, in some of its recent publications and activities, has taken the major role in developing principles, guidelines and requirements for involvement of minority groups, particularly indigenous groups (World Bank 1993).

Many of the ideas and directions suggested in this chapter are appropriate to these cultures, however, they need to be applied with even greater scrutiny. As an example, community surveys are not part of indigenous or ethnic cultures. Such instruments are foreign and are not appropriate for use. Public meetings can meet with the fate of either no attendance, or being used for local politics and grand-standing. Again, the western form of public meetings are not appropriate to other cultures. However, if techniques are used that 'fit' with the relevant indigenous or ethnic culture, such as small group sessions, one-on-one sessions, and potlatches, the results can be very positive, valuable, and very strong. Coupled with using people from these cultures to implement the process, the results are even stronger. The message is the same as with any culture or community: use approaches which are familiar to that community. Ask in advance how the members of the community want to be involved and what they believe is on the agenda for discussion and negotiation. Finally, each and every involvement process must be undertaken in a time frame that suits that culture.

A company or government agency with a fixed agenda, whether it be a project or a new policy, faces defeat from the very beginning because many indigenous communities are looking for land and resource claims to be resolved, or for broader political issues to be raised. These broader concerns will almost always be placed on the agenda for discussion and negotiation, even if they fall outside of the scope of the current activity, or the mandate of the agencies and companies involved. Public participation facilitators need to be able to deal with these issues in an informed and compassionate manner so that discussion can progress to the more immediate issues raised by the current project.

Stakeholder Accountability

A recent trend in public processes is the focus on stakeholder representation or multi-stakeholder processes. Representatives of public interest groups are usually invited to discuss and negotiate key issues and recommendations, or even to take responsibility for implementing results. In some cases, this process works exceptionally well, however, in other cases, it is unclear whether these stakeholders represent their organisation, a number of organisations or only themselves. Of even greater concern is whether they actually communicate the results of their discussions and negotiations with their own membership(s). It is now a requirement in some processes that there be a method for determining that information is getting back to the membership of the stakeholder group.

Organisations undertaking public involvement processes need to realise that stakeholder processes are one technique in the broader public involvement process. Contrary to what some organisations see as hype and others as pressure, stakeholder processes are not public processes. A stakeholder process is only one part of a larger public process. If an organisation bases its actions totally on a stakeholder process, there is a strong likelihood that the general public will want to know why it was not consulted before a decision was reached. The public and interest groups may even turn to the courts or elected officials to block implementation. Involving stakeholders does not remove the need to involve other groups throughout the public process. No one group represents all of the interests.

PRECEDENT–SETTING GUIDELINES

In 1993, the World Bank released their environmental assessment sourcebook update entitled, *Public Involvement in Environmental Assessment: Requirements, Opportunities and Issues*, which established precedent-setting requirements for public consultation in environmental assessments (EA) for certain projects. Its intent was to improve understanding about the potential impacts of proposed projects and to establish procedures for project development. It recognised that consultation can also identify contentious issues and generate alternative sites, designs and mitigation measures while creating a sense of local ownership of, and commitment to, projects, as well as mutual accountability during project implementation. The Update requires consultation with affected groups in the very early stages of project design and planning, and then again when the draft EIA is prepared. This is seen as an absolute minimum, however, and many organisations are developing creative and innovative ways to involve the public.

Identifying affected and relevant groups is essential to the World Bank's consultation requirements. Groups directly affected should be more involved than other interested parties and the Update requires that directly affected groups should be involved early and extensively in the project cycle. These groups include the intended beneficiaries of a project, 'at-risk' groups, and key stakeholders in physical proximity to the project or who fall within its area of influence. Organisations must try to predict a project's impact on each participating group including the extent, intensity, duration, and reversibility of the potential impacts.

A recognised consultation framework should be established that defines the 'who, what, where, when and how' of the consultation process. The World Bank stresses that clear agreement on the framework at the start of the consultation process promotes respect and trust among participants. Consultation on the draft EA report for certain World Bank project types is mandatory and is considered one of the most important elements of the EA process. The World Bank recognises that public consultation during EA preparation helps clarify misconceptions and enhances a project's social acceptability.

The World Bank believes effective consultation is characterised by the development of a clear consultation framework and wide distribution of information before the consultation process begins. The consultation process must permit and encourage project modification as a result of public consultation. It requires the use of two-way communication through a wide sampling of affected people. It also requires feedback mechanisms to inform participants of the results of their participation.

Public participation is not a World Bank requirement, except in certain circumstances. Public participation is required for projects that affect indigenous people, involve involuntary resettlement, or depend on local responsibility for their success. While the Update does not mandate public participation, the World Bank states that public participation in decision making further strengthens local ownership of and accountability for projects and can achieve positive developmental results.

For those working in the international arena, indeed even within domestic settings, Directive 4.01 is setting the tone for projects funded by the World Bank. Now, projects outside World Bank authority, including those undertaken by the private sector, are also coming under the influence of this directive. Thus, this directive is having a far greater influence than anticipated in requiring, not only consultation, but also participation in many project and policy development exercises, and especially where cultural and indigenous populations may be affected.

THE FUTURE OF PUBLIC INVOLVEMENT

As stated earlier, public involvement is not a fad or a fashionable, politically correct practice. It is true that in both the public and corporate sector, public involvement processes have been abused, and publics have misinterpreted consultative processes to be something more than they are. These have placed into question the validity of public involvement mechanisms. However, the evolution of public and corporate institutions in society has created a lasting place for public involvement in decision making, one that is being confirmed each time a successful consultation takes place.

The public today is more informed about issues that affect it than ever before. Information and communications technology play a substantial role in creating an educated public, and in conveying the demands of the public to those who are making the decisions. Specialised interest groups are more agile, more easily mobilised, and frequently possess technical information which is superior to that of decision makers (Montgomerie 1994). Therefore, in dealing with a public which is increasingly more aware, there are often very good reasons to seek the participation of the public in decisions which affect them.

Governments across the globe are undergoing a fundamental change in their orientation to the business of governing. Pressured by the sheer enormity of the task of governing, they are moving away from large bureaucratic delivery structures toward smaller-scale policy-oriented roles in which governments "steer more than they row" (Osborne 1991). Governments are now becoming more focused upon clients instead of the bureaucracy, upon outcomes instead of inputs, and upon persuasion instead of commands. To be successful in these strategies, the public must be involved. With downsizing and flattened organisations, governments must rely more heavily on public input to generate public policy. Greater empowerment of communities and more articulate interest groups require that governments be more sensitive to the needs of the public. Therefore, public involvement is now an increasingly important component of public policy, and a component which is recognised for its positive contribution.

The corporate sector is also changing its view of the role of consultation in corporate decision making. There is a growing ability of an informed and active public to block corporate decisions that were once made behind closed doors. With corporations more accountable to the public for their actions, it is in their interests to involve the public to increase ownership in those decisions and an understanding of their implications. There is also a growing awareness in the corporate sector that huge investments of time, money and effort made in the planning of major projects can be lost when those plans are challenged. Projects which are developed with the support and involvement of the community have a far greater chance of succeeding. Implementation costs are also substantially reduced by working with a public that is involved from the beginning. Corporations are also being judged more frequently by both stockholders and customers on their image as good corporate citizens. Public involvement practices can have an enormous impact on creating a very positive presence in the community or in the country.

Consultation has typically taken place on a project-by-project basis, especially in the environmental field. This is now shifting dramatically to include ongoing consultation and participation in the development of policy, legislation, regulations, and even program delivery in fields such as health, education, and social services. Many of the approaches to public involvement were developed in the EIA field and are being transferred to the policy and program areas of government with much success.

We are witnessing a slow change in the role that the public plays in the decisions that affect them toward greater involvement in decisions of all sorts. Governments are moving rapidly toward greater public participation in the development of public policy both for reasons of necessity and because they are being pressured to do so. In the private sector, whether for reasons of strategic success, reduced implementation costs, or developing a positive corporate image, public involvement has an expanding role in the future of corporate decisions. However, the greatest guarantee of a continuing growth of public involvement is the public itself. Having once been allowed in to participate in decision making, it is difficult to stand on the sidelines the next time. If the public is not invited to participate, people will demand it.

THE CHALLENGE OF PUBLIC INVOLVEMENT

As societies and peoples become more informed about their environments and organise into more complex structures, the current demand by the public to be involved in decision making will continue to increase. People expect and demand to be more involved. They live with the consequences of decisions and expect to share and be responsible for making them. To be successful in meeting these demands, organisations will need to be more proactive, less reactive. They will need to meet the public in communities and on the street. They will need to find a common language, to learn to listen, and to consider and incorporate what is being said. But most of all, they will need to learn that if it is well organised, open and honest, public involvement can be more than just a means to an end. It can be an end in itself, a permanent dialogue that will benefit the organisation for many years.

ACKNOWLEDGMENTS
Assistance in writing this article was provided by Nancy Marshall, Andrea Matishak, Ian Montgomerie, and Patrick Lewis, all of Praxis, Inc.

REFERENCES

Benn, S.I. 1979. "The problematic rationality of political participation." *Philosophy, Politics and Society*. Peter Laslett and James Fishkin, eds. Oxford: Basil Blackwell, 292–312.

Bregha, F.J. (nd) *Public Participation in Planning Policy and Programme*. Toronto: Ontario Ministry of Community and Social Services.

Bregha, F.J. 1978. "Further directions for public participation in Canada." *Involvement and Environment: Proceedings of the Canadian Conference on Public Participation*. Barry Sadler, ed. Edmonton: Environment Council of Alberta.

Burdge, R.J. 1990. "The benefits of social impact assessment in third world development." *Environmental Impact Assessment Review* 10(1/2): 123–134.

Connor, D.M. 1985. *Constructive Citizen Participation, A Resource Book* (rev edn). Victoria BC: Development Press.

Craig, D. 1990. "Social impact assessment: Politically oriented approaches and applications." *Environmental Impact Assessment Review* 10(1/2): 37–54.

Creighton, J.L. 1980. *Public Involvement Manual: Involving the Public in Water and Power Resources Decisions*. Washington DC: US Department of the Interior, Water and Power Resources Service.

Creighton, J.L. 1984. *Public Participation: A Manual for EEI Member Companies*. Saratoga CA: Edison Electrical Institute.

Creighton, J.L., J.L. Chalmers, and K. Branch. 1983. "Integrating planning and assessment through public involvement." In *Public Involvement Techniques: A Reader of Ten-Years Experience at the Institute for Water Resources*, (IWR Report-82-R1). James L. Creighton, Jerry Delli Priscoli, and C. Mark Dunning, eds. Fort Belvoir VA: US Army Corps of Engineers' Institute for Water Resources.

Draper, J.A. 1978. "Evolution of citizen participation in Canada." *Involvement and Environment: Proceedings of the Canadian Conference on Public Participation*. Barry Sadler, ed. Edmonton: Environment Council of Alberta.

Farrell, G.M., J.P. Melin, and S.R. Stacey. (nd) *Involvement: A Saskatchewan Perspective*. Regina: Deptartment of the Environment, Government of Saskatchewan.

Gougeon, R.B. 1986. Modifying the Impact of Resource Development: Externalities and Public Involvement in Parks Canada. Unpublished MA thesis, Department of Geography, University of Ottawa.

Gudbranson, W. 1994. "The textual revolution for information power." A special supplement to the *Financial Post Magazine*, April: D33.

Hampton, W. 1977. "Research into public participation in structure planning." In *Public Participation in Planning*. W.R.D. Sewell and J.T. Coppock, eds. London: John Wiley and Sons.

Montgomerie, I. 1994. Public Consultation and Public Policy. Unpublished Ph.D. dissertation, University of Alberta, Edmonton.

Praxis. 1988. *Manual on Public Involvement Environmental Assessment: Planning and Implementing Public Involvement Programs*. Calgary: Federal Environmental Assessment Review Office.

World Bank, Environment Department. 1993. *Public Involvement in Environmental Assessment: Requirements, Opportunities and Issues*. Environmental Assessment Sourcebook Update. Washington DC: World Bank.

ADDITIONAL SOURCES

Arnstein, S.R. 1969. "A ladder of citizen participation." *American Institute of Planning Journal* 35(4): 216–224.

Bleiker, A-M. and H. Bleiker. 1981. *Citizen Participation Handbook: for Public Officials and Other Professionals Serving the Public* (4th edn). Wyoming: Institute for Participatory Planning.

Burdge, R.J. 1990. "Social impact assessment and the public involvement process." *Environmental Impact Assessment Review* 10(1/2): 81–90.

Canadian Petroleum Association. 1989. Public Guidelines for the Canadian Petroleum Industry: Planning Implementing, Evaluating. Calgary: Canadian Petroleum Association.

Chess, C. 1989. *Improving Dialogue with Communities: A Short Guide for Government Risk Communication.* New Brunswick: New Jersey Department of Environmental Protection, Division of Science and Research.

Creighton, J.L., J. Delli Priscoli, and C.M. Dunning, eds. 1983. *Public Involvement Techniques: A Reader of Ten-Years Experience at the Institute for Water Resources,* (IWR Report-82-R1). Fort Belvoir VA: US Army Corps of Engineers' Institute for Water Resources.

Dale, D. 1978. *How to Make Citizen Involvement Work: Strategies for Developing Clout.* Amherst: University of Massachusetts (Citizen Involvement Training Project).

Ducksik, D.W. 1982. Edison Electrical Institute's Workshop on Utility Experience with Advance Public Participation in Planning—Proceedings. Washington DC: Edison Electrical Institute.

Environment Canada. 1992. *Consultations and Partnerships: Working Together with Canadians.* Transition Team Steering Committee on Consultations and Partnerships. Ottawa: Environment Canada.

Fisher, R. and W. Ury. 1981. *Getting to Yes.* Markham: Penguin.

Fraser, B. 1981. *Public Involvement Handbook.* Victoria BC: Information Services, Ministry of Forests, Province of British Columbia.

Garcia, M. 1975. *A Workshop Manual for Public Participation.* Tucson: University of Tucson.

Holtz, S. 1988. "Perspectives on Federal and National Consultations," a paper presented to the Ontario Environment Network Workshop on Public Consultation, at Scarborough College, Toronto.

Howell, R.E., M. Olsen, and D. Olsen. 1987. *Designing a Citizen Involvement Program: A Guidebook for Involving Citizens in the Resolution of Environmental Issues.* Corvallis: Western Rural Development Centre, Oregon State University.

Langton, S., ed. 1979. *Citizen Participation Perspectives: Proceedings of the National Conference on Public Involvement.* Medford MA: Lincoln Filene Center for Citizenship and Public Affairs, Tufts University.

Mater, J. 1977. *Citizens Involved: Handle with Care! A forest industry guide to working with the public.* Toronto: Timber Press.

Niagara Institute. 1989. *Public Participation Handbook: How to Make Better Decisions by Involving Interested People and Groups in your Planning and Implementation Process.* Ontario: Niagara-on-the-Lake.

Olsen, B. 1976. The Role of Public Participation in Parks Canada Planning. ME Des (Urbanism) degree project, University of Calgary.

Ontario Ministry of the Environment. 1989. *Public Consultation: A Resources Kit for Ministry Staff.* Toronto: Environment Ontario.

Osborne, D. and T. Gabler. 1991. *Reinventing Government.* Reading MA: Addison Wesley Publishing.

Praxis. 1990. Public Involvement and Community Relations in Integrated Resource Planning. Prepared for Resource Planning Branch, Alberta Forestry Lands & Wildlife. Vol.II.

Sadler, B., ed. 1978. *Involvement and the Environment—Proceedings of the Canadian Conference on Public Participation.* 2 vols. Edmonton AB: Environment Council of Alberta.

Sewell, W.R.D. and J.T. Coppock, eds. 1977. *Public Participation in Planning.* London: John Wiley and Sons.

Wallace, R.R. (Dominion Ecological Consulting). 1985. Public Input to Government Decision Making. Occasional Paper No. 13, Ottawa: Federal Environmental Assessment Review Office.

Young, C., G. Williams, and M. Goldberg. 1992. *Evaluating the Effectiveness of Public Meetings/Workshops: A New Approach for Improving Department of Energy Public Involvement.* Argonne: Argonne National Laboratory.

Part III:

TOOLS FOR THE FUTURE

Chapter 11
Climate Impact Assessment[1]

ROSLYN TAPLIN
Macquarie University, Australia
ROCHELLE BRAAF
Macquarie University, Australia

INTRODUCTION

With the emergence of widespread scientific concern about the greenhouse effect over the last 10 years, the field of climate impact assessment has changed in role from that of a challenging interdisciplinary endeavour to being a lynchpin in policy debate about climate change. During this period, considerable climate impact assessment work has been attempted by the international scientific community and, in particular, the Intergovernmental Panel on Climate Change (IPCC). Many substantial policy issues and debates have arisen in association with the broad dissemination of climate change science—some of which are discussed in Taplin (1994a, 1994b, in press) and Braaf et al. (1994). Climate impact assessment is of fundamental significance in clarifying arguments about the greenhouse effect and accordingly is an important area of impact assessment that is currently experiencing growth and change.

Carter et al. (1994: 1) defined climate impact assessment as "a sequential set of activities designed to identify, analyse, and evaluate the impacts of climatic variability and climatic change on natural systems, human activities, and human health and well-being, [and] to estimate the uncertainties surrounding these impacts." The technical difference in emphasis between today's mode of thinking about climate impact assessment and that of a decade ago is that climate impact assessment then focussed on climate variability, rather than climate change due to the enhanced greenhouse effect. Current thinking is exemplified by Carter et al. (1992: 2) who assert that the "ultimate objective [of climate impact assessment] is to provide the general public and policy-makers with estimates of the extent to which climate change may affect the environment and human activities and result in changes in social and economic welfare."

The change in the political status of climate impact assessment in the last 10 years is well demonstrated by the differences in approach of two guides to the subject: *Climate Impact Assessment: Studies of the Interaction of Climate and Society* (Kates et al. 1985) and the Intergovernmental Panel on Climate Change's *Technical Guidelines for Assessing Climate Change Impacts and Adaptations* (Carter et al.

1994). Both of these publications were the result of input from dozens of experts and reviewers in the climate impact assessment field. In Kates et al. (1985), the intellectual challenge of climate impact assessment is stressed: "Scientists in all parts of the world who recognize the practical necessity of informed adjustment to climate are invited to take up the challenge of integrated assessment, and to share in the intellectual adventure of developing theories and methods on the borders of current thought" (Kates et al. 1985: 33). In comparison, Carter et al. (1994) focus on human-induced climate change and emphasise the political and policy implications of the assessment process. They say:

> The general responsibility of science is to expand the [climatic impact assessment] knowledge base for the common benefit. This should be achieved by developing the research methodology for assessment, collecting information on trends in the environment and in society, developing predictive tools for evaluating impacts, forging scientific links across disciplinary, institutional and political boundaries and communicating results objectively to other scientists, decision-makers and the public. Policymakers require climate impact assessments to provide them with the necessary scientific information for policy decisions. . . . They could also provide a basis for negotiating global and transnational protocols for addressing climatic change issues, which lie outside the jurisdiction of individual policymakers. (Carter et al. 1994: 1).

Climate impact assessment thus has an important role in both domestic and international policy development.

CLIMATE IMPACT ASSESSMENT AND THE IPCC

The change of focus of climate impact assessment was initiated by scientists from 29 nations at a conference in Villach, Austria, in October 1985.[2] At that conference, the atmospheric science community made an unprecedented statement, saying, "Some warming of climate now appears inevitable due to past [human] actions" (WMO 1986, cited in Lowe 1989:4). They also emphasised that government actions with respect to energy conservation, use of fossil fuels, and control of greenhouse gas emissions would profoundly affect the rate and degree of future warming and thus the degree of impact on the global climate. Their landmark statement, known as the Villach Statement, has been referred to as a "clarion call to the political world [by scientists] to take some serious account of the problem" (Lowe 1989:4).

In 1988, two and a half years after the Villach Conference, the Intergovernmental Panel on Climate Change (IPCC), an epistemic community of scientists, was established by the World Meteorological Organization (WMO) and the United Nations Environment Programme (UNEP) as a response to growing international

---

[2] Organised under the auspices of the World Meteorological Organization, the United Nations Environment Programme, and the International Confederation of Scientific Unions.

concern about the greenhouse effect. The IPCC was the first multilateral vehicle for scientific and policy treatment of the issue. Its mandate, which included climate impact assessment, was to involve scientists from many nations in—

    (a) Assessing the scientific information that is related to the various components of the climate change issue such as emissions of major greenhouse gases and modification of the Earth's radiation balance resulting therefrom, and that needed to enable the environmental and socioeconomic consequences of climate change to be evaluated; and

    (b) Formulating realistic response strategies for the management of the climate change issue (Government of Australia 1988: 2).

Currently, scientists and social scientists from more than 80 countries are involved in the IPCC process.

The IPCC decided in 1988 to focus its efforts on (I) assessment of available scientific information on climate change; (II) assessment of environmental and socioeconomic impacts of climate change; and (III) formulation of response strategies. This work was taken up by Working Groups I, II, and III respectively. The specific terms of reference for the IPCC's climate impact assessment research were that it—

> . . . should view the environmental and socioeconomic impacts of climate change in an integrated manner, . . . emphasise evaluation of the impacts on a regional scale as they affect agriculture, forestry, health, water resources and floods, desertification, energy, and other sectors, . . . [and] consider the impact of [a] continuously changing climate (Government of Australia 1988: 5-6).

The three working groups' reports of the IPCC's First Assessment Report (IPCC 1990a; 1990b; 1990c) were released progressively in the latter half of 1990 in preparation for the Second World Climate Conference. Stephen Schneider, a climatologist and head of the Interdisciplinary Climate Systems Section at the US National Center for Atmospheric Research in Boulder, Colorado, in reviewing these reports in the journal *Environment*, aptly commented that the reports become "decreasingly quantitative and increasingly nonspecific as they move from the physical and biological sciences to impact assessment and, ultimately, policy. . . [reflecting] the increasing divergence of opinions" (Schneider 1991a: 26).

In November 1992, the IPCC was reorganised and a plan was agreed upon for IPCC work towards a Second Assessment Report to be produced by mid-1995. Working groups II and III were assigned new research agendas. Working group II took on the combined tasks of the old working groups II and III—climate impact assessment and development of adaptation and mitigation response strategies. The new working group III was given the role of handling general issues affecting all groups and, in particular, assessment of future scenarios of greenhouse gas emissions and socioeconomic issues. The IPCC also took on the task of producing a special report (completed and approved by the IPCC in November 1994) to provide information for the first Conference of the Parties to the Climate Convention in

Berlin in March 1995.[3] These tasks have placed a heavy burden on the greenhouse research community.

Working group I currently is assessing available information on the science of climate change. Major tasks for working group I have been to develop a methodology for the preparation of national inventories of anthropogenic emissions by sources and removals by sinks of greenhouse gases, and to establish estimates of past and present regional net emissions of greenhouse gases. These tasks are consistent with the aims of the convention and information compiled will feed into impact assessments and policy development. More emphasis on research at the regional level has also been a focus of working group I as a result of a request from the Intergovernmental Negotiating Committee for the Climate Convention. However, as Bert Bolin (the chair of the IPCC) has pointed out, "Reliable regional scenarios are not possible yet" (cited by Nilsson 1992: 212). This was affirmed at an IPCC Regional Climate Evaluation Workshop held at the Climatic Impacts Centre, Macquarie University, in February 1994.

The reorganised working group II is working on assessing impacts of climate change and developing common methodologies including guidelines for national assessments of impacts of climate change, assessing vulnerability to climate change, and assessing technology options. The working group is comprised of four subgroups. Subgroup A is examining energy issues, industry, transportation, urban issues including related human settlement issues, air quality and health, and waste management and disposal. Subgroup B is looking at impacts on the small islands and coastal zones, oceans and marine ecosystems, tropical cyclones, storm surges, and sea level change. Subgroup C is studying unmanaged resources and terrestrial ecosystems, mountain regions, the cryosphere,[4] hydrology, and terrestrial impacts of climate events such as floods. And finally, subgroup D's mandate is to look at impacts relating to desertification, droughts, agriculture, forests, land use including various forms of human settlements, health, and management of water resources. This broad range of topics currently being researched gives an indication of the mammoth task being attempted by the IPCC in assessing the impacts of climate change.

---

[3] The UN Framework Convention on Climate Change came into force on 26 March 1994 and the first Conference of the Parties to the Convention is to be held in Berlin in March 1995. Currently, national strategies for greenhouse mitigation and adaptation are being developed and refined by nations that have ratified the Convention.

[4] The cryosphere refers to the Earth's snow and ice cover, both land-based and oceanic. The terrestrial components of the cryosphere that the IPCC is investigating with respect to climate change include seasonal snow cover, mountains, glaciers, ice sheets, frozen ground including permafrost, and seasonally frozen ground.

SCIENTIFIC AND TECHNICAL ISSUES AND THEIR POLICY RAMIFICATIONS

In general, climate impact assessments focus on regional interactions of climate, environment, and society. Climate impact assessment is an area of impact assessment that is exceptionally broad in its scope and thus exceptionally challenging. It involves integration of input from both the physical sciences and social sciences; accordingly, as well as climate information, ecological, social, cultural, political, and economic dimensions are critical components of a climate impact assessment. An indication of the challenges involved in climate impact assessment can be gained by listing its incursions into other impact assessment areas. These include social impact assessment, technology assessment, policy assessment, development impact assessment, risk assessment, demographic impact assessment, economic and fiscal impact assessment, and ecological impact assessment—which are all complex fields in themselves. In general, knowledge resources from these more well-developed areas of local environmental assessment appear to have been underutilised to date in globally orientated climate impact assessment endeavours. This is possibly due to a need for more interaction between the communities of experts involved in local environmental assessment and those involved in climate impact assessment.

Papers published from the World Bank–International Association for Impact Assessment symposium, *Environmental Assessment and Development*, demonstrate this need (Goodland and Edmundson 1994). The foreword states, "Protecting the global environment will require greater levels of effort and probably restraint within each country and unprecedented levels of information sharing and cooperation between countries. There are many daunting problems that require. . . unconventional ways for dealing with the new generation of development and environment challenges" (El-Ashry 1994: vii); and yet the contributions in the volume do not address the issue of climate change and development projects. Climate impacts due to accumulated development projects potentially may become the most severe of all impacts from any development project. Fostering collaboration and knowledge sharing between those involved in the many subfields of environmental assessment and experts carrying out integrated climate impact assessments should result in better assessment work overall.

Integrated climate impact assessment attempts to operationalise a full interactive model of climate and society (Kates 1985; Henderson-Sellers 1991; Carter et al. 1994; Mendelsohn and Rosenberg 1994). That is, a systems approach is employed to identify the full range of structures and activities within a region which are vulnerable to climate, and, given a sufficient understanding of how they are interconnected, the impacts of climatic change on natural and human systems may be determined (Riebsame and Magalhaes 1991: 416). Impact studies to date have used a variety of methods, making it difficult to provide consistent and comparable results. Ideally, a common set of guidelines for conducting climate impact studies should be employed such that national and international comparisons between regions are possible. The IPCC's new report, *Technical Guidelines for Assessing Climate Change Impacts and Adaptations*, is an attempt to provide such uniform guidelines. To this end, in 1993 the UNEP climate unit initialised a two-year project focusing on 10 pilot country studies to test these IPCC impact assessment guidelines (UNEP 1993: 7).

Predictions of the impacts of climate change are based on general circulation model (GCM) output data which are based on scenarios of future atmospheric concentrations of greenhouse gases. GCMs are complex compuer models that simulate the global climate. Meehl (1992: 555) explains that "each successive generation of supercomputers has expanded the limits of computing power and has allowed longer runs with more complicated models." As Schneider (1992: 17) has commented, "The validation of the predictions of such models becomes a chief concern." In 1992 reports, the IPCC's updated scientific assessment produced GCM scenarios with CO_2 doubling times in the range of 60 to 100 years, and a globally averaged mean increase in surface air temperature of $1.5\,^{\circ}C$ to $4.5\,^{\circ}C$ (IPCC 1992). Bert Bolin, chair of the IPCC, has suggested that greenhouse gas accumulation may lead to significant climate change impacts in the next 20-30 years, that is in one generation's time (Bolin 1994: 26). Bolin (1994: 26) also has said, "It is clear from IPCC assessments that climate change cannot be avoided unless emissions of CO_2 are reduced." Countries with environments at risk acknowledged in the United Nations Framework Convention on Climate Change (UNFCCC 1992: 4) include small island countries, countries with low-lying coastal areas, countries with arid and semi-arid areas, countries with areas prone to natural disasters, countries liable to drought and desertification, countries with high urban atmospheric pollution, and countries with areas with fragile ecosystems.

There are two fundamental problems with GCMs from the perspective of climate impact assessment researchers: (1) uncertainties about the timescale of climatic change, and (2) climate change predictions at the regional level are not yet available (rectangular grids of GCMs are currently of the order of 300 km by 500km). Several attempts have been made to provide a better base for impact assessment, such as: development of limited area models (LAMs) that have spatial resolutions as fine as 60 km (Gordon et al. 1994; Fowler et al. 1992); nesting of LAMs within GCMs; and production of multiple scenarios based on a variety of GCM outputs and historical (analogue) climate data (Climate Change Newsletter, May 1994). However, even given this research, predictions of climatic impacts for regions cannot be reliably based on climate modelling. This means that knowledge of the timescale of the impacts of climatic change and the degree of severity of their impacts at the regional level is uncertain.

Nonetheless, scientific concern about greenhouse has spurred policymakers internationally to address seriously the issue even in the face of scientific uncertainty. According to the IPCC science working group, "The unequivocal detection of the enhanced greenhouse effect is not likely for a decade or more" (IPCC 1992: 5). However, the hundreds of leading scientists involved in the IPCC assessment undertaking are arguably involved because of their concerns that policy action should be taken to avert the potential impacts of climatic change. As Schneider (1991b: 48) has said, "The public policy dilemma is how to act even though we will not know in detail what will happen. . . . Public policymakers will have to address how much information is 'enough' to warrant action. . . we will have to rely on the intuition of experts. " The timescale of greenhouse predictions indicates that decision makers need to act now to mitigate emissions of greenhouse gases and to put into place adaptation policies even though greenhouse impacts are in the timescale of decades. Bolin said recently, "The risk is there. It is not certain that there is a major [hu]man-induced change of climate, but there is a risk that it might happen. Person-ally, I think that there will be a change. The question is rather: how quickly it will occur, what it will look like, and how serious will it be" (van Zijst 1994: 5).

Uncertainty highlights a further time phenomenon with regard to greenhouse; that is, political moves to develop policy—at least symbolically[5]—both at inter-national and domestic levels, are running ahead of developments in scientific knowledge about climate change and climate impacts. The existence of the Framework Convention on Climate Change testifies to this. The convention's ultimate aim as stated in Article 2 (UNFCCC 1992: 5) is to achieve—

> . . . stabilization of greenhouse gas concentrations in the atmosphere at a level that would prevent dangerous anthropogenic interference with the climate system. Such a level should be achieved within a timeframe to allow ecosystems to adapt naturally to climate change, to ensure that food production is not threatened and to enable economic development to proceed in a sustainable manner.

The convention does not have definite targets for reduction of greenhouse gas emissions but does include reporting process requirements. These are legally weak in terms of compliance but interesting from a policy perspective. Developed country parties are required to make commitments to aim to stabilise emissions by the end of the century at 1990 levels and have these commitments reviewed periodically by the Conference of the Parties. The convention states in Article 4.2(a) and (b):

> Each of the Parties shall adopt national policies. . . and take corre-sponding measures on the mitigation of climate change. . . [and] communicate. . . periodically. . . detailed information on its policies and measures. . .with the aim of returning individually or jointly to their 1990 levels of these anthropogenic emissions (UNFCCC 1992: 8-9).

The climate convention's requirements, of course, are dependent on climate impact assessment information. Difficulties undoubtedly exist for impact assessors in terms of estimating the degree of compliance that parties to the convention may adopt. The convention process thus exemplifies the particular type of environmental

---

[5] Symbolic policymaking takes place when governments formulate policies and decisions as a pretence at responding to concerned interest groups or mass publics. These policy responses are often deliberately designed to appease public concerns but are not intended to be effectively implemented or to represent a fundamental change to existing practices and procedures (Edelman 1964).

decision-making process that Godard (1992: 242) describes as evolving "in controversial contexts, [with] an association of ignorance, uncertainty, potential gravity, and irreversibility."

Australia is a useful case example of a developed nation attempting to implement the climate convention. Taplin (1994b) discusses Australian greenhouse policy formulation and implementation to date and concludes that there are many barriers to achieving the climate convention's stabilisation goal for Australia, including: scientific uncertainty, the long-term climate impacts horizon of decades, lack of political will, Australia's economic dependence on the energy sector, the tensions of Australia's federal system of government, bureaucratic inertia and obstruction, and technical problems in assessing greenhouse gas emissions. These types of barriers are not unique to Australia and exemplify why implementation of the convention will not be straightforward.

A further challenge in the climate impact assessment field is associated with the context within which assessment takes place. The social, cultural, political, ecological, and economic concerns of developed and developing countries differ vastly, and climate impact assessment research in differing nations thus necessarily needs to be sensitive to differing societal values and goals. Accordingly, climate impact assessors from developed nations involved in climate impact assessment work in developing nations need to be careful not to contribute further to the myth "that industrial countries have many answers to environmental dilemmas that need only to be shared with the South" (MacDonald 1994: 29).

In the IPCC and Climate Convention negotiation processes, North-South tensions are a noticeable undercurrent and have arisen explicitly on several occasions. For example, in the early days of the IPCC process, frustration was felt by developing nations about the lack of comprehension displayed by developed nations about the expertise base in developing countries. Mexico at one stage complained that "with the plethora of working groups and subgroups of the IPCC, it had to employ more people to keep tabs on the IPCC than to monitor climate change itself" (IPCC 1990d: 4). A more recent example is the current tension between developed and developing nations with respect to proposals for joint implementation of the climate convention. The concept of joint implementation is that "it can be cost-effective for one nation to meet its commitment to reduce greenhouse gas emissions through financial, technical, or other support to another nation which results in emissions reductions in that second nation" (Granich and Kelly 1994: 1). In practice, this involves partnerships between developed and developing nations with the investment by developed countries being rewarded by some form of 'climate credit'.

Some developing nations are unhappy with the joint implementation concept, however, because they perceive that it may be a cheap way for the rich industrialised countries to shirk their obligations to reduce their own greenhouse gas emissions. Taking these North-South tensions into account, climate impact assessment ideally should be an initiative arising from within nations and should be carried out in close conjunction with relevant national and regional institutions.

POLICY ISSUES

Scientists and social scientists involved in the climate impact assessment field are in an unenviable position—governments want impacts information as input to their policymaking on greenhouse and yet the state of the art in climate change impact assessment is in its infancy. That is, precautionary measures cannot be formulated and implemented without an information base.

Climate impact assessment serves two major purposes in the policy context. First, and more obviously, climate impact assessment is necessary to establish the degree of potential impact of climatic change on natural environments, social systems, communities, economic structures, and infrastructure for particular regions. Also a knowledge of impacts provides justification for mitigating the impacts of climate change via the reduction of greenhouse gas emissions. A second policy-relevant purpose of climate impact research is to establish what kind of changes in ecological and human systems will potentially result from climatic change, so as to guide the development and implementation of adaptation policy responses to minimise adverse effects.

The need by policymakers for climatic impact information was emphasised recently at a climate impacts workshop in Melbourne, Australia. A senior Australian environment bureaucrat emphasised the need for quantification of the social and economic costs of climate change as input to policy formulation (Carruthers 1994: 13). He said:

> We need to draw together skills from across many scientific disciplines in order to understand key climate change impacts. Of the various disciplines, I'd like to highlight the neglect of the integration of our knowledge and understanding of climate change impacts with the socio-economic analysis. As a result, we don't yet have a sound understanding of the dimensions of the economic costs and social implications of climate changes, especially where there are impacts on human systems.

However, the incorporation of economic modelling methods in the climate impact assessment process to predict costs is another area which is far from being straightforward. Most economic models include some processes of substitution, supply and demand adjustment, and trade and other market factors that are based on various assumptions. As Adams et al. (1990) and Rosenberg (1990) have observed in relation to climate impact assessment, broader mitigation or adaptation policy considerations or cultural adjustment factors are difficult to take into consideration due to the problem of quantifying such processes. Also economic models may not be robust beyond certain limits of change (Riebsame and Magalhaes 1991: 423). A final point about economic assessment as a subset of climate impact assessment is the dilemma of how can economists sensibly attempt to estimate costs if the physical impacts of climate change are uncertain?

Ravetz (1986: 417) said, in reference to the global environmental problems that confront governments, "Some of our ideas about science and its applications will have to change. The most basic of these is the assumption that science can, indeed, be useful for policy. . . and can provide 'the facts' unequivocally. . . we must cope with the imperfections of science, with radical uncertainty, and even with ignorance, in forming policy decisions for the biosphere." More recently, Pielke (1994: 316) has said: "Although some scientists and policymakers have understood that scientific information is necessary but not sufficient for making policies related to global changes, the conception that scientific information is sufficient for policymaking persists." He points out that it is an easy course of action for decision makers to procrastinate in the expectation of scientists clarifying information and says that science should be a "component to be incorporated with the broader decision process" (Pielke 1994: 317).

Certainly, attempts to incorporate the *precautionary principle*[6] in the formulation of environmental policy where science is uncertain and knowledge is incomplete are a positive development. In the case of the climate convention, inclusion of the precautionary principle is very significant, notwithstanding its critiqued socioeconomic context caveat.[7] The convention states:

> The Parties should take precautionary measures to anticipate, prevent or minimize the causes of climate change and minimise its adverse effects. Where there are threats of serious or irreversible damage, lack of full scientific certainty should not be used as a reason for postponing such measures, taking into account that policies and measures to deal with climate change should take into account different socio-economic contexts, be comprehensive, cover all relevant sources, sinks and reservoirs of greenhouse gases and adaptation, and comprise all economic sectors (UNFCCC 1992: 5).

---

[6] The *precautionary principle* first emerged in its present form in Germany in domestic policy in the early 1980s. By the early 1990s, at the time of the Climate Convention drafting, the precautionary principle had been taken up by many nations as a policy concept and had been incorporated in numerous international environmental agreements. The precautionary principle is incorporated in Australian environmental policy in the National Strategy for Ecologically Sustainable Development (Government of Australia 1992: 8) and the Intergovernmental Agreement on the Environment (IGAE 1992: 13-14). In Australian policy, it is defined thus: "Where there are threats of serious or irreversible environmental damage, lack of full scientific certainty should not be used as a reason for postponing measures to prevent environmental degradation" (IGAE 1992: 13). With respect to implementation of the precautionary principle, Australian policy guidelines are that "public and private decisions should be guided by: (i) careful evaluation to avoid, wherever practicable, serious or irreversible damage to the environment; and (ii) an assessment of the risk-weighted consequences of various options" (IGAE 1992: 14).

[7] The qualification in the Climate Convention that the precautionary principle should take into account different socio-economic contexts is arguably a weakening of the endorsement of the need for precautionary action and has been critiqued by those who would like to see a stronger Convention.

However, whether the precautionary principle will go beyond being just a symbolic political device in the climate convention, or for that matter in domestic policy, is questionable. Rayner (1986: 433) perceives that the problem with uncertain science is the lack of "permeability of intellectual boundaries maintained by intellectual and political institutions." Certainly, political institutions seem to be looking towards science in the climatic impact assessment field in a traditional sense—as in Beer's (1973) discussion of technocratic politics and Haas' (1990) epistemic community perspective—to provide 'facts' on which to base policies before they will move beyond 'no regrets' greenhouse policies, that is, policies that have no net negative economic impacts. This, of course, is contrary to the concept of the precautionary principle!

The uncertain status of the knowledge associated with climate change thus means that the manner in which greenhouse policy is developed ultimately may be a test of the viability of the precautionary principle, and thus a test of how effectively policymakers at the domestic and international level can deal with global environmental risk.

THE LIKELY FUTURE OF THE FIELD

Henderson-Sellers and Blong (1989: 181) have said, "We can develop strategies for prevention, mitigation, and adaptation of all the greenhouse effects. It is already too late for prevention of the effects themselves, but not for prevention of some of the impacts." Certainly, policymakers should not wait another decade for further modelling research developments before going beyond 'no regrets' policies. Climate change is an environmental problem that scientists have identified and warned the global community about. Their warnings need to be heeded and adaptation and mitigation policy action should be taken.

As indicated earlier in this chapter, the environmental assessment technique of climate impact assessment has become a lynchpin in the greenhouse policymaking process. Much scientific and technical effort is being dedicated in carrying out research in the area but there are difficulties associated with climate model validation at the regional scale. Wigley (1994), in a presentation to an IPCC working group meeting, stated that "a climate change forecast (even a bad one) is better than no forecast, similarly a policy based on this forecast is better than no policy— perhaps it is not necessary to minimise the uncertainties and the policy should be made ahead of the science, as policy is more complicated than science." Policy is being made ahead of science at the international level; however, it is unlikely that policy will be developed beyond 'no regrets' measures at the domestic level with uncertain climate impact assessment data as a basis.

One area that exhibits potential in the climate impact assessment field that has not been seriously addressed to date is the identification of thresholds for regions. Cohen et al. (1994:9) have noted its analytic capabilities, saying, "Physical, biological, or economic systems may exhibit threshold behavior, experiencing sharp changes in response as driving factors cross thresholds. This could happen, for example, as species become extinct, or as increases in sea level overwhelm coastal defences." One advantage of a thresholds approach is that by thoroughly investigating and identifying current-day thresholds, and also considering the best available GCM information, adaptation policies may be able to be developed to avoid severe climatic impacts. The utility of the thresholds concept has also been alluded to by Pearce (1991:18) and Henderson-Sellers (1994).

Therefore, if social, economic, environmental, and political analysts can identify what current threshold limits exist after which severe adverse impacts may be experienced for particular systems and regions, then climate modellers can suggest how often such threshold limits potentially will be reached or exceeded in conditions of changing climate. This kind of information can be generated from GCMs at the present time for the global outlook and in a more limited way for regions.

With respect to the application of thresholds, it should be noted that the draft report of the IPCC Special Workshop on Article 2 of the United Nations Framework Convention on Climate Change held at Fortaleza, Brazil (IPCC 1994: 19), asserts:

> A critical level is not necessarily a physical property of the system or the valued attribute under study. A few exposure effect relationships will display a sufficient non-linearity that might enable a critical level to be determined. For most systems, however, a critical level may not be so easily identified. Furthermore, where a system responds to many variables over time there may be more than one critical level applicable to the same system. The term 'critical level' should, as far as possible, be avoided by scientists so as not to mislead policymakers.

This denial of the utility of the thresholds concept for policymakers does not take into account the complexities that are considered in policy analysis. For example, Braaf et al. (1994) in a study of adaptation responses to climate change in Australia have ascertained that threshold levels for particular outcomes will differ from region to region and that within regions different groups and individuals will have different threshold limits. To illustrate, Braaf et al. (1994) point out that older adults in northern Queensland, Australia, may be more susceptible to ill health effects from heatwaves than younger, more able-bodied adults, and acknowledge that threshold limits may result from a combination of factors (for example, hot days with high humidity over a protracted period of time will create incidences of heat stress in older adults, who suffer chronic illness and who do not have access to air-conditioning). Thus, Braaf et al. (1994) argue that the identification of threshold levels for particular regions and vulnerable groups should serve as a general guide to administrators and decision makers to aid adaptive planning for climatic change.

Accordingly, development of knowledge about regional and sectoral thresholds for ecological systems, human systems, or economic sectors should improve the utility of climate impact assessments to policymakers. In association with the assemblage of such knowledge, it is essential that assessment of the potential social, cultural, and political constraints that may be involved in implementing climate change policy at the regional level be a component of the climate impact assessment procedure. For example, policy to address the impacts of tropical cyclones on a coastal region of a developed nation such as Australia would undoubtedly be framed differently from that developed for a Pacific Island developing country due to differing social, cultural, and political contexts. Climate impact assessments thus need to incorporate such region specific information.

CONCLUSION

Climate impact assessment has evolved from intellectual endeavour to an important policy tool. Unfortunately, to date, the uncertainties in global climate modelling have been translated to, and exacerbated in, climate impact assessment, resulting in much climate impact assessment research being unusable in the policy process. As such, work needs to be done in developing an improved methodological approach for climate impact assessment. The reliance of policymakers on this rational assessment technique means that those scientists and social scientists involved in the field have a responsibility to attempt to be innovative and to explore the boundaries beyond dependency on climate modelling information. It is suggested first, that adoption of a thresholds approach, and second, further collaboration and information sharing with the broader environmental impact assessment community, as discussed in this chapter, are directions climate impact assessors could beneficially move towards.

REFERENCES

Adams, R.M. et al. 1990. "Global climate change and US agriculture." *Nature* 345: 219-224.

Beer, S. 1973. "The modernization of American federalism." *Publius* 3(2): 74-79.

Bolin, B. 1991. "The intergovernmental panel on climate change." In *Climate Change: Science, Impacts and Policies*—Proceedings of the Second World Climate Conference, pp. 19-21. J. Jager, H.C. Ferguson, eds. Cambridge: Cambridge University Press.

Bolin, B. 1994. "Science and policy making." *Ambio* 23(1): 25-29.

Braaf, R. et al. 1994. A Study of Adaptation Responses to Climatic Change for Australia: Final Draft Report. Prepared for the Department of Environment, Sport and Territories, Government of Australia.

Carruthers, I. 1994. "Opening Address—Integrated Climate Change Impact Assessment: Quantifying the Costs of Climatic Change as Input to Policy Formulation." Paper presented at CSIRO division of atmospheric research workshop—Towards an Integrated Approach to Climate Impact Assessment. Melbourne. 26 April.

Carter, T.R. et al. 1992. Preliminary Guidelines for Assessing Impacts of Climate Change. Oxford: Environmental Change Unit and Ibaraki: Centre for Global Environmental Research. For the Intergovernmental Panel on Climate Change.

Carter, T.R. et al. 1994. Technical Guidelines for Assessing Climate Change Impacts and Adaptations. Oxford: Environmental Change Unit and Ibaraki: Centre for Global Environmental Research. For the Intergovernmental Panel on Climate Change.

Climate Change Newsletter. May 1994. Special Issue: *Developments in Understanding Climate Change.*

Cohen, D. et al. 1994. "The Global Climate Policy Evaluation Framework." Proceedings of the 1994 A&WMA (Air & Water Management Association) Global Climate Change Conference, Phoenix, 5–8 April.

Edelman, M. 1964. *The Symbolic Uses of Politics.* Urbana: University of Illinois Press.

El-Ashry, M.T. 1994. "Foreword." In *Environmental Assessment and Development.* R. Goodland and V. Edmundson, eds. Washington DC: World Bank. World Bank-IAIA Symposium: vii-viii.

Fowler, A.M. et al. 1992. Regional Impact of the Enhanced Greenhouse Effect on New South Wales: Annual Report 1991-92. Melbourne: CSIRO (Commonwealth Scientific and Industrial Research Organisation)

Godard, O. 1992. "Social decisionmaking in the context of scientific controversies: The interplay of environmental issues, technological conventions and economic stakes." *Global Environmental Change.* September: 239-249.

Goodland, R., and V. Edmundson, eds. 1994. *Environmental Assessment and Development.* Washington DC: World Bank. World Bank-IAIA Symposium.

Gordon, H.B., B.J. McAvaney, and J.L. McGregor. 1994. "Modelling of Climate Change." Paper presented at Greenhouse '94: An Australia-New Zealand Conference on Climate Change, Wellington, 9-14 October.

Government of Australia. 1988. Australian Delegation Report on World Meteorological Organization–United Nations Environment Programme Intergovernmental Panel on Climate Change, First Session, Geneva, 9-11 November.

Government of Australia. 1992. *National Strategy for Ecologically Sustainable Development.* Canberra: Australian Government Publishing Service.

Granich, S., and M. Kelly. 1994. "Towards joint implementation." *Tiempo* 13: 1-3.

Haas, P.M. 1990. "Obtaining international environmental protection through epistemic consensus." *Millennium: Journal of International Studies* 19(3): 347-363.

Henderson-Sellers, A. 1991. "Policy advice on greenhouse-induced climate change: The scientist's dilemma." *Progress in Physical Geography* 15(1): 53-70.

Henderson-Sellers, A. 1994. "An Adaptation Approach to Climatic Change: Its Future Role in Oceania." Paper presented at Greenhouse '94: An Australia-New Zealand Conference on Climate Change, Wellington, 9-14 October.

Henderson-Sellers, A., and R. Blong. 1989. *The Greenhouse Effect: Living in a Warmer Australia.* Kensington: New South Wales University Press.

Intergovernmental Agreement on the Environment. 1992. Text. Canberra: Government of Australia.

IPCC (Intergovernmental Panel on Climate Change). 1990a. *Climate Change: The IPCC Scientific Assessment.* Oxford: Oxford University Press.

IPCC (Intergovernmental Panel on Climate Change). 1990b. *Climate Change: The IPCC Impacts Assessment.* Canberra: Australian Government Publishing Service.

IPCC (Intergovernmental Panel on Climate Change). 1990c. *Climate Change: The IPCC Response Strategies.* Washington DC: Island Press.

IPCC (Intergovernmental Panel on Climate Change). 1990d. IPCC First Assessment Report. Geneva.

IPCC (Intergovernmental Panel on Climate Change). 1992. *Climate Change 1992: The Supplementary Report to the IPCC Scientific Assessment.* Cambridge: Cambridge University Press.

IPCC (Intergovernmental Panel on Climate Change). 1994. IPCC Special Workshop on Article 2 of the United Nations Framework Convention on Climate Change: Informal working documents. Fortaleza, Brazil, 17–21 October.

Kates, R.W. 1985. "The interaction of climate and society." In *Climate Impact Assessment: Studies of the Interaction of Climate and Society*. R.W. Kates, J.H. Ausubel, and M. Berberian, eds. Chichester: John Wiley and Sons.

Kates, R.W., J.H. Ausubel, and M. Berberian, eds. 1985. *Climate Impact Assessment: Studies of the Interaction of Climate and Society*. Chichester: John Wiley and Sons.

Lowe, I. 1989. *Living in the Greenhouse*. Newham: Scribe.

MacDonald, M. 1994. "What's the difference: A comparison of EA in industrial and developing countries." In *Environmental Assessment and Development*. R. Goodland and V. Edmundson, eds. Washington DC: World Bank. World Bank-IAIA Symposium.

Meehl, G. 1992. "Global coupled models: Atmosphere, ocean, sea-ice." In *Climate System Modeling*. K. Trenberth, ed. Cambridge: Cambridge University Press.

Mendelsohn, R., and N.J. Rosenberg. 1994. "Framework for integrated assessment of global warming impacts." *Climatic Change* 28(1-2): 15-44.

Nilsson, A. 1992. *Greenhouse Earth*. Chichester: John Wiley and Sons.

Pearce, D. 1991. "Evaluating the socio-economic impacts of climate change: An introduction." In *Climate Change: Evaluating the Socio-economic Impacts*. Paris: Organisation for Economic Cooperation and Development.

Pielke R.A. 1994. "Scientific information and global change policymaking." *Climatic Change* 28(4): 315-319.

Rayner, S. 1986. "Commentary." In *Sustainable Development of the Biosphere*. W. Clark, R. Munn, eds. Cambridge: Cambridge University Press.

Ravetz, J. 1986. "Usable knowledge, usable ignorance." In *Sustainable Development of the Biosphere*. W. Clark, R. Munn, eds. Cambridge: Cambridge University Press.

Riebsame, W.E., and A. Magalhaes. 1991. "Assessing the regional implications of climate variability and change." In *Climate Change: Science, Impacts and Policy:* Proceedings of the Second World Climate Conference, pp. 415-430. J. Jager, H.L. Ferguson, eds. Cambridge: Cambridge University Press.

Rosenberg, N.J. 1990. Processes for Identifying Regional Influences of and Responses to Increasing Atmospheric CO_2 and Climate Change: The MINK project working paper 1— background and baselines. Washington DC: Resources for the Future.

Schneider, S. 1991a. "Three reports of the intergovernmental panel on climate change." *Environment* 33(1): 25-30.

Schneider, S.1991b. "Prediction of future climate change." In *Energy and the Environment in the 21st Century:* Proceedings of the conference, 26-28 March 1990, pp. 41-54. J. Tester, D. Wood, N. Ferrari, eds. Cambridge MA: Energy Laboratory, Massachusetts Institute of Technology.

Schneider, S. 1992. "Introduction to climate modeling." In *Climate System Modelling*. K. Trenberth, ed.. Cambridge: Cambridge University Press.

Taplin, R. 1994a. "International policy on greenhouse and the Island South Pacific." *The Pacific Review* 7(3): 271-281.

Taplin, R. 1994b. "Greenhouse: An overview of Australian policy and practice." *Australian Journal of Environmental Management* 1(3): 142-155.

Taplin, R. Forthcoming. "Climate science and politics: The road to Rio and beyond." In *Climate Change, People and Policy: Developing Southern Hemisphere Perspectives*. A. Henderson-Sellers and T. Giambelluca, eds. Chichester: John Wiley and Sons.

UNEP (United Nations Environment Programme). 1993. *Climate Change Bulletin* 1(1).

UNFCCC (United Nations Framework Convention on Climate Change). 1992. Text. Geneva: United Nations Environment Programme/World Meteorological Organization Information Unit on Climate Change.

van Zijst, P. 1994. "Politicians deal with politics, scientists better not get mixed up with that." Interview with Professor Dr. Bert Bolin, *Change* 18: 4-7.

Wigley, T. 1994. "UCAR (University Corporation for Atmospheric Research) : Development of High Resolution Scenarios." Paper Presented at IPCC working group I regional evaluation workshop, Climatic Impacts Centre, Macquarie University, Sydney, 7-9 February.

WMO (World Meteorological Organization). 1986. Report of the International Conference on the Assessment of the Role of Carbon Dioxide and of Other Greenhouse Gases in Climate Variations and Associated Impacts. Geneva: World Meteorological Organization. Report No. 661.

Chapter 12

Development Impact Assessment:
Impact assessment of aid projects in nonwestern countries[1]

JANICE JIGGINS
Consultant, The Netherlands

INTRODUCTION

The attempt to assess development impacts is an unfinished story of only modest achievement. There is still no commonly accepted protocol among development practitioners defining 'impact', no common methodology for measuring whatever it is, and unending dispute over how to interpret whatever results from an impact assessment exercise. This is perhaps not surprising for development interventions; even 'bounded' constructions such as a bridge or roadway interact with conditions that are complex, diverse, contingent, and, to varying degrees, unpredictable. The diversity of approaches also signals the fundamental dilemma, that the result of the analysis is foreshadowed by the choice of method. Simple 'fitness for function' rules are often ignored; impact assessment can be rapidly overloaded by questions which, though interesting and even important in themselves, are not germane to the management, investment, or resource allocation issue that requires clarification.

Nonetheless, the activities of monitoring, evaluation, and impact assessment have become locked into the formal procedural requirements of aid givers and aid receivers, each activity with its own distinctive rituals and expertise. Maddock (1993) provides a brief but useful review of the institutionalisation of monitoring, evaluation, and impact assessment within aid agencies. But since development impacts are inseparable from the flow of time and contingent history, these technical distinctions tend to disappear in the hands of the development manager and field-based professional looking for answers to specific problems and queries. The difference in perspective can be stated thus: while the development agency tends to see impact assessment as the crowning activity of a linear process and of an investment of limited duration, development practitioners tend to see monitoring, evaluation, and impact assessment as codependent, iterative, and cyclic.

There is growing appreciation that the tools of technical assessment (including environmental impact assessment) commonly used in industrialised countries at the project, company, or enterprise level, are not wholly suitable for development

[1]*Environmental and Social Impact Assessment* - Edited by F. Vanclay and D.A. Bronstein. Copyright © 1995 by the International Association of Impact Assessment. Published in 1995 by John Wiley & Sons Ltd. A version of this chapter will appear in *Impact Assessment,*the quarterly journal of IAIA.

impact assessments, chiefly on three counts. First, the conditions of a bounded activity, to which objective criteria can be applied within conscious management strategies, rarely apply. Second, assessments of possibility and probability are usefully reliable mainly where the intervention is the major source of variance, which is rarely the case in development. And third, the consequences of an intervention are—in the development case—unlikely to be uniform, specific, or entirely proximate. Thus the question, *Do development impact assessments give value for money?* cannot be answered, since their influence can scarcely be tracked, measured, or quantified.

The balance between *ex ante* and *ex post* assessments has tended to shift over time. There is increasing focus on methods for *ex ante* assessments as development funding becomes tighter, partly in order to guide resource allocation more tightly and to estimate returns to investment more keenly. Another factor may be the apparent lack of influence that *ex post* impact assessments have had on funding allocations or project design (Maddock 1993).

It is particularly notable that while the adoption and impact of technical innovations in agriculture are by far the most thoroughly researched development topics in terms of number of studies over a period of more than 30 years, the results have had remarkably little influence on the staffing, design or organisation of agricultural research, development and extension activities (Röling 1988). The coherent, integrated management of agricultural knowledge and information, personnel, resource flows and institutional capacity remains the exception, and disjointed, fragmented systems the rule. For example, as Collinson and Tollens (1994) point out, the impact of the Consultative Group for International Agricultural Research (CGIAR) on agricultural research is essentially dependent on how effectively the rest of the research and development sequence operates—a sequence over which the CGIAR has no control. Further, such influence as impact studies in agriculture have had could be considered in a broad sense, malign. Diffusion research for example focuses on *ex post* assessment of successful diffusion processes; the lessons have been used to justify an expectation that *ex ante* effort to promote diffusion can be used to disseminate new agricultural technologies.

It would be wrong to conclude either that conditions in nonwestern countries are so different from those in industrialised countries that none of the familiar tools of impact assessment apply, or that, however conducted, assessment of development impacts has no value. The state of the art merely reflects continuing difficulties in defining 'development', and in assessing a somewhat haphazard process rather than a controlled event.

Development impact assessment is used, and is useful, both for addressing generic questions in development and for comparative questions, both *ex ante* and *ex post*. This chapter reviews the key clarifications, difficulties, assumptions, and philosophic traditions underlying development impact assessment, including the somewhat neglected issues of culture and local knowledge, then examines more closely recent innovations in the use of gender-sensitive tools and participatory methods, concluding with comments on future directions.

EX ANTE ASSESSMENTS

Alongside the more formal and technocratic methods, guidelines and checklists have been constructed to provide first-order screening tools. In 1988, for example, the evaluation unit of OXFAM constructed a simple procedure for identifying 'who wins, who loses' among groups of people in the Third World experiencing the impact of changes in 'rich world' agricultural policy. Carruthers (1977), addressing the problem of multiple objectives in function analysis and linear programming, advocated the wider use of the decision matrix as an alternative solution to either the constraint approach or the construction of weighted objective functions. Today, decision matrices, once mainly used by advisers and experts, are as often found in the tool kits of field workers, who construct the matrices in a more participatory manner together with project beneficiaries. Others such as Gass and Biggs (1993) propose the participatory use of methods such as institutional mapping, interest group analysis, and 'pay-off' matrices to assess otherwise hard to handle areas such as power dynamics and institutional context in *ex ante* technology assessment.

Other interesting and highly practicable new directions in the agricultural sector include the construction of models for *ex ante* assessment of priorities in agricultural research in terms of natural and economic resource endowments (Ravnborg 1993). Some procedures, such as the estimate of Total Factor Productivity, focus more directly on the question of sustaining the productivity of natural resources in conditions of technical and economic change in agriculture. They promise much, yet are so demanding of data, they remain in the desirable but impractical class. Further, as with other tools embedded in neo-classical economics and reliant on historical data, they are least informative where the clearest guidance is needed, that is, the likelihood of unforeseen future thresholds and interactions.

Teledetection, remote sensing and GIS techniques, and their accompanying data bases and computer software, for their part, are enhancing capacity to observe, monitor, quantify, and relate environmental impacts on an unprecedented scale (CEC 1992), yet major problems of interpretation and meaning remain (Fedra 1994). Their utility as a tool of *ex ante* impact assessment is substantially enhanced when married to rigorous 'ground truthing' and discussion with local inhabitants. ICLARM (International Centre for Living Aquatic Resource Management, based in the Philippines), for example, is developing computerised capacity to match GIS data with farmers' maps and diagrams of resource flows and environmental trends on their farms and surrounding landscapes, as a tool for assessing farm-level innovations.

In addition, computer-based aids for testing the environmental and economic consequences of management options in farm system development, biological, climatic and statistical simulation models of the interactions governing plant growth and crop yield, and expert systems, are expanding rapidly our understanding of the effects of people's interaction with their environment (Fresco et al. 1994). Whether they contribute as much to people's capacity to manage their environment sustainably over the longer term is less obvious. I return to this question in the final section. Meanwhile others are searching for a rigorous but more practical approach to sorting out the social effects of natural resource management decisions and policies (Freeman and Scott-Frey 1990–91).

EX POST ASSESSMENTS

With regard to *ex post* assessments, useful distinctions typically are made between: (1) the efficiency of resource use, which is fairly easy to identify and measure; (2) immediate to medium-term effects, which are often difficult to identify, measure, and assess; and (3) longer term impacts. However, experience has led to the conclusion that impacts are almost impossible to identify, measure, or assess to any satisfactory extent (Patton 1986) because perturbation and complex interactive effects play a progressively larger role over time. Nonetheless, the effort to assess impacts continues unabated as a means, however imperfect, of maintaining accountability, and to legitimate funding (Collinson and Tollens 1994).

A more practicable distinction is sometimes made between: (1) efficiency, as a measure of fitness for function; (2) effectiveness, as a measure of whether the approach, technology, or project does in fact do the job it was designed to do; and (3) efficacy, as a measure of whether the activity is in fact the right thing to be doing to achieve stated objectives. Bilateral and multilateral funding agencies in particular devote considerable resources to addressing these three questions, sometimes through dedicated agencies such as the Dutch Inspectorate for Development Cooperation at the Field Level, partly as a means of assuring, however inadequately, public scrutiny of the development lending process and financial accountability. However, as the timing and focus of studies conducted under this heading typically are driven more by the concerns of the lending agency than by recipients, the results tend to be used, if at all, mainly to improve administrative and regulatory procedures within the agency. Bureaucratic inertia, political expediency, and over-reliance on consultants for project and program execution also may contribute to the slowness of change in institutional behaviours which might lead to greater development efficiency, effectiveness or efficacy (Jiggins 1988).

Whatever the purpose or focus of *ex post* assessments, general methodological problems persist. Often no baseline is available for comparison, or no time frame or measure has been specified initially for the achievement of the stated objective(s). Even where baseline information exists, questions may still be raised as to the validity of the chosen starting point and the boundaries drawn around the information sets. Typically there are only weak links between the specified objective(s), the proposed activities, and actual activities.

INNOVATIONS BORROWED FROM MANAGEMENT

One response has been to improve the formal procedures for project design and program planning. For example, the ZOPP method—Zentrale Orienterte Projekt Plannung (objective-oriented project planning) (see GTZ 1987)—has been adapted from management literature and from USAID's introduction of the logical framework into development planning in the 1970s (Morss 1976; Waterston 1976; Cohen and Uphoff 1977). The intended purpose is to strengthen the link between the information input, stated objectives, and project activities, within specified time horizons, financial parameters, and sets of actors or agencies, accompanied by identified and measurable indicators of progress and achievement.

As with other management tools borrowed from industrial and commercial settings, the ZOPP procedure makes four unhelpful assumptions: (1) there is a necessary causal relationship between inputs and outputs; (2) everything that it is necessary to know can be determined in advance; (3) the pathways identified in the ZOPP exercise will in fact be those by which the project develops; and (4) the intervention is the major source of variance in terms of impact. Experience suggests that the more that development agencies seek to formalise development interventions in order to get a better grasp of and assurance of desired impacts, the more the intervention locks into the starting situation, missing the dynamic of unfolding history and the larger sources of variance.

Another management tool borrowed from industrial settings is utilisation research, or evaluation of impacts as a problem-solving process, with 'just-in-time delivery' of information inputs. However, the design and maintenance of channels for the timely and reliable feedback of information instrumental to problem solving, has proven almost impossible on any scale in the context of overly bureaucratic and hierarchical organisations, large status (and gender) differences, dispersed populations, and poor physical and communication infrastructures.

CONTINUING DOMINATION OF NEO-CLASSICAL ECONOMIC TOOLS

Although management innovations have made an important contribution, it would be fair to say that neo-classical economic analysis continues to dominate *ex post* assessment of development impacts. (Standard references include: Norton and Solis 1982; Schuh and Tollini 1979; Gittinger 1972; Cernea 1979; Bell and Hazell 1980). Economists have evolved increasingly sophisticated methods for application at the enterprise level or at the aggregated level of a sector, industry, or nation, but the methods remain largely based on partial equilibrium models, handicapped by assumptions about necessary causal relationships between inputs and outputs. Although considerable fine-tuning has occurred (for example, to derive more accurate distributional weights and to price values appropriate to semi-subsistence economies), problems remain (Hardaker et al. 1984). For example, there are continuing difficulties with regard to determining appropriate dimensions for scale- and time-dependent effects, the weight to give to interpretations of the political economy and institutional issues, and the ideological, disciplinary, or other biases of the assessor (see, for example, Tendler 1982, on urban bias).

Anthropologists such as Goodell (1983) and others (Wagemans 1987; Wijeratne 1988) meanwhile have documented the almost insurmountable barriers that 'external' assessors face in checking the degree to which reality conforms to what was supposed to happen, and in interpreting what might appear to be the facts. For example, extension agents may 'inhabit' the village houses built for them under Training and Visit extension programs only when inspection teams are on tour, living the rest of the year in town houses closer to schools and other amenities. Agricultural agents and farmers may collude to establish demonstration plots of a cash crop, on the condition that the agents divert the fertiliser designated for cash crop production to the farmers' food crops (Röling 1994, quoting Von der Luhe).

Other problems arise with respect to what might be termed hidden or invisible effects. Hazell and Ramsamy (1991) recently revisited a district in India to assess in terms of equity the long-term impacts of the 'green revolution' (GR) on agricultural producers and rural communities. The GR strategy, as it unfolded through the decades following its introduction in the 1960s, was based on the development and introduction of improved varieties of major food crops, supported by innovations in irrigation; inorganic fertilisers; chemical control of pests, diseases, and weeds; and mechanisation. Although successful in terms of stimulating a vast increase in the supply of cheap food staples to commercial markets, the strategy's net effect on incomes, labour, and employment, especially among poorer or smaller producers, and the long-term consequences for the environment, remain in dispute (see Lipton and Longhurst 1989).

Hazell and Ramsamy's study is one of the most thorough assessments so far made of the GR's impacts on different classes and social categories at the community level. They concluded that over time, adoption of the package of improvements had become to all practical purposes universal among producers, and thus that the effects were indeed 'scale neutral'. However, they found no solution to the common methodological problem, that those with less access to the new technology or employment opportunities, with weaker incomes or farming land of lower inherent potential, over time tend to be pushed out of farming and leave the rural area. They are no longer available for impact studies. Other fundamental concerns have to do with the continuing blindness of researchers to gender-related effects and impacts (Jiggins 1986).

GENDER ANALYSIS

The United Nations Decade for Women (1971–1981) gave powerful impetus to measures to overcome gender blindness in development impact assessment. The measures were principally of three kinds: (1) the creation of gender-sensitive instruments (e.g., checklists, guidelines) for use within development agencies for screening and adjusting the process and content of project design, planning, monitoring, and evaluation; (2) the commissioning of special studies to assess impacts on women, and the concomitant development of appropriate tools for doing so; and (3) the development of tools for practitioners, specific to particular disciplines, sectors, or types of activity.

Gender-sensitive Instruments

Examples range from simple checklists of do's and don'ts (FAO 1982; Netherlands Ministry of Foreign Affairs 1988), to more elaborate guidelines (CEPAL 1982; Weekes-Vagliani 1985), through to assessments of the impact of gender-sensitive impact studies (UNFPA 1986).

There is an emerging consensus that these instruments form a useful component of wider strategies designed to bring about a more gender-sensitive development process. Their success lies not so much in the fact that their prescriptions have been honoured (often they have not), but that in the hands of advocates for greater gender sensitivity, they form a powerful lever for changing attitudes and performance criteria within aid bureaucracies and their counterpart agencies in developing countries. They give a voice to the hitherto ignored or undervalued.

Special Studies

One expression of such change has been the commissioning of special impact studies to provide deeper insight into how project design and implementation may fail to improve or may even worsen women's status and access to benefits and resources, and to consider what might be needed to ensure greater gender equity. These studies encompass reviews of projects designed specially to benefit women (van Vliet 1984; United Nations 1985), projects where women could be expected to be major beneficiaries because of the nature of the project activity (Hazzard 1987), and studies of the effects and impacts on women of sectoral investment programs (IFAD 1985; Carloni 1987; Poats 1991).

A number of general lessons emerge from these studies highlighting the reasons why the gender question remains difficult. They include: (1) confusion between women's advancement and gender analysis, and between gender analysis and gender staffing issues; (2) continuing bias in favour of technocratic approaches, with insufficient attention to issues such as social relationships, power structures, and institutional factors; (3) lack of contact between women in the community and service agency personnel and/or project personnel; (4) lack of commitment to gender issues among senior professionals and planners; (5) marginalisation of gender questions in the form of special projects for women rather than integration of gender issues into all projects; and (6) a lack of mechanisms and tools for ensuring that gender questions are dealt with adequately at all stages and levels of development.

Tools for the Field

Partly in response to the challenge to devise more effective mechanisms and tools for gender analysis, recent years have seen the elaboration of a range of *ex ante* screening and assessment tools appropriate to various development sectors. For example, Feldstein and Jiggins (1994) present a collection of practitioners' experiences in applying gender-sensitive tools to farming systems research and extension activities from initial diagnosis through to evaluation and assessment of effects. PROWESS/UNDP (1991) reports on the findings of a 'forward-looking assessment' of the effect of involving women in water and sanitation development, in particular the effects of community-based approaches to participatory monitoring, evaluation, and research. Thomas-Slayter et al. (1993) provide a handbook of gender-sensitive field methods for sustainable resource management and evaluation tools. The Department of Health Services Studies of the Tata Institute of Social Sciences, Bombay, produces a newsletter, *Qualitative Research Methods*, that shares methodological experiences with regard to women's reproductive health and health-seeking behaviours. OXFAM's gender and development unit produces the *GADU Newspack* twice a year. Each issue includes detailed methodological guidelines and bibliographic information on selected issues (e.g., Issue 15: June 1992 focuses on women's health). Others have focused on the planning function and development training, providing a range of analytic case studies to demonstrate the impact on women of planning with and without adequate prior gender analysis and gender-sensitive implementation (for example, Illo 1988; Joseph 1987).

In general, the effort to make development impact assessment more gender-sensitive, and thereby increase the likelihood that women will in fact benefit from development interventions and processes, may form the main exception to the conclusion drawn earlier, that impact assessments have had little influence on practice. The reason surely is that there is an increasingly well-organised network of concerned women (and men) working as advocates and champions of change, and ready to act on the findings of such assessments. There are mechanisms, often informal but nonetheless powerful, to ensure accountability, and which have an activist energy to translate the feedback from experience into future policy and practice.

OTHER NEGLECTED ISSUES: CULTURE AND LOCAL KNOWLEDGE

The neglect or under-valuation of gender issues can be seen as a special case of a broader concern. Both the cultural aspects of impact assessments and the role of local knowledge in defining assessment criteria and forming judgements have been seen as important for many years (Bennis et al. 1961, 1970; Chambers 1979)—but both continue to be 'neglected' in the sense that the issues raised by earlier studies have had little apparent influence on the practices and attitudes of the major development assistance bureaucracies, or even of the many local development bureaucracies which have swallowed wholesale the evaluation and assessment methodologies derived from experience and practice in other contexts. Nor do many of the consultants employed to provide 'objective' assessments of project and program performance show much sign that they have absorbed the voluminous anthropological literature and studies from the communication field demonstrating the importance of cross-cultural communication, culturally embedded interpretation and bias, and local knowledge (Jiggins 1988; Warren et al. 1989; Hoffman 1991; Long and Long 1992). In this respect, they stand in contrast to the considerable effort that multinational business puts into training staff in cross-cultural communication and in the importance of local knowledge in defining and exploiting local markets (Röling and Kuiper 1990).

More recently there has been renewed effort to devise ways to link impact assessment to local people's capacities in ways that are less dependent on outsiders' insights and empathy, and more reliant on partnerships and negotiated procedures and conclusions, thus reflecting the biases and cultures of both parties (Box 1990). The ideal, achievable perhaps among individuals, still tends to fall foul of institutional incompatibilities. The demands of an external finance organisation such as the World Bank or FAO, for example, may be so far removed from the institutional expectations and behaviours of a local bureaucracy that formal evaluation may be reduced to a ritualistic sham. Attempts to design and implement evaluation procedures that are meaningful to the management objectives and staff working patterns of a local agency, on the other hand, may not generate the data and information feedback demanded by senior bureaucrats far removed from the action, whether located in the capital city or in Washington or Rome (Röling and Talug 1993).

INNOVATIONS IN PARTICIPATORY METHODS

Participatory methods have been used for many years for both *ex ante* and *ex post* impact assessment. Ascroft et al. (1973), for example, promoted the use of farmer panels under the Special Rural Development Program of the Kenyan government in order to assess the impact on small farmers of the 'claims and benefits' of technologies in terms of money and labour days, while Ashby (1987) has developed procedures for reviewing current technology use by different farmer categories in order to assess the impact of proposed new technologies.

A new impetus was given to participatory methods in the mid-1980s by wide-ranging reviews of aid agencies' overall development assistance record by, among others, the World Bank, FAO, and USAID (see WRI 1989, 1991). In 1985, the World Bank reviewed the short-term results of more than 1000 projects it had funded. It concluded that against a range of 'hard' criteria (such as financial rate of return), the failure rate in agriculture and rural development projects was high, and that the highest failure rate was recorded by agricultural projects in Africa, particularly in the livestock sector—a conclusion shared by the FAO's review of its own investments over a 15-year period (WRI 1989). A 1984 USAID review of AID-assisted projects found that no large-scale plantation or small village woodlot project in Africa could be judged successful (WRI 1991). The common inference drawn was that, in addition to policy, institutional and cultural differences among donors and recipients, the introduction of unfamiliar technology and interventions requiring major changes in a way of life were major causes of failure. There had been a profound misreading of social and economic factors on the part of project planners, implementers, and evaluators.

A part of the problem undoubtedly lay in undue reliance on inadequate quantitative data and statistical instruments in both *ex ante* and *ex post* assessments. As Gill (1993: 13) cogently argues:

> There is no way in which rigour in analysis can compensate for an unknown and unknowable degree of inaccuracy in the measurement of independent variables. Modern statistical analysis can handle sampling errors, but non-sampling errors of a type arising either from the inherent difficulties. . . or in the operational environment. . . give a totally different picture.

Statistical enquiry and formal modelling have their place, as noted in earlier sections; however, there is an increasing realisation that these instruments need to be informed by other complementary methods lest inappropriate quantification and formalism lead to a false sense of security, or, indeed, present too harsh a picture.

Another part of the problem has to do with a failure to select assessment methods appropriate to their intended use. The capacities and objectives in information terms of formal, quantitative methods compared to rapid assessment methods, for example, are complementary but nonsubstitutable (Hursh-Cesar 1988).

Chambers and Jiggins (1987) argued the need for including users in the design and assessment of new technology and in the consideration of its potential and actual impacts as a way of overcoming the kinds of problems which the World Bank, FAO and AID reviews had highlighted. They pointed to various participatory methods, drawn from a variety of disciplines and sectors, which would make such an approach possible. Building on earlier experimentation with Rapid Rural Appraisal methods, the last seven years has seen a veritable explosion of methodological creativity and the elaboration of user guides (e.g., Jiggins and de Zeeuw 1992) and trainers' manuals (e.g., PROWESS/UNDP 1992). Participatory assessment methods have been extended beyond agriculture to the fields of health, nutrition, low-cost buildings and infrastructure development, water and sanitation, soil and water conservation, and wildlife and biodiversity conservation (see Shah et al. 1991 for an example of a participatory impact assessment in a soil and water conservation program).

Such methods are often strongest where formal, quantitative instruments such as questionnaire surveys are weakest, as in assessment of community organisations and social capacity (Fowler 1988). They utilise local people's strengths, such as accuracy of memory, familiarity with and observation of trends over time and of diversity and variability over space and time, and the ways in which people interact with their social and physical landscape.

RISKS IN PARTICIPATORY APPROACHES

Whatever benefits participation has, participatory methods carry their own risks and shortcomings. For example, 'participatory' in the hands of the gender-blind practitioner does not ensure that women are more adequately included than under any other approach (Palmer 1981). In other words, method cannot substitute for commitment to principle. Nor can method safeguard against the use of participation for extracting information or for other exploitative purposes.

Then again, if technocratic approaches assume technically competent, neutral expertise focused on management and policy options, participatory approaches assume skills of perception, listening, communication, cooperation, and conflict management focused on the practical pathways forward. Both sets of assumptions about capacity may be considered unrealistic, though undoubtedly there are individuals who approximate either type. Further, while technocratic approaches assume authority and instrumentality, participatory approaches assume willingness to participate, partnership in learning, and shared responsibility. Again, both sets of assumptions may be unrealistic or too facile in a given context, successful application requiring prior effort in both cases to establish rules of mutual accountability, an effort which is all too rarely made. Checkland and Scholes (1990) offer convincing examples, however, that participatory techniques for managing 'soft' systems by formalising the linkages between information feedback, perception of goals and system behaviour, can inculcate among stakeholders a sense of shared responsibility for the process and the outcome.

It is both significant and potentially of great consequence that pressure to strengthen shared responsibility and accountability is a major emerging theme in development impact assessment. At the national and international levels, the demand for accountability is reflected in the growing NGO presence at both technical/sectoral conferences as well as policy fora, including in recent years such global discourse as the UN Conference on the Environment and Development (1992 at Rio de Janiero). At the level of technical assistance, there is increasing pressure for development consultants to advance financial guarantees against their assurance that specified results will be reached.

Finally, participatory process gives rise to the danger of creating 'philosopher-kings' in the place of 'scientific experts'. The participation of stakeholders in information-sharing, interpretation, and the negotiation of meaning, judgements, and decisions, does not in itself remove inequalities in power, status, and what might be termed wisdom. The conscientious facilitator of participatory impact assessment thus has an obligation to establish, by mutual agreement at the start of the process, rules of fairness and accountability and an open negotiation process; but anecdotally, the record again suggests that practice falls short of the ideal.

CONTRASTING ASSUMPTIONS AND TRADITIONS

The technocratic and participatory traditions are based on strongly contrasting differences in philosophic underpinnings and outlook (see Table 1).

Participatory traditions deliberately seek to include the human, social, political, cultural and contextual elements involved in human endeavour. They are based on the understanding that reality is not a given actuality waiting to be discovered by the detached scientist, but a constructed understanding—an informed perception—

developed by those who are engaged in the activity under scrutiny. Assessments of impact in the participatory tradition are not descriptions, measurements or models of the way things are, but representations of actors' attempts to understand the situation in which they act. Diversity and disagreement among actors are embraced as enriching the picture which is constructed and analysed. The product is not a conclusive judgement so much as a negotiated agreement on future action with regard to both resolved and unresolved claims, concerns, and issues (van Beek 1991).

Table 1. Contrasting traditions

Key Words

Technocratic	Participatory
Universalistic ethic	Pluralistic ethic
Deductive	Inductive
Uses only externalised knowledge	Uses externalised and tacit knowledge
Technocratic working method	Democratic working method
Objective	Subjective
Factual	Normative, interpretive
Data-based	Stories, narrative
Reductionist	Holistic, naturalistic
Positivist	Constructivist
Outsiders' perspectives	Stakeholders' perspectives
Expert	Facilitator
Judgement-oriented	Negotiation-oriented
Excludes or subsumes values, principles	Includes, makes explicit participants' values, principles
Assumes a controlled, experimental framework	Assumes a political process

Source: Author's summary, drawing on Guba and Lincoln 1989, Abma 1993.

PARTICIPATORY APPROACHES AND THE ENVIRONMENTAL CHALLENGE

The traditions and potential of participatory approaches are precisely the reasons why, notwithstanding their hazards and risks, they are likely to assume increasing prominence in the future. As two leading philosophers of science have noted (Funtowizc and Ravetz 1990), the environmental challenge requires that decisions with profound consequences are taken with a degree of urgency yet in conditions of uncertainty. The stakes are large but the probability is high that decisions cannot wait until more or better quality information is at hand. Moreover, many of the key relationships are inherently indeterminate and therefore unknowable in detail.

Further, in a divided world, the negotiation of who is to carry the burdens of any specific adjustment cannot be made without recourse to discussion of contrasting values and principles. A retreat toward narrower and deeper specialisation in industrial, scientific, and technical understanding of environmental relationships and effects carries the danger of increasing knowledge among the elite few while contributing little to the development and management of the options for change in societal terms. This is the nub of the concern that, for example, greater investment in GIS and computer simulation systems cannot substitute for the development of societal capacity to guide and implement change in directions informed by competing values and principle.

The only way to ensure that good quality decisions are made in such circumstances, is to pool information, experience and understanding among different fields of knowledge and diversities of people. Further, if action is required on a societal scale and within a limited timeframe, the way to achieve significant but voluntary change in behaviour within and among societies is to involve the people affected in assessing the situation and developing and reviewing the options.

FUTURE DIRECTIONS

The future of development impact assessments are thus likely to move in three directions. First, there is likely to be stronger effort toward the socialisation of the process of problem definition, and the development and review of options. However, this is not to argue that assessments will become wholly 'participatory' in approach and method. There is growing realisation that meeting the challenge of large scale, rapid societal change under conditions of environmental hazard requires better management of meaning at the interface between 'hard' and 'soft' systems. Thus, and this is the second conclusion, there is likely to be increasing emphasis on the ways in which the two systems can be 'coupled' to achieve environmentally sustainable futures (Röling 1994). Third and in turn, these efforts are likely to require the creation and support of new networks, as channels for exchange and discussion of information and meaning and as social institutions capable of deriving agreement on action and carrying it through.

REFERENCES

Abma, T.A. 1993. *Beyond the Technocratic Orientation in Evaluation*. Institute for Health Care Policy and Management. Rotterdam: Erasmus University.

Ascroft, J.R. et al. 1973. "Extension and the Forgotten Farmer." *Bulletin van de Afdelingen Sociale Wetenschapen* No. 37.

278PART III: TOOLS FOR THE FUTURE

Ashby, J.A. 1987. "The effects of different types of farmer participation on the management of on-farm trials." *Agricultural Administration and Extension* 24: 235–252.

Bell, C.L.G. and P. Hazell. 1980. "Measuring the indirect effects of an agricultural investment project on its surrounding region." *American Journal of Agricultural Economics* 62(1): 75–86.

Bennis, W.G., K.D. Benne, and R. Chin. 1961. *The Planning of Change*. New York: Holt, Rinehart and Winston.

Bennis, W.G., K.D. Benne, and R. Chin. 1970. *The Planning of Change* (2nd edn). London: Holt International.

Box, L., ed. 1990. *From Common Ignorance to Shared Knowledge Networks in the Atlantic Zone of Costa Rica*. Studies in Sociology 28. Wageningen: Agricultural University.

Carloni, A. 1987. *Women in Development: AID's Experience, 1973–1985.*" Vol. I: Synthesis Paper. Washington DC: United States Agency for International Development.

Carruthers, I.D. 1977. "Applied project appraisal: The state of the art." *ODI Review* 2: 12–28.

CEPAL. 1982. *Women and Development Guidelines for Programme and Project Planning*. Santiago: United Nations.

Cernea, M.M. 1979. *Measuring Project Impact: Monitoring and evaluation in the PIDER rural development project–Mexico*. Staff Working Paper No. 332. Washington DC: The World Bank.

Chambers, R., ed. 1979. "Rural development: Whose knowledge counts?" *IDS Bulletin* 10: 2.

Chambers, R. and J. Jiggins. 1987. "Agricultural research for resource—poor farmers, part I: Transfer of technology and farming systems research; Part II: A parsimonious paradigm. *Agricultural Administration and Extension* 27: 35–52 and 27: 109–128.

Checkland, P. and J. Scholes. 1990. *Soft Systems Methodology in Action*. Chichester: John Wiley and Sons.

Cohen, J.M. and N.T. Uphoff. 1977. *Rural Development Participation: Concepts and Measures for Project Design, Implementation and Evaluation*. Ithaca: Cornell University.

Collinson, M. and E. Tollens. 1994. "The impact of the international agricultural centres: Measurement, quantification and interpretation." *Experimental Agriculture* (in press).

Commission of the European Communities. 1992. *TREES—Tropical Ecosystem Environment Observations by Satellites*. Brussels: ESA/CEC.

FAO. 1982. *Guidelines, Women in Land and Water Development*. Rome: FAO Land and Water Development Division.

Fedra, K. 1994: *GIS and Environmental Modelling*. RR-94-2. Laxenberg, Austria: International Institute for Applied Systems Analysis. February.

Feldstein, H. and J. Jiggins, eds. 1994: *Tools for the Field: Gender Analysis in Farming Systems Research and Extension*. West Hartford: Kumarian Press

Fowler, A. 1988. "Evaluating Development Interventions: Towards Community-based Comparative Evaluation (CBCE)" and "Guidelines and Field Checklist for Assessments of Community Organisations." Care International Regional Workshop on the Agro-Forestry Monitoring and Evaluation Methodology Programme, May. Nairobi: The Ford Foundation.

Freeman, D. and R. Scott Frey. 1990–91. "A modest proposal for assessing social impacts of natural resource policies." *Journal of Environmental Systems* 20(4): 375–404.

Fresco, L.O. et al., eds. 1994. *The Future of the Land: Mobilising and integrating knowledge for land use options*. Chichester: John Wiley and Sons.

Funtowicz, S.O. and J.R. Ravetz. 1990. "Global Environmental Issues and the Emergence of Second Order Science." EUR 12803 EN. D-G Science, Research and Development. Luxembourg: Commission of the European Communities.

Gass, G.M. and S.D. Biggs. 1993. "Rural mechanisation: A review of processes, policies, practice and literature." *Project Appraisal* 8 (3): 157–187.

Gill, G.J. 1993. "O.K., the data's lousy, but it's all we've got (Being a critique of conventional methods)." Gatekeeper Series No. 38. London: IIED.

Gittinger, J.P. 1972. *Economic Analysis of Agricultural Projects.* Baltimore: John Hopkins University Press.

Goodell G.E. 1983. "The administrators' difficulty in obtaining accurate feedback." *Agricultural Administration* 13(1): 40–55.

GTZ. (Deutsche Gesellschaft für Technische Zusammenarbeit). 1987. *Planification des projets par objectifs (ZOPP).* 3/87 6S. Eschborn: GTZ.

Guba, E. and Y. Lincoln. 1989. *Fourth Generation Evaluation.* Beverly Hills: Sage.

Hardaker, J.B., J.R. Anderson and J.L. Dillon. 1984. "Perspectives on Assessing the Impacts of Improved Agricultural Technologies in Developing Countries." Invited paper to 28th Annual Conference of the Australian Agricultural Society. Sydney: 7–9 February.

Hazell, P.B.R. and C. Ramsamy. 1991. *The Green Revolution Reconsidered: The impact of high-yielding rice varieties in South India.* Baltimore MD: Johns Hopkins University Press.

Hazzard, V. 1987. *UNICEF and Women, The Long Voyage, A Historical Perspective.* UNICEF history series monograph VII. New York: UNICEF.

Hoffman, V. 1991. *Bildgestutzte Kommunikation in Schwarz-Afrika.* Tropical Agriculture 7. Weikersheim, Germany: Verlag Josef Margraf.

Hursh-Cesar, G. 1988. "The Context for Using Rapid, Low-Cost Methods for Monitoring and Evaluating Health Services Delivery." Prepared for the Center for Development Information and Evaluation, USAID, November. Washington DC: Intercultural Communication Inc.

IFAD. 1985. *Rural Women in Agricultural Investment Projects 1977–1984*, A Report of Findings Based on an Assessment Study of IFAD Experience Relating to Rural Women." Nairobi: IFAD.

Illo, J.F.I. 1988. "Irrigation in the Philippines: Impact on women and their households." Women's Roles and Gender Differences in Development: Cases for Planners, Asia: 2. Bangkok: The Population Council.

Jiggins, J. 1986. Gender-related Impacts and the Work of the International Agricultural Research Centres. CGIAR Study Paper No. 17. Washington DC: The World Bank.

Jiggins, J. 1988. "Beware the Greeks bearing gifts: Reflections on donor behaviour and cooperative performance." In *Cooperative Revisited*, H. Hedlund, ed. Seminar Proceedings No. 21. Uppsala: Scandinavian Institute of African Studies.

Jiggins, J. and H. de Zeeuw. 1992. "Participatory technology development in practice: Process and methods." In *Farming for the Future*, A. Waters-Bayer, ed. London: Macmillan.

Joseph, R. 1987. "Worker, Middlewomen, Entrepreneur: Women in the Indonesian Batik Industry." Women's Roles and Gender Differences in Development: Cases for Planners, Asia: 1. Bangkok: The Population Council.

Lipton, M. and R. Longhurst. 1989. *New Seeds and Poor People.* London: Unwin Hyman.

Long, N. and A. Long, eds. 1992. *Battlefields of Knowledge: The interlocking of theory and practice in research and development.* London: Routledge.

Maddock, N. 1993. "Has project monitoring and evaluation worked?" *Project Appraisal* 8(3): 188–192.

Morss, E.R. et al. 1976. *Strategies for Small Farmer Development.* Boulder: Westview.

Netherlands Ministry of Foreign Affairs. 1988. *Women in Agriculture.* Sector papers No.1. The Hague: Directorate General for Development Cooperation, Institute of Cultural Affairs .

Norton, R.D. and L. Solis. 1982. *The Book of CHAC: Programming studies for Mexican agricultural policy.* Baltimore: John Hopkins University Press.

Palmer, I. 1981. "Women's issues and project appraisal." *IDS Bulletin* 12(4): 32–39.

Patton, M.Q. 1986. *Utilisation-focused Evaluation* (2nd ed.). Newbury Park: Sage.

Poats, S.V. 1991. *The Role of Gender in Agricultural Development.* Issues in Agriculture 3. Washington DC: CGIAR.

PROWESS/UNDP. 1991. *A Forward-Looking Assessment of PROWESS.* New York: PROWESS/UNDP.

PROWESS/UNDP. 1992. *Tools of Community Participation: A Manual for Training Trainers in Participatory Techniques.* New York: UNDP.

Ravnborg, H.M. 1993. *Planning and Priority Setting in Agricultural Research. The Case of Tanzania.* CDR Working Papers 93.5. Copenhagen: Centre for Development Research.

Röling, N. 1988. *Extension Science: information systems in agricultural development.* Cambridge: Cambridge University Press.

Röling, N. 1994. "Platforms for decision-making about ecosystems." In *The Future of the Land: Mobilising and integrating knowledge for land use options.* L.O. Fresco et al., eds. Chichester: John Wiley and Sons.

Röling, N. 1994. "The Changing Role of Agricultural Extension." Keynote paper to Workshop on Agricultural Extension in Africa organised by CTA. Yaounde: 24–28 January.

Röling, N. and C. Talug. 1993. Monitoring and Evaluation in TYUAP: Staff Manual for the Province. Department of Evaluation and Coordination. Ankara: Ministry of Agriculture and Rural Affairs.

Schuh, E. and H. Tollini. 1979. *Costs and Benefits of Agricultural Research: The State of the Arts.* Staff Working Paper No. 360. Washington DC: The World Bank.

Shah, P., G. Bharadwaj, and R. Ambastha. 1991. *Participatory Impact Monitoring of a Soil and Water Conservation Programme by Farmers, Extension Volunteers and AKRSP.* RRA Notes 13: August. London: IIED.

Tendler, J. 1982. *Rural Projects through Urban Eyes.* Staff Working Papers No. 532. Washington DC: The World Bank.

Thomas-Slayter, B., A.L. Esser, and M.D. Shields. 1993. *Tools of Gender Analysis: A Guide to Field Methods for Bringing Gender into Sustainable Resource Management.* Worcester: Clark University.

UNFPA. 1986. *Monitoring the Impact of UNFPA Basic Needs Assessment on Women's Projects.* Special Unit for Women and Youth. New York: UNFPA.

United Nations. 1985. *United Nations Development Fund for Women, Development Co-operation with Women: The Experience and Future Directions of the Fund.* New York: United Nations.

van Beek, P. 1991. "Using a Workshop to Build a Rich Picture: Defusing the Ponded Pasture Conflict in Central Queensland." Paper presented to workshop on Managing Complex Issues in Uncertain Environments: Systems Methodologies in Queensland Agriculture. 26–29 August. Brisbane: Department of Primary Industry.

van Vliet, M. 1984. *Vrouwen in de Marge: Een inventarisatie van 17 vrouwenprojekten ex. Cat. III-c.* The Hague: Directorate-General for Development Cooperation.

Wagemans, M. 1987. *Voor de Verandering.* Published Ph.D. dissertation. Wageningen: Agricultural University.

Warren, D.M., L.J. Slikkerveer, and S.O. Titiola, eds. 1989. *Indigenous Knowledge Systems: Implications for Agricultural and International Development.* Studies in Technology and Social Change 11. Ames: Iowa State University

Waterston, A. 1976: *Managing Planned Agricultural Development.* Washington DC: Governmental Affairs Institute.

Weekes-Vagliani, W. 1985. *The Integration of Women in Development Projects.* Development Centre Papers. Paris: OECD.

Wijeratne, M. 1988. *Farmer, Extension and Research in Sri Lanka: An Empirical Study of the Agricultural Knowledge System with Special Reference to Rice Production in Matara District.* Published. Ph.D. dissertation. Wageningen: Agricultural University.

World Resources Institute. 1989. *World Resources 1988–89.* Washington DC: WRI, 82-84.

World Resources Institute. 1991. *World Resources 1990–91.* Washington DC: WRI, 91-92.

Chapter 13
Environmental Auditing[1]

RALF BUCKLEY
Griffith University, Australia

INTRODUCTION

Development, Types, and Definitions of Environmental Audit

In its broadest sense, environmental audit is simply a check on some aspect of environmental management (Buckley 1990a, 1991a: 121–164, 1991b). From the first uses of the term over a decade ago, a number of somewhat different activities have all been referred to as environmental audits. For example, these included checking performance of environmental monitoring equipment, checking corporate compliance with environmental legislation, and checking the accuracy of environmental impact predictions (e.g., Oates 1982; Levin 1983; Harrison 1984; Henz 1984; Bisset 1985; Blakeslee and Grabowski 1985; Hall 1985; Schaeffer et al. 1985). Some of these arose in response to the various requirements of bodies such as the US Environmental Protection Agency (US EPA) and the US Securities and Exchange Commission (SEC). Others arose as practitioners attempted to evaluate the success of environmental impact assessment (EIA) and related environmental planning tools. Others again arose from the internal management practices of transnational corporations, as they began to compare environmental aspects of their operations between different nations (Brown 1994).

The development of environmental audit has been rapid. Training courses and handbooks were available in the USA by the mid-1980s (Harrison 1984; Blakeslee and Grabowski 1985), and in other countries by the late 1980s (Buckley 1989a; IBC 1990). These were at a rather general level and encompassed a wide range of different types of audit. By 1991, there were handbooks for specific users, such as the international development assistance organisations (Buckley 1991b). And in the last few years, environmental audit has become incorporated in legislation in a number of countries and jurisdictions; accreditation systems for environmental auditors, both private and government-sponsored, have become widespread; and audits have become a standard prerequisite or component of a range of business transactions such as financing, insurance, and corporate restructuring. Training courses and manuals, in consequence, have become much more detailed and specific (Brown 1994; IHEI 1994).

Several classifications of environmental audit have been proposed, based on various criteria. Tomlinson and Atkinson (1987) proposed a terminology for

[1]*Environmental and Social Impact Assessment* - Edited by F. Vanclay and D.A. Bronstein. Copyright © 1995 by the International Association of Impact Assessment. Published in 1995 by John Wiley & Sons Ltd.

environmental audits associated with development control processes, based on the stage in development. Buckley (1989a, 1991a,b) proposed a broader classification based on subject matter, differentiating audits of environmental compliance, monitoring programs, impact predictions, plant and equipment performance, physical risks and hazards, financial risks and liabilities, products and markets, baselines and benchmarks, management programs, management structures, planning procedures and legislative frameworks. Detailed descriptions of each of these are given by Buckley (1991a,b).

Audits of environmental monitoring programs, for instance, check whether the monitoring program is actually capable of detecting changes in relevant parameters due to impacts from the development concerned, with sufficient precision and in sufficient time to be able to manage such impacts effectively. This covers aspects such as selection of parameters, sampling frequency and spatial layout, replication and statistical analysis (Buckley 1989a, 1990a, 1991a,b). Audits of equipment performance test whether equipment, e.g. for environmental monitoring, is meeting specifications. Audits of physical risks are analogous to environmental risk analyses in EIA, but in much more detail since they can identify, for example, the precise physical location and quantity of all hazardous substances on-site. Audits of financial risks examine all potential pathways to financial loss associated with the environmental management aspects of corporate operations. Environmental product and market audits examine the ways in which environmental aspects of the company's product lines affect sales and prices in different actual and potential market sectors. They can include aspects such as sources of raw materials, wastes generated during manufacturing, recyclability and recycled content, packaging and labelling, and so on. Audits of environmental management programs, management structures, planning procedures and legislative frameworks test how well each of these actually performs its intended environmental management task.

Alternatively, environmental audits may be classified by origin (internal or external), scale, timeframe, purpose, or some combination of such criteria (Buckley 1989a, 1990a, 1991a,b). A recent industry handbook, for example (Brown 1994) lists environmental management audits, compliance, technical and process, merger-acquisition–divestment (MAD), environmental impairment liability (EIL), environmental marketing and environmental impact audits. Many of these correspond directly to the subject-based categories outlined above. Overall, however, this is principally a purpose-based classification: what will the results of the audit be used for?

As practices have changed, so have terminologies. Some of the activities listed above are now rarely referred to as environmental audit, and other uses of the term have been amalgamated. Environmental product and market audits, for example, would now fall under the rubrics of environmental purchasing (BIE 1993), life-cycle analysis (BIE et al. 1993), and environmental marketing (Coddington 1993). Again, environmental baseline and benchmark audits have been known at times as state-of-the-environment reports; and in some countries they still are. In many countries, however, environmental reporting now refers to public reports of environmental performance by operating corporations (DTTI et al. 1993; Elkington 1994; Larderel

1994; UNEP 1994). Snapshot descriptive reports of environmental parameters throughout a defined geographic area, therefore, are once again being referred to as environmental audits.

Currently, there is a broad distinction between environmental audits carried out ultimately to protect corporate shareholders from potential financial liabilities associated with poor environmental management, and those carried out ultimately to improve public environmental planning and management processes. The former are commonly called corporate environmental audits. If the term environmental audit is used without other qualification, in most cases this will now be the meaning intended. This is also the meaning used in most environmental audit handbooks and much, though not all, recent government literature on environmental audits. Typically, corporate environmental audits involve an objective review of a corporation's environmental operations including policies, management structures and practices, past and present legal compliance, and physical and financial risks. They are generally instigated by and carried out on individual private companies, by company staff or private-sector consultants, and the results are often kept confidential.

Audits carried out to improve public environmental planning and management processes may perhaps be referred to as planning environmental audits or environmental planning audits (Buckley 1992a). Typically, they span a number of individual developments, are carried out by the public sector, and results are made publicly available. Audits of environmental impact predictions, environmental planning and EIA processes, and legislative frameworks would fall in this category.

Although the distinction between corporate and planning environmental audits is useful, there is still some overlap. Corporate environmental audits, for example, may include tests of impact predictions for the company or plant under audit, and though the purpose and scale differ from broad-scale public audits of environmental impact predictions, the data and techniques involved are the same. Public-sector agencies may conduct or require legal-compliance, equipment-performance, physical-risk or even financial-risk environmental audits on individual corporations as an environmental policy tool, and again, the only real difference between these and voluntary corporate environmental audits is in origin and final reporting requirements. In addition, corporate environmental audits may be carried out on public-sector corporate bodies, such as electricity or water supply utilities. So a classification based on subject matter, although no longer commonly used, is still a valuable adjunct to any broad distinction such as the above.

Given the multiplicity and mutability of the term environmental audit as outlined above, is there now any generally accepted definition? Not really. One commonly quoted definition is that of the International Chamber of Commerce, "a management tool comprising a systematic, documented, periodic, and objective evaluation of how well environmental organisation, management, and equipment are performing with the aim of helping to safeguard the environment by. . . facilitating management control of environmental practices [and] assessing compliance with company policies, which includes meeting regulatory requirements" (Graham-Bryce 1989). This is not universally accepted, and clearly refers only to corporate environmental audits in the sense outlined above.

A more focused, albeit strangely phrased, definition is used by the Environmental Auditors' Registration Scheme in the United Kingdom. As quoted by Brown (1994), it is: "a systematic process of objectively obtaining and evaluating evidence to determine the reliability of an assertion with regard to environmental aspects of activities, events and conditions, as to how they measure to established criteria, communicating the results to the client" *[sic]*. In principle at least, this definition could cover planning as well as corporate environmental audits. It also captures the essential components of environmental audit (Buckley 1989a, 1990a, 1991a,b), as follows. It incorporates three steps: (1) assessing the current status of the environmental management parameters under audit; (2) testing them against some predetermined norm; and (3) attesting or testifying to the outcome. And it involves three parties: the auditor, the auditee, and third parties who require verification of the outcome. This essential structure applies even if all three parties belong to the same organisation.

Environmental Audit and Environmental Impact Assessment

How is environmental audit related to environmental impact assessment (EIA)? Clearly, it depends on the type of environmental audit referred to, and the definition adopted. Arguably, the various types of environmental planning audit are means to improve the overall public policy, regulatory and administrative framework for environmental management, including EIA. In a sense, these types of environmental audit are extensions of EIA.

Corporate environmental audits are intended to ensure that the corporations concerned are complying with that framework, including requirements specified in EIA documents or subsequent environmental management plans (Buckley 1994a). Corporate environmental audits are an extension of environmental compliance commitments or requirements, but not EIA as such. Even so, there are many links between EIA and corporate environmental audit. Corporate environmental audits are generally concerned with assessing actual past impacts, for example, whereas EIA is generally concerned with assessing potential future impacts; and audits are generally carried out on operating facilities, whereas EIA is generally carried out before a project commences.

Both EIA and corporate environmental audit, however, commonly describe the current status of defined environmental parameters at a given site; EIA as a baseline, audit as a benchmark. Both review past impacts as a means of predicting future ones; EIA to gain development approval, audit to assess risks or compliance. Some EIAs are not carried out until after development approval has already been granted. Equally, some audits are conducted as part of an approval process, either by environmental regulatory agencies for operating licences or insurances; or by corporate regulatory agencies for corporate prospectuses, mergers, acquisitions, divestments and share reissues.

Similarly, corporate environmental audits generally involve some form of objective verification, usually to a third party; and EIA documents are also subject to external scrutiny by government agencies and the public. In some jurisdictions, individuals who conduct environmental audits have a statutory liability for their content, and may be penalised heavily for inaccuracies. Such statutory liability is much less common for individuals who prepare EIA documents. Both environmental auditors and EIA practitioners, however, typically have professional responsibilities and duties of care to their clients under civil law. Overall, though environmental audit and EIA are certainly different processes, there are many similarities. It is therefore not surprising that many EIA practitioners, particularly those with an engineering background, also conduct environmental audits. Note, however, that environmental audits are also carried out by many individuals and firms which do not generally conduct EIA; environmental law and accounting firms in particular.

In many respects, the development of environmental audit has parallelled that of EIA, but lagging it by a decade or more. Experienced EIA practitioners are numerous and there are international professional organisations such as the International Association for Impact Assessment (IAIA), and national bodies such as the Environment Institute of Australia. Although there are many EIA manuals (e.g., Goodland et al. 1991), most EIA practitioners do not need to refer to them since they are already familiar with the techniques described. Techniques continue to improve, but the marginal rate has slackened. In most developed nations at least, although the standard of EIA is still poor (Buckley 1989b,c, 1990b,c,d; Warnken and Buckley in prep., cited in Anderson 1994), this is perhaps due mainly to cost constraints, lack of basic scientific knowledge in some cases, and legislative and administrative shortcomings in EIA frameworks (Buckley 1989b, 1990b, 1991a:180–196), rather than to any deficiencies in the average abilities of EIA practitioners.

Experienced environmental auditors, in contrast, are still comparatively few; training courses are still well attended, and manuals highly sought after. Despite this, the number of accredited environmental auditors in Victoria, Australia, for example, has grown rather slowly (Brown 1994 and previous editions, para. 2265). There are controversies over issues such as accreditation and legal liabilities of auditors. The experience of many practitioners is still relatively narrow and the standard of individual audits may differ considerably. The involvement of professional bodies such as IAIA, and the application of relevant national and international standards such as BS7750, ISO9000, and ISO14000 for environmental management systems, from the United Kingdom and the International Standards Organisation respectively, are thus important for environmental audit to mature as a discipline. There is also a need for audits themselves to be subject to review and evaluation, in the same way that EIA has been.

ENVIRONMENTAL PLANNING AUDITS

Audits of Environmental Legislation and Regulatory Frameworks

An audit of environmental legislation and regulatory framework in any jurisdiction is a check on how well they achieve the environmental policy goals they are intended to serve. Relevant issues include: who has standing to bring actions under the legislation concerned; how much ministerial discretion it provides, and to which ministers; what policy instruments it uses for each major policy goal; and how it interacts with other rights and laws, including those relating to property, health and safety, crime, tort, nuisance, and contract (Buckley 1991a,b). In the specific case of EIA legislation, for example, this would involve checking how well the legal provisions for EIA, wherever and however they are couched, actually serve the policy intentions of EIA (Buckley 1991a,b).

Is EIA provided for in a single piece of legislation in the jurisdiction concerned, or through multiple references in different laws? If the latter, is EIA defined and interpreted in the same way if it is triggered by different laws? (Buckley 1992b). Does it include broad strategic issues (Buckley 1994b) and cumulative impacts (Buckley 1994c), or is it very narrowly circumscribed? There is a somewhat infamous instance in the State of Queensland, Australia, for example, where EIA law had to be amended because a judge held that under previous law, the impacts of a proposed tourist development on a turtle colony were irrelevant to a development application since, in his view, the definition of environment did not include turtles (Buckley and McDonald 1991).

Audits of Environmental Agencies and Decisions

Planning and regulatory agencies are rarely free to make arbitrary decisions. In addition to the electoral accountability of the government concerned, and their own policy goals, agencies are generally bound by legally enforceable statutory requirements. They are generally required to consider a predefined set of factors, often including environmental factors, in their decision-making processes; and their decisions may be subject to legal challenge if these considerations clearly have been ignored. Such challenges are usually made to individual decisions, but if public dissatisfaction with agency performance becomes electorally threatening, governments may appoint independent commissions to audit a regulatory agency's procedures and decision-making processes as well as its actual decisions.

Besides testing whether the agency has followed statutory requirements, such audits may examine, for example, whether the agency's procedures operate smoothly, without frequent ministerial intervention; whether they provide adequate public participation; and whether they provide equitable mechanisms for resolving disputes (Buckley 1992a). For example, the Australian International Development Assistance Bureau (AIDAB), Australia's bilateral aid agency, conducted a detailed environmental audit of its operations three years ago (Ovington 1992). The Australian Commonwealth Environment Protection Agency has also conducted a review of its operations and enabling legislation recently, employing consultants to examine specific issues, and circulating discussion papers widely for public criticism and comment (CEPA 1994).

Audits of Environmental Planning Procedures

Environmental legislation is only as good as the procedures used to implement and enforce it. In addition, many environmental laws are written in quite broad language, establishing government agencies and their overall powers, but leaving them to articulate their own enabling legislation through administrative guidelines and procedures. Particularly in jurisdictions where there is no avenue for third-party court challenges to agency decisions, it is often the administrative procedures rather than actual legislation that determines how well environmental policy goals, such as effective EIA, are achieved in practice. Audits of environmental planning procedures are therefore an important tool in improving environmental management at a national or regional scale. Audits of this type can be applied to any aspect of the institutional frameworks for environmental planning (Buckley 1990b,d, 1991c).

What permits, licences, approvals, notices, and consents are required for developments of different types in different industry sectors in different types of land tenure, for example? And as in audits of environmental agencies, what opportunities are there for public involvement, what provisions for resolving disputes, and who has standing to bring disputes before the courts? In the case of EIA, for example, do relevant administrative procedures ensure that all developments likely to produce a significant impact on the environment are subject to EIA, whereas those unlikely to do so are not? (Buckley 1990b,c,d, 1992b,c). Do they lead to correct identification and accurate prediction of major impacts, and to design modifications and permitting conditions which ensure that impacts are minimised? And do they establish monitoring programs and feedback mechanisms? (Buckley 1990d, 1991c; Warnken and Buckley in prep., cited in Anderson 1994).

Audits of Environmental Impact Predictions

The ability to predict environmental impacts with reasonable accuracy is central to effective EIA (Buckley 1989b, 1990b,c,d, 1991a: 93–120, 1991d, 1992c). Auditing predictions in environmental impact statements simply means testing how accurate they were by comparing them with actual monitored impacts once the projects concerned are under way. There were at least 25 published references to this type of audit up to 1990 (Buckley 1991a: 93–120), but only five actual audits of more than one project had been carried out to that date, and only one for an entire country (Buckley 1989c, 1991a,d). Bisset (1985) identified 791 predictions from four projects in the UK, finding that 77 of these were testable and 57 were 'probably accurate'. Knight (1985) examined seven projects in Victoria, Australia; only one testable prediction had been made, and this proved to be 'reasonably accurate'. Henderson (1987) found 122 predictions from two projects in Canada, of which 70 were testable and 54 'substantially correct'. Culhane (1987) examined 29 projects in the USA, finding 239 predictions of which less than 25 percent were quantified, and less than 30 percent of these 'unqualifiedly close to forecast'. And Buckley (1989c, 1991a,d) reviewed more than 1000 projects in Australia, finding that both testable predictions and the monitoring data to test them were available for only 181 predictions from 19 projects. For 28 percent of these, actual impacts were more severe than predicted.

The average accuracy of impact predictions, expressed as a log ratio between actual and predicted impacts (Buckley 1989c, 1991a,d) was 44 percent $\pm$ 5 percent standard error. Predictions where actual impacts proved more severe than expected were on average significantly (p < 0.05) less accurate (33 percent ± 9 percent) than those where they proved as or less severe (53 percent ± 6 percent). An audit of environmental impact predictions for all tourist developments in Australia over the past 15 years is currently nearing completion (Warnken and Buckley in prep., cited in Anderson 1994).

CORPORATE ENVIRONMENTAL AUDITS

Incentives for Corporate Environmental Audit

Corporations may conduct environmental audits for a wide range of reasons. For example, these may include the development of environmental policies and training programs; improvement of resource utilisation and waste management, selection of suppliers and contractors; and identification of new markets. Most commonly, however, corporations conduct environmental audits as a core component in risk management; to obtain insurance, to check compliance with environmental laws, or to ascertain potential environmental liabilities during mergers, acquisitions, and divestments.

The legal and financial risks which industry and commerce face from poor environmental management have increased greatly and are continuing to do so (Buckley 1991a: 165–179, 1991e, 1992d, 1993). They include costs associated with: fines for breaching regulations; other statutory penalties, such as closure of plant or site, injunctions to stop particular activities, forfeiture of assets, compulsory actions, or statutory compensation; penalties for individuals, including fines, convictions, and gaol terms; recovery of costs, expenses, loss, and damages by public authorities; restraining orders over property, and charges against such property; compulsory control, prevention and mitigation measures; clean-up, repair and rehabilitation costs; compensation claims, citizens' lawsuits and class actions; temporary closure by regulatory agencies or court injunctions; upgrading, retrofitting or replacing equipment to more stringent standards; delays in approvals for future projects; lost market share from poor public image or product boycotts; falls in share prices; reduced credit from suppliers; higher insurance premiums; delayed and reduced cash flows; and losses in value of assets and securities. Penalties and other costs can also be transferred to financiers, insurers, partners, and shareholders through corporate mergers and acquisitions, contractual arrangements between operating corporations, equity holdings in operating corporations, loan agreements and other financing arrangements, and insurance contracts and their legal interpretation by the courts (Buckley 1992d, 1993).

Corporate Environmental Audits and Environmental Management Systems

In many countries, the emphasis is moving from environmental audit as an isolated activity, to environmental audit as part of an overall environmental management system (EMS). There are now several national and international standards for EMS, notably BS7750 in the UK and ISO14000 from the International Standards Organisation. These are essentially standards for quality assurance in EMS. Organisations applying for accreditation of their EMS under BS7750 must have the EMS inspected by an accredited certification body, and must undertake environmental audits and environmental management reviews.

Organisations in countries other than the UK may also be accredited under BS7750. In Australia, for example, a new certified environmental management (CEM) scheme run by a subsidiary of Standards Australia can offer certification to BS7750, as can private firms such as SGS Australia Pty Ltd and Det Norske Veritas, using its so-called International Environmental Rating System (IERS). The IERS offers 10 levels of achievement, named S1 to S5 and then A1 to A5, with A5 the highest. The S5 level qualifies for certification under BS7750, and also fulfils the requirements of the European Communities' eco-management and audit scheme (Brown 1994).

Public Reporting of Corporate Environmental Audits

In many countries, there is increasing pressure, and in some cases legal insistence, for public reporting of corporate environmental performance, generally based on regular environmental audits. The European Community's eco-management and audit scheme (ECC 1993) is a good example. Promulgated in June 1993, this program has been adopted by many European corporations. It is a voluntary, site-based scheme, and an organisation can register some or all of its operating sites. To do so, it must establish an environmental policy, carry out a comprehensive environmental audit, and issue a public statement, including supporting data, of environmental management policies, targets and actual performance. The audit can be carried out by company personnel or an external auditor, but the public statement must be validated by an accredited 'environmental verifier' and submitted to the relevant national authority, designated by each of the EC nations. If the accredited statement is accepted, the company is then authorised to display one of the eco-management and audit logos established under the scheme. Perhaps as a result of this, public environmental performance claims by European companies are commonplace. *Environment Strategy Europe*, for example, an annual publication advertising corporate environmental responsibility, is well sponsored: more than 60 corporations are represented in the 1994/95 edition.

Internationally, Chapter 30 of *Agenda 21* encourages private corporations to 'report annually on their environmental records' (Larderel 1994), and the United Nations Environment Programme carried out a progress survey in 1992 in conjunction with the UK-based consultancy firm, SustainAbility (DTTI et al. 1993; Elkington 1994; Larderel 1994; UNEP 1994). Industry associations have also developed reporting guidelines, such as the Public Environmental Reporting Initiative, PERI (Balta 1994). Perhaps one of the strongest pressures for public environmental reporting comes not from government initiatives, but direct market pressures. The major international business magazine, *Fortune*, published its first rankings of corporate environmental performance in the 26 July 1994 edition (Elkington 1994). Irrespective of the source, public environmental reporting requires routine audits of environmental performance as well as environmental management systems, and is likely to be a major factor in the continued growth of environmental auditing.

Industry-specific Protocols for Corporate Environmental Audit

Different types of industrial operation produce different environmental impacts. As with scoping checklists and guidelines for EIA, environmental audit protocols need to be tailored to individual industry sectors or types of development, if they are not to be extremely generalised. To date, expertise in auditing specific sectors has been treated as proprietary knowledge by individual auditors or auditing consultants, and guarded accordingly. However, in some sectors at least, sector-specific environmental management manuals incorporating audit protocols are now being published by industry associations. There is a good example in the tourism sector, where the International Hotels Environment Initiative has produced a 200-page manual, *Environmental Management for Hotels: The Industry Guide to Best Practice* (IHEI 1994), which though not strictly an audit protocol, includes sufficient information for hotels both to conduct their own environmental audits and to take appropriate management action subsequently. Its 13 chapters deal with environmental culture and policy, waste management, energy and water conservation, water quality, product purchase, indoor air quality, external air emissions, noise, stored fuel, poly-chlorinated biphenyls, pesticides and herbicides, hazardous materials, and asbestos.

Special-purpose or Subsidiary Environmental Audits

As the term environmental audit comes to encompass more and more of an organisation's environmental performance and management, and hence require more and more time, resources and expertise, there is a trend for some organisations to audit different aspects of their operations at different times. This has led to a variety of special-purpose or subsidiary environmental audits, such as energy audits, hazardous-waste audits, materials-consumption audits, and so on. In particular, many corporations are now embarking on life-cycle analysis, which is an attempt to quantify and characterise all the environmental impacts associated with the production, transport, packaging, marketing, sale, use, recycling, and ultimate disposal of a manufactured item (BIE 1993; BIE et al. 1993; Coddington 1993). This is essentially a form of detailed product audit.

CURRENT ISSUES AND DEBATES

Accreditation of Environmental Auditors

There are active debates at present in a number of countries and jurisdictions over mandatory or voluntary accreditation and registration schemes for environmental auditors, either as individuals or consulting firms. This issue is particularly significant where environmental audit is referred to in legislation. The State of Victoria, Australia, for example, may require environmental audits as a condition of permitting and licensing, and such audits must be carried out by an auditor appointed under the Environment Protection Act 1970—essentially, a government-accredited auditor. Only individuals can be appointed as auditors under this legislation, and there are different categories for different types of audit. Appointment as an auditor confers responsibilities and potential liabilities as well as privileges, and relatively few individuals have chosen to accept it.

In other areas, legislation may refer to environmental audit without requiring that it be carried out by auditors appointed under law. This applies in several other states of Australia, for example. In the United Kingdom, the privately run, nonprofit Environmental Auditors Registration Association (EARA) has three levels of membership: associate environmental auditor, environmental auditor, and principal environmental auditor. Although it has been referred to as an accreditation scheme (Brown 1994, para. 1330), it describes itself as a registration scheme, and is essentially a professional association with entrance requirements, such as the Institute of Biology, or the Institute of Mining and Metallurgy in the UK, and their equivalents in other countries.

In the USA, there are state registration schemes in California and New Jersey. The privately run national Registry of Environmental Professionals also offers registration as a certified environmental auditor (Brown 1994, para. 1330). The much-debated eco-management and audit scheme of the European Community Council (1993) is a voluntary scheme and does not provide for statutory accreditation of environmental auditors.

Even in jurisdictions that refer to environmental audit in legislation, there is, of course, nothing to stop an organisation commissioning a voluntary audit by any person, accredited or not. Such an audit would simply not be considered as an audit within the meaning of the relevant legislation; but equally, it would be exempt from any requirements (e.g., as regards disclosure) applying specifically to statutory audits. Similarly, even in jurisdictions where environmental audit is not mentioned in statutes, there are still advantages in commissioning auditors whose competence is externally certified in some way. Even if not immediately contemplated, there is always the possibility that an organisation may wish to rely on an environmental audit either in legal proceedings or as a marketing tool, and in either case the credibility of the auditor is critical.

Liability of Environmental Auditors

Registration, certification, or accreditation as an environmental auditor generally constitutes a declaration, by the individuals concerned, that they consider themselves professionally competent to practice in this discipline. In most jurisdictions, all professionals and consultants have an implied duty of care towards their clients, and in most cases, a higher level of certification would allow a client to require a higher standard of performance. If an organisation suffered loss because it relied on an environmental audit which proved to be faulty or incompetent, it could well sue the auditors. In defence, the auditors would generally have to show that they were as skilled as others in the profession and that they had used their skills to the best of their professional ability, that is, that they were (a) capable and (b) neither malicious, reckless, nor negligent.

In addition to professional liabilities towards commercial clients, environmental auditors may have legal liabilities established by statute. In Victoria, Australia, for example, statutory (i.e., government-accredited) environmental auditors who provide false or misleading information, or conceal relevant information, can be fined up to Aust$250,000 (US$175,000 as of January 1995) and gaoled for up to two years.

Statutory References to Environmental Audit

If environmental audits are to be referred to in legislation, as they already are in some countries and states, the precise legal definition of environmental audit may become critical. There clearly would be advantages if similar definitions were used in different laws and jurisdictions, although this has not always happened in regard to terms used in other legislation. Currently, the State of Victoria, Australia, defines environmental audit as:

> A total assessment of the nature and extent of any harm or detriment caused to or the risk of any possible harm or detriment which may be caused to, any beneficial use made of any segment of the environment by any process or activity, waste, substance (including any chemical substance) or noise.

Mining law in Queensland, Australia, also refers to environmental audit reports, but rather than provide a definition, the relevant government agency has drawn up detailed guidelines for questions to be answered in such reports. Other jurisdictions, such as New South Wales, Australia, appear to be adopting the International Chamber of Commerce definition given earlier, or some similar wording. This is quite a different definition from that used in Victoria.

By using a term in legislation without defining it, the legislature effectively leaves its definition to the courts. This need not necessarily be a problem: EIA is widely used in many jurisdictions, for example, but defined in few. It may, however, produce unexpected outcomes, as in the case referred to earlier (Buckley and McDonald 1991) where a court decided that environment did not include fauna. It could also lead to unnecessary and expensive litigation, as seems quite possible in regard to the definition of environmental management plans (EMPs) in Australia (Buckley 1994a).

Environmental Audits for Insurers, Investors, and Shareholders

Many commercial transactions and public policy measures depend on the public availability of reasonably accurate financial information on public companies. It is for this reason that national bodies such as the Securities and Exchange Commission in the USA, or its equivalent in other nations, specify reporting and disclosure requirements for company prospectuses, takeovers, and share offers, etc. As environmental management issues assume an ever more significant role in a company's overall financial performance, accurate information on a company's environmental opportunities and liabilities becomes correspondingly important to insurance underwriters, investors and investment managers, and shareholders. Environmental audits provide this information, and that is why they have become standard components in obtaining environmental impairment liability (EIL) insurance, in obtaining loan finance, and increasingly, in corporate reporting.

All of these also provide public policy benefits. A viable environmental insurance industry provides a commercial incentive for improved environmental management, and a pool of funds for rehabilitation of pollution, contamination or other environmental damage (Buckley 1991a: 165–179, 1991e). Avoiding losses to financiers, investors, and shareholders, through penalties or claims for environmental damage, as for any other reason, helps to maintain investment confidence nationally as well as for the companies concerned, contributing to a strong currency and a healthy balance of payments.

In addition, regulatory systems that effectively pass the onus for checking environmental performance from government agencies to accredited auditors or verifiers may improve economic efficiency, as long as their effectiveness is assured through appropriate penalties for negligence or fraud. This, perhaps, is what the system of accredited environmental auditors in Victoria, Australia, is intended to achieve, and to a lesser degree, since it is voluntary, the EC's eco-management and audit scheme. These attempts may not always work, as the history of the environmental insurance industry in the USA shows clearly (Buckley 1993). But whoever takes ultimate responsibility for the results, environmental audit is an increasingly important tool for the public as well as the private sector.

Confidentiality of Corporate Environmental Audit Documents

For companies that have undertaken to publish an independently verified environmental performance statement, as under the EC eco-management and audit scheme, confidentiality of environmental audits is unlikely to be a major issue. In many nations, however, including the USA and Australia, companies are concerned that environmental audits undertaken in order to check compliance, improve management, or obtain finance or insurance, might be used by regulatory agencies in support of prosecutions. In recent years, this has been one of the most significant policy issues in regard to environmental audit, especially in the USA, since without some form of protection for the results, many companies may choose not to initiate environmental audits (Buckley 1991f, 1992e, 1993; McDonald 1994).

From a policy perspective, it is important that companies should not be discouraged from undertaking voluntary audits through concerns over confidentiality; but it is equally important that regulatory agencies should not be hampered from prosecuting breaches of environmental laws simply because relevant data have been included in an environmental audit report. In other words, there is a balance between public rights to monitor compliance and prosecute breaches of environ-mental laws, and private rights against self-incrimination.

In more detail, precisely what type and degree of protection should be available, from what kinds of audit, for what kinds of information, held in what form? what types of enquiry should it be protected against? and how should protection depend on other sources for the information concerned? To address these questions, distinctions must be drawn between: (a) powers of search and seizure, cf. discovery in court proceedings; (b) power to seize documents, cf. the right to demand information; (c) the use of audit documents to initiate proceedings, cf. their use as evidence in proceedings started on other grounds; and (d) access to information, cf. its admissibility as evidence in court.

There are four main potential mechanisms to protect the confidentiality of an environmental audit report. The first is to keep its existence completely secret. This is difficult in practice, especially if the report is to be used as a basis for manage-ment or remedial action. The second is through a statutory protection for such reports in the jurisdiction concerned. The state of South Australia, for example, provides such protection though S58 of the Environment Protection Act 1993 (SA). To obtain such protection, the person or corporation concerned must apply for it before starting the audit, with a description of the audit program proposed. Protection may be subject to conditions on the content, format, and lodgement of the report. Subject to such conditions, protected reports are then exempt from seizure and inadmissible as evidence in court proceedings.

Similarly, the states of Oregon and Arizona, USA, have adopted statutory self-evaluative privileges during 1994 (McDonald 1994). Nationally, the USEPA has a declared policy not to demand access to internal corporate environmental audits except: as evidence of *mens rea*, mental state, in criminal prosecutions; in relation to risks or accidents which are already under investigation for other reasons; or if a corporation puts forward its environmental management system as an affirmative defence, for example, as an element of due diligence (McDonald 1994). Similar policy statements have been made by the states of Queensland and New South Wales, Australia. Such policy statements, however, do not have the same force as statutory protection.

In default of statutory protection, a company may try to protect audit reports by claiming either a privilege against self-incrimination, or legal professional privilege. These are known by different legal terms in different countries. In the USA, for example, there are two forms of legal professional privilege, applying respectively to client-attorney communications and attorney work products. The privilege against self-incrimination is a common law privilege derived from the British system of justice. In Australia, the High Court recently decided that the privilege against self-incrimination is available only to individuals, and not to corporations (McDonald 1994). Note that in some jurisdictions, such as the state of Victoria, Australia, environmental legislation provides statutory rights to claim the privilege against self-incrimination, and this would presumably survive the High Court's decision regarding the common-law right (McDonald 1994).

In the UK, Australia, and the USA, therefore, corporations will generally have to rely on legal professional privilege (LPP) if they wish to protect the confidentiality of their environmental audit reports. LPP can only be claimed under quite restricted circumstances (Buckley 1991f, 1993; McDonald 1994). In general, LPP applies only to communications between a lawyer and client whose sole purpose is either to provide legal advice, or in anticipation of future litigation; and which has been kept confidential in practice. In the UK, it is sufficient for the above to be the dominant rather than the sole purpose. Communications can include letters and memoranda, investigative reports and interview transcripts, all of which may form part of an environmental audit. Whether LPP applies to communications between the auditor and the client or lawyer depends on whether the auditor is an agent of one or the other, or an independent contractor, and in each case is subject to a variety of legal tests which may differ between jurisdictions (McDonald 1994).

The sole (or dominant) purpose test applies to the purpose for which the document was created, not to its subsequent use. The non-legal content of a document, though not attracting LPP, may be used to indicate the purpose of the document as a whole. This might apply, for example, if management recommendations were included in an environmental audit report providing legal advice on non-compliance. In determining whether a communication was made in anticipation of pending litigation, courts have considered issues such as whether litigation would automatically ensue if a breach were detected; and whether any breach would automatically be revealed by regulatory reporting requirements (McDonald 1994).

Legal professional privilege may apply only to communications that have been kept confidential in practice. Courts have interpreted this to mean that: active measures must have been taken to protect confidentiality; information must not have been disclosed by directors, executives or authorised employees; and information must not have been disseminated beyond employees who need to know it. Authorised publication of relevant information by the corporation will generally constitute a voluntary waiver of privilege. Disclosure to directors does not constitute publication; and illegal publication does not waive privilege. If there are alternative public sources for relevant information, however, this may be taken as waiving any right to privilege. It is possible to waive privilege to only part of a communication, however; for example in concurrent civil and criminal cases.

Legal professional privilege can protect only documents, not the information they may contain or the people who may know it; and it can only protect the legal content of such documents. LPP can provide protection against search and seizure by regulatory authorities, unless relevant statutes specify expressly and unambiguously that search and seizure rights override privilege. In general, it can also provide protection against demands under freedom-of-information legislation, and against discovery in either civil or criminal proceedings. Overall, therefore, it is clear that the availability of legal professional privilege is generally too restricted and uncertain to give companies the confidence to report actual breaches of environmental laws in their environmental audit reports, if such breaches are in fact occurring.

A workable policy framework which balances the competing interests of public and private rights, encouraging self-evaluation without self-incrimination, might include the following components (Buckley 1993; McDonald 1994). Voluntary audits would be exempt from search and seizure, but discoverable in cases initiated on other grounds; as under current USEPA policy. Regulatory agencies would have the authority, but not the obligation, to undertake not to demand audit documents. Privilege would be available, as at present, to protect documents rather than witnesses. Any statute intended to override privilege should have to specify this expressly and unambiguously, as at present. The rights of regulatory authorities to undertake compliance checks would remain unaffected. And finally, mandatory self-audits of compliance could be required only as provided for under statute, licence conditions or other equivalent regulatory instrument.

CONCLUSION

During the past decade or longer, environmental audit has become a central component in a wide range of public policy measures and private-sector commercial transactions in many nations. It is probably the fastest-growing type of environmental assessment at present, and one with major financial implications. Its focus is different from that of EIA, but much of the expertise required is similar; it is not surprising that many EIA consultants have turned to environmental audit. Environmental audit, however, is less mature as a discipline than EIA: people still argue about terms, techniques and expectations. Professional associations such as IAIA can help by establishing and promulgating standards for professional environmental auditors, and providing a mechanism for them to compare skills and experiences, as it has already done for other forms of environmental assessment.

There have been many different definitions and classifications of environmental audit at different times during its brief history. It may be useful to draw a broad distinction between environmental planning audits and corporate environmental audits. The former include audits of environmental baselines and benchmarks, legal and regulatory frameworks, agencies and decisions, planning procedures, and impact predictions. These individual types are usually separate. The latter includes corporate environmental policies, management structures and practices, legal compliance, and physical and financial risks. These may be audited separately or together. Alternatively, all these aspects may be audited for a subset of corporate operations, as in the case of energy audits or life-cycle analyses.

There are strong trends toward increasing routine use of corporate environmental audits in financing, insurance and other inter-corporate transactions, as well as in permitting and licensing. Coupled with this are trends toward more widespread and detailed accreditation procedures for environmental auditors, increasing statutory provision for environmental audit, and increasing statutory liabilities for environmental auditors; and a much weaker trend toward statutory provisions for protection of corporate environmental audits undertaken voluntarily.

Within the corporate sector, there are trends towards public reporting of corporate environmental audits; industry-specific and issue-specific environmental audits; and corporate environmental management systems which include environmental audits as one component. Through such environmental management systems, some companies in some industry sectors in some countries are now approaching the goal of integrated environmental planning and management, which has been propounded for nearly a decade (e.g., Buckley 1987). Whilst industry leaders may be approaching this goal, however, there is a long way to go before it becomes the norm for industry world-wide. It is the industry leaders and their customers who have the greatest opportunities to promote improved environmental management throughout their industry sectors as a whole, and corporate environmental audits are one of their main tools in doing so.

In the public sector, although institutional frameworks for environmental planning and management have evolved considerably over the past decade in many countries, they are still ineffective in many instances; and planning environmental audits is one of the main tools in improving them. Audits of environmental impact predictions and EIA processes, for example, urgently need to evolve from occasional academic initiatives to routine aspects of environmental policy.

ACKNOWLEDGEMENTS AND DISCLAIMER
I thank Associate Professor Janet McDonald for information on legal aspects of environmental audit. Please note that the author is not a lawyer and this chapter is not intended to be used as legal advice.

REFERENCES

Anderson, I., ed. 1994. "Ecotourism: opportunism or opportunity." *New Scientist supplement* 26.11.94: 7

Balta, W.S. 1994. "PERI: guidelines for credible reporting." *Environment Strategy Europe 1994/95*, 123–124. London: Camden Publishing Limited.

BIE. 1993. *Buying into the Environment: Guidelines for Integrating the Environment into Purchasing and Supply*. London: Business in the Environment.

BIE et al. 1993. *LCA Sourcebook: A European Business Guide to Life-Cycle Analysis*. London: Business in the Environment, SustainAbility and Spold.

Bisset, R. 1985. "Post-development audits to investigate the accuracy of environmental impact predictions." *Zeitschrift Umweltpolitik* 4/84: 463–484.

Blakeslee, H.W. and T.M. Grabowski. 1985. *A Practical Guide to Plant Environmental Audits*. New York: Von Nostrand Reinhold.

Brown, G.A. 1994. *Environmental Audit Guidebook*. Melbourne: Law Book Group.

Buckley, R.C. 1987. "Critical problems in environmental planning and management." *Environmental and Planning Law Journal* 5: 206–225.

Buckley, R.C. 1989a. *Environmental Audit: Course Handbook* (1st edn). Gold Coast: Bond University.

Buckley, R.C. 1989b. "What's wrong with EIA?" *Search* 20: 146–147.

Buckley, R.C. 1989c. *Precision of Environmental Impact Predictions: First National Environmental Audit*. Canberra: Australian National University.

Buckley, R.C. 1990a. "Environmental audit: review and guidelines." *Environmental and Planning Law Journal* 7: 142–146

Buckley, R.C. 1990b. "Shortcomings in current institutional frameworks for environmental planning and management." *Bulletin of Public Administration* 62: 50–56.

Buckley, R.C. 1990c. "Environmental science and environmental management." *Search* 21: 14–16.

Buckley, R.C. 1990d. "Adequacy of current legislative and institutional frameworks for environmental impact audit in Australia." *Environmental and Planning Law Journal* 7: 127–141

Buckley, R.C. 1991a. *Perspectives in Environmental Management*. Heidleberg: Springer.

Buckley, R.C. 1991b. *A Handbook for Environmental Audit*. Canberra: Australian International Development Assistance Bureau.

Buckley, R.C. 1991c. "Environmental planning legislation: court backup better than ministerial discretion." *Environmental and Planning Law Journal* 8: 250–257.

Buckley, R.C. 1991d. "How accurate are environmental impact predictions?" *Ambio* 20: 161–162.

Buckley, R.C. 1991e. "Environmental management as an insurance problem." *Risk* 7(2): 24–29.

Buckley, R.C. 1991f. "Environmental audit and legal professional privilege." *Environmental and Planning Law Journal* 8: 338–340.

Buckley, R.C. 1992a. Environmental audit: planners' perspectives. *Planning for Sustainable Development*. Royal Australian Planning Institute, Auscript, Canberra.

Buckley, R.C. 1992b. "Triggers, thresholds and tests for EIA." *Environmental Policy and Law* 22: 146–149.

Buckley, R.C. 1992c. "Environmental impact predictions in science and planning." *Ambio* 21: 323–324.

Buckley, R.C. 1992d. "Environmental opportunities and risks in finance." *Environmental Management and Health* 3(2): 22–25.

Buckley, R.C. 1992e. "Confidentiality of corporate environmental audit documents: policy issues." *Environmental Planning and Law Journal* 9: 297–298.

Buckley, R.C. 1993. "Privilege and environmental audits." In Business Law Environment Conferences, *Environmental Liability Law*, 111–129. Melbourne: Longman Cheshire.

Buckley, R.C. 1994a. "Cumulative environmental impacts: problems, policy and planning law." *Environmental and Planning Law Journal* 11: 317–320.

Buckley, R.C. 1994b. "Strategic environmental assessment." *Environmental and Planning Law Journal* 11: 166–168.

Buckley, R.C. 1994c. "Defining environmental management plans." *Environmental and Planning Law Journal* 11: 355–356.

Buckley, R.C. and McDonald, J. 1991 "Science and law: the nature of evidence." *Search* 22: 94–95.

CEPA. 1994. *Public Review of the Commonwealth Environment Impact Assessment Process*. Canberra: Commonwealth Environment Protection Agency.

Coddington, W. 1993. *Environmental Marketing's Relationship to Overall Corporate Environmental Management*. Sheffield: Greenleaf.

Culhane P J. 1987. "The precision and accuracy of US environmental impact statements." *Environmental Monitoring and Assessment* 8: 217–238.

DTTI et al. 1993. *Coming Clean: Corporate Environmental Reporting.* London: Deloitte Touche Tohmatsu International, International Institute for Sustainable Development and Sustainability Ltd.

Elkington, J. 1994. "Reporting on corporate performance." In *Environment Strategy Europe 1994/95*, 127–130. London: Camden Publishing Limited.

European Community Council (ECC). 1993. *Eco-management and Audit Scheme.* Regulation EEC 1836/93. Luxembourg: ECC.

Goodland, R. et al. 1991. *Environmental Assessment Sourcebook.* 3vv. Washington DC: The World Bank.

Graham-Bryce, I. (chair). 1989. *ICC Position Paper on Environmental Auditing.* Paris: International Chamber of Commerce.

Hall, W.N. 1985. "Environmental audits: a corporate response to Bhopal." *Environment Forum* 4: 36.

Harrison, L.L., ed. 1984. *The McGraw-Hill Environmental Auditing Handbook: A Guide to Corporate Environmental Risk Management.* New York: McGraw-Hill.

Henderson, L.M. 1987. "Difficulties in impact prediction auditing." *Environmental Impact Assessment Worldletter* May/June 1987: 9–12.

Henz, D.J. 1984. "Safeguarding confidential business information: environmental audits and litigation." *Pollution Engineering* 16: 30.

IBC. 1990. *Environmental Audit Conference.* Sydney: International Business Conferences Pty Ltd.

IHEI (International Hotels Environment Initiative). 1994. *Environmental Management for Hotels: The Industry Guide to Best Practice.* Oxford: Butterworth-Heinemann.

Knight, M. 1985. *Review of Seven Environmental Assessments Carried Out between 1975 and 1982.* Melbourne: Victoria Ministry of Planning and Environment.

Larderel, J.A. de. 1994. "The agenda for corporate responsibility." *Environment Strategy Europe 1994/95*, 121–122. London: Camden Publishing Limited.

Levin, M.H. 1983. "An EPA response on confidentiality in environmental auditing." *Environment Law Reporter* 13.

McDonald, J. 1994. "Corporate confidentiality after Caltex: how safe is your audit?" *Environmental and Planning Law Journal* 11: 193–210.

Oates, J.A.H. 1982. "Use of an environmental audit procedure." *Clean Air* 12: 3.

Ovington, J.D. 1992. *1992 Audit of the Environment in the Australian International Development Cooperation Program.* Canberra: Australian Government Publishing Service.

Schaeffer, D.J. et al. 1985. "The environmental audit. I. Concepts." *Journal of Environmental Management* 9: 191–198.

Tomlinson, P. and S.F. Atkinson. 1987. "Environmental audits: proposed terminology." *Environmental Monitoring and Assessment* 8: 187–198.

United Nations Environment Programme. 1994. *Company Environment Reporting: A measure of the progress of business and industry towards sustainable development.* Nairobi: UNEP.

Chapter 14
Environmental Sustainability[1]

ROBERT GOODLAND
Environment Department, the World Bank
HERMAN DALY
University of Maryland, USA

ECONOMY AND THE ENVIRONMENT

The paramount importance of sustainability arose partly because the world is recognising that current patterns of economic development are not generalisable. Present levels of OECD per capita resource consumption cannot possibly be generalised to all currently living people, much less to future generations, without liquidating the natural capital on which future economic activity depends. Sustainability thus arose from the recognition that the profligate and inequitable nature of current patterns of development, when projected into the not-too-distant future, lead to biophysical impossibilities. The transition to sustainability is urgent because the deterioration of global life-support systems—the environment—imposes a time limit. We do not have time to dream of creating more living space or more environment; we must save the remnants of the only environment we have, and allow time for, and invest in the regeneration of what we have already damaged.

The widely accepted definition of sustainability in economics is: 'maintenance of capital', sometimes phrased as 'nondeclining capital'. Historically, at least as early as the Middle Ages, the merchant traders used the word 'capital' to refer to human-made capital. The merchants wanted to know how much of their trading ships' cargo sales receipts could be consumed by their families without depleting their capital. Of all the forms of capital (discussed below), environmental sustainability refers to natural capital. So environmental sustainability is defined by two further terms, namely 'natural capital' and 'maintenance of' or at least 'nondeclining' natural capital. These terms are amplified below.

NATURAL CAPITAL

Natural capital is basically our natural environment, and is defined as the stock of environmentally provided assets (such as soil, atmosphere, forests, water, wetlands), which provide a flow of useful goods or services. The flow of useful goods and

[1]*Environmental and Social Impact Assessment* - Edited by F. Vanclay and D.A. Bronstein. Copyright © 1995 by the International Association of Impact Assessment. Published in 1995 by John Wiley & Sons Ltd.

services from natural capital can be renewable or nonrenewable, and marketed or nonmarketed. Sustainability means maintaining environmental assets, or at least not depleting them. 'Income' is sustainable, at least in terms of the generally accepted definition of economics given by Nobel laureate, Sir John Hicks (1946) as "the maximum value a person can consume during a week, and still expect to be as well off at the end of the week as at the beginning." Any consumption that is not sustainable cannot be counted as income. Prevailing models of development treat consumption of natural capital as income, and therefore are unsustainable. Consumption of natural capital is liquidation, the opposite of capital accumulation.

Natural capital is distinguished from other forms of capital, namely human or social capital (people, their capacity levels, institutions, cultural cohesion, education, information, knowledge),[2] and human-made capital (houses, roads, factories, ships). From the mercantilists until very recently, capital referred to the form of capital in the shortest supply, namely human-made capital. Investments were made in response to the limiting factor, such as sawmills and fishing boats, because their natural capital complements—forests and fish—were abundant. That idyllic era has ended.

Now that the environment is so heavily used, the limiting factor for much economic development has become natural capital, rather than human-made capital. Fish have become limiting, rather than fishing boats. As natural forests and fish populations become limiting, we begin to invest in plantation forests and fish ponds. This introduces a hybrid category that combines natural and human-made capital—a category we may call 'cultivated natural capital'.[3] While cultivated natural capital is often a worthwhile investment, it does not avoid entirely the problem of complementarity and the limiting role of natural capital. For example, in a plantation forest, human-made capital can be used for planting, disease control, etc., but ultimately the natural capital proper of sunlight, rainfall, soil, and the tree gene pool, will prove limiting to the growth of trees. Also a plantation forest is a poor habitat for most other species and converting natural capital to cultivated natural capital imposes a cost in the form of lost biodiversity.

Natural Capital Is Now Limiting

In an era in which natural capital was considered infinite relative to the scale of human use, it was reasonable not to deduct natural capital consumption from gross

[2] Human capital formation, by convention, is left out of the national accounts for various reasons, one of which is that, if it is truly productive, it will eventually be reflected, through enhanced productivity, in a higher GDP. Realisation of the values of education and administration, for example, are lagged, and are conventionally assumed to be equal to their costs. The loss of natural capital, if not recorded, as largely is the case today, may take some time before it will reflect itself in income and productivity measurements.

[3] The subcategory of marketed natural capital, intermediate between human capital and natural capital, is 'cultivated natural capital' such as pond-bred fish, cattle herds, and plantation forests.

receipts in calculating income. That era is now past. Today the limiting factor in development is more often remaining natural capital than extra human-made capital. The fish catch is limited by the remaining fish population, not by fishing boats; timber is limited by remaining forests, not by sawmills; petroleum is limited by geological deposits and atmospheric capacity to absorb CO_2, not by refining capacity.

The goal of environmental sustainability is the conservative effort to maintain the traditional meaning and measure of income in an era in which natural capital is no longer a free good, but is more and more the limiting factor in development. At a conceptual level, the justification for making environmental sustainability a *sine qua non* for project eligibility could not be stronger or more conservative. The difficulties in applying the concept arise mainly from operational problems of measurement and valuation of natural capital, as emphasised by Ahmad et al. (1989), Lutz (1993), and El Serafy (1991, 1993).

ENVIRONMENTAL SUSTAINABILITY

Sustainability can be conceived as having three levels, weak, strong, and absurdly strong, depending on how strictly one elects to hew to the concept of maintenance or nondeclining capital (Daly and Cobb 1989). Weak sustainability is maintaining total capital intact. Thus oil may be depleted as long as the receipts are invested in sustainable (e.g., energy-producing, human capital development) activities elsewhere. The sustainable receipts are income. This generously assumes that human-made and natural capital are to a large extent substitutable. Strong sustainability requires maintaining both human-made and natural capital intact separately. Receipts from depleting oil should be invested in sustainable energy production, rather than in any asset. This assumes that natural and human-made capital are not really substitutes but complements in most production functions. A sawmill (human capital) is worthless without the complementary natural capital of a forest. Economic logic requires us to invest in the limiting factor which now is increasingly natural capital, rather than the human-made capital, which was limiting yesteryear. Absurdly strong sustainability would never deplete anything. Nonrenewable resources—absurdly—could not be used at all; for renewables, only net annual growth rates could be harvested, in the form of the overmature portion of the stock.

Investing in natural capital (nonmarketed) is essentially an infrastructure investment on a grand scale, that is the biophysical infrastructure of the entire human niche. Investment in such 'infra-infrastructure' maintains the productivity of all previous economic investments in human-made capital, public or private, by rebuilding the natural capital stocks that have come to be limitative. Operationally, this translates into: encouraging the growth of natural capital by reducing our level of current exploitation; investing in projects to relieve pressure on natural capital stocks by expanding cultivated natural capital, such as tree plantations to relieve pressure on natural forests; and increasing the end-use efficiency of products (improved cookstoves, solar cookers, hay-box cookers, wind pumps, solar pumps, manure rather than chemical fertiliser).

CRITERIA FOR ENVIRONMENTAL SUSTAINABILITY

From the above 'maintenance of natural capital' approach to environmental sustainability (EnvSus), we can draw practical rules-of-thumb (figure 1) to guide the design of economic development.[4] As a first approximation, the design of new projects should be compared with the input/output rules (figure 1) in order to assess the extent to which a project is sustainable. At the next level of detail, specific indicators of environmental sustainability can be used (Goodland et al. 1993).

Figure 1. **Rules-of-thumb for environmental sustainability**

---

1. **Output rule**
    Waste emissions from a project should be within the assimilative capacity of the local environment to absorb without unacceptable degradation of its future waste absorptive capacity or other important services.

2. **Input rule**
    a. *Renewables:* harvest rates of renewable resource inputs would be within the regenerative capacity of the natural system that generates them.
    b. *Nonrenewables:* depletion rates of nonrenewable resource inputs should be equal to the rate at which renewable substitutes are developed by human invention and investment. Part of the proceeds from liquidating nonrenewables should be allocated to research in pursuit of sustainable substitutes.*

*For a theoretical development of this idea, see El Serafy (1991, 1993).

The implications of implementing environmental sustainability are immense: renewable resources must be managed for the long term; waste and pollution need to be reduced; energy and materials should only be used with scrupulous efficiency; solar energy and other renewable energy sources in all their possible forms need to be developed; and investment must be made in repairing the damage done to the earth in the past. Environmental sustainability needs enabling conditions in order to

---

[4] Environmental sustainability is a 'constraint' in economic terms, rather than a maximising criterion, such as profit or discounted utility, or present value of a future income stream. It is important to distinguish sharply between the environmental sustainability constraint, and the largely unrelated profit maximising criterion of 'sustained-yield'. Other misconceptions about environmental sustainability are discussed later in this chapter.

be developed: not only economic and social sustainability (figure 2), but also democracy, human resource development, empowerment of women, and much more investment in human capital than common today (i.e., increased literacy, especially ecoliteracy, Orr 1992). (See also Figure 3 for comparisons.)

Figure 2. **Objectives of social, economic, and environmental sustainability**

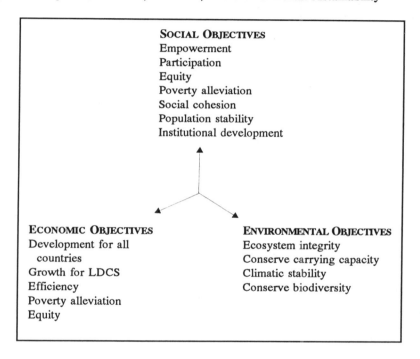

CONSTRAINTS TO ACHIEVING ENVIRONMENTAL SUSTAINABILITY

A worldwide view of 'business-as-usual' is making it ever more difficult to change course. The sooner the world starts to approach environmental sustainability, the easier it will become. Technology and education make big differences: the demographic transition from high birth and death rates to low birth and death rates took a century in Europe, but only a decade in Taiwan. But the longer the delay, the worse the eventual quality of life (e.g., fewer choices, fewer species, more risk). The prime evidence that the world is hurtling away from environmental sustainability at present (Simonis 1990; Meadows et al. 1992, Hardin 1993, Brown et al. 1994, Pimentel 1994) is that the global economy is being temporarily maintained only through the exhaustion and dispersion of a one-time inheritance of natural capital, such as topsoil, groundwater, tropical forests, fisheries, and biodiversity.

Figure 3. Comparison of social, economic, and environmental sustainability

SOCIAL
SUSTAINABILITY (SOCSUS)

Socio-cultural stability is the social scaffolding provided through networks of people's organisations that empower self-control and self-policing in peoples' management of natural resources (cf. Cernea 1993). Resources should be used in ways that increase equity and social justice, while reducing social disruptions. Human rights, education, employment, women's empowerment, transparency of decision making, fiscal accountability, and participation seem to be integral to SocSus. SocSus will emphasise qualitative improvement of social organisation patterns and community well-being over quantitative growth of physical assets; and cradle-to-grave pricing to cover full costs, especially social. It will be achieved only by systematic community participation. Social cohesion, cultural identity, diversity, sodality, sense of community, love, tolerance, compassion, humility, patience, forbearance, fraternity, fellowship, institutions, pluralism, commonly accepted standards of honesty, laws, discipline, etc., constitute the part of social capital that is least subject to rigorous measurement, but probably most important for SocSus. This 'moral capital', as some have called it, requires maintenance and replenishment by shared values and equal rights, and by community, religious and cultural interactions. Without this care it will depreciate just as surely as will physical capital. Human capital investments in education, health and nutrition of individuals are now accepted as part of economic development, but social capital, as needed for social sustainability, is not yet adequately recognised.

ECONOMIC
SUSTAINABILITY (ECONSUS)

Economic capital should be stable. The widely accepted definition of economic sustainability is 'maintenance of capital', or keeping capital intact, and has been used by accountants since the Middle Ages to enable merchant traders to know how much of their sales receipts they and their families could consume without reducing their ability to continue trading. Thus Hicks' (1946) definition of income—'the amount one can consume during a period and still be as well off at the end of the period'—can define economic sustainability, as it involves the consumption of interest rather than capital. We now need to extrapolate the definition of Hicksian income from sole focus on human-made capital and its surrogate (money) to embrace the other three forms of capital (natural, social, and human). Economics has rarely been concerned with natural capital (e.g. intact forests, healthy air) because until relatively recently it had not been scarce. This new pattern of scarcity, that of scarce natural capital, arose because the scale of the human economic subsystem has now grown large relative to its supporting ecosystem. To the traditional economic criteria of allocation and efficiency, must now be added a third, that of scale. This scale criterion would constrain throughput growth—the flow of material and energy (natural capital) from environmental sources to sinks, via the human economic subsystem. Economics prefers to value things in money terms, and it is having major problems valuing natural capital, intangible, intergenerational, and especially common access resources, such as air, etc. Too, environmental costs used to be 'externalised', but only now are starting to be internalised through sound environmental policies and valuation techniques. Because people and irreversibles are at stake, economics needs to use anticipation and the precautionary principle routinely, and should err on the side of caution in the face of uncertainty and risk.

ENVIRONMENTAL
SUSTAINABILITY (ENVSUS)

Although environmental sustainability is needed by humans and originated because of social concerns, EnvSus itself seeks to improve human welfare by protecting the sources of raw materials used for human needs and ensuring that the sinks for human wastes are not exceeded, in order to prevent harm to humans.

Humanity must learn to live within the limitations of the biophysical environment. Natural capital should be maintained, both as a provider of inputs (sources), and as a sink for wastes (Serageldin 1993a). This means holding the scale of the human economic subsystem to within biophysical limits of the overall ecosystem on which it depends.

On the sink side, this translates into holding waste emissions within the assimilative capacity of the environment without impairing it.

On the source side, harvest rates of renewables must be kept within regeneration rates.

Nonrenewables cannot be made fully sustainable, but quasi-EnvSus can be approached for nonrenewables by holding their depletion rates equal to the rate at which renewable substitutes can be created (see El Serafy 1991, 1993).

The rapid depletion of these essential resources, coupled with the degradation of land and atmospheric quality, show that the human economy, as currently configured, is already inflicting serious damage on global supporting ecosystems, and is probably reducing future potential biophysical carrying capacities by depleting essential natural capital stocks (Daily and Ehrlich 1992). In spite of the massive global consumption of capital inheritance (as opposed to utilisation of income), this has been largely a result of the peoples of the OECD countries: most of the world consumes at barely subsistence levels. Even where there is environmental damage in developing countries (e.g., tropical deforestation, mining), it is often done by multinational corporations or national governments (often in order to make debt repayments) with minimal benefits to local communities.

Can humanity attain a more equitable standard of living which does not exceed the carrying capacity of the planet? The transition to environmental sustainability *will* inevitably occur. However, the ability of nations to plan for an orderly and equitable transition to environmental sustainability, rather than allowing biophysical limits to dictate the timing and course of this transition, remains in doubt.

It seems unlikely that the planet can sustain a doubling of the scale of the world economy if the industrial and energy components of the economic activity remain constant or continue to grow. Increasing the throughput growth (i.e., materials and energy) is not the way to reach environmental sustainability; we cannot 'grow' our way into sustainability, but we can 'develop' toward that goal. The global ecosystem, which is the source of all the resources needed for the economic subsystem, is finite and has now reached a stage where its regenerative and assimilative capacities have become very strained. It looks inevitable that the next century will witness double the number of people in the human economy, depleting sources and filling sinks with their increasing wastes. A single measure—population times per capita resource consumption—encapsulates what is needed to achieve environmental sustainability. This scale of the growing human economic subsystem is judged large or small relative to the finite global ecosystem on which it so totally depends, and of which it is a part.

The global ecosystem is the source of all material inputs feeding the economic subsystem, and is the sink for all its wastes. Population times per capita resource consumption is the total flow—throughput—of resources from the global ecosystem to the economic subsystem, then back to the global ecosystem as waste, as dramatised in figure 4. The 'empty world' case, where the scale of the human economic subsystem is small relative to the large, but nongrowing global ecosystem, is long gone. In the lower diagram, the 'full world' case, the scale of the human economic subsystem is large and still growing, relative to the finite global ecosystem. In the full world case, the economic subsystem has already started to interfere with global ecosystemic processes, such as altering the composition of the atmosphere (e.g., climate change, acid rain, and *waldesterben* or forest death), or the now nearly global damage to the ozone shield.

Figure 4. **Relationship between the economic and the environmental systems: 3 views**

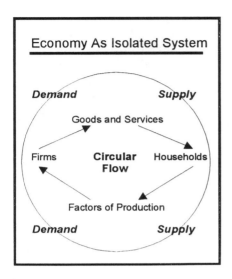

← **A. The economy as an isolated system**

B. Linking the economic and environmental systems →

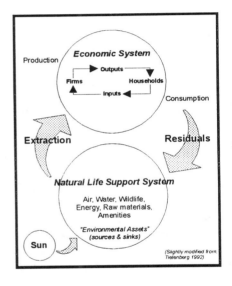

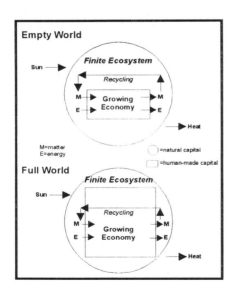

← **C. The economy dependent on the environment**

GROWTH COMPARED WITH DEVELOPMENT

To grow means to increase in size by the assimilation or accretion of materials. *To develop* means to expand or realise the potentialities of; to bring to a fuller, greater or better state. This is a definitional distinction useful in conceptualising sustainability, in that development is sustainable, and throughput growth is not. Thus 'sustainable growth' is a bad oxymoron. This is unrelated to the unhelpful 'development vs environment' perception of a policy tradeoff. True, growth refers to added value, but sustainability demands that we disaggregate what part of value added increase is due to quantity change (throughput) and what part to qualitative improvement.

When something grows, it gets quantitatively bigger; when it develops, it gets qualitatively better or at least different. Quantitative growth and qualitative improvement follow different laws. Our planet develops over time without growing. Our economy, a subsystem of the finite and nongrowing earth must eventually adapt to a similar pattern of development without throughput growth. The time for such adaption is now.

It is neither ethical nor helpful to the environment to expect poor countries to cut or arrest their development, which tends to be highly associated with throughput growth. Therefore, the rich countries, which are responsible for most of today's environmental damage, and whose material well being can sustain halting or even reversing throughput growth, must take the lead in this respect. Poverty reduction will require considerable growth, as well as development, in developing countries. But global environmental constraints (atmospheric CO_2 accumulation, ozone shield damage, acid rain, etc.) are real, and more growth for the South must be balanced by negative throughput growth for the North. Future Northern growth should be sought from productivity increases in terms of throughput (e.g., reducing energy intensity of production).

Development by the North must be used to free resources (source and sink functions of the environment) for growth and development so urgently needed by the poorer nations. Large-scale transfers to the poorer countries also will be required, especially as the impact of economic stability in North countries may depress terms of trade and lower economic activity in developing countries. Higher prices for the exports of poorer countries, as well as debt relief, therefore will be required. Most importantly, population stability is essential to reduce the need for growth everywhere, but especially where population growth is highest, i.e., in the poor countries (Goodland et al. 1994).

ENVIRONMENTAL IMPACT DISAGGREGATED

Carrying capacity is a measure of the amount of renewable resources in the environment in units of the number of organisms these resources can support. It is thus a function of the area and the organism: a given area could support more lizards than birds with the same body mass. Carrying capacity is difficult to estimate for humans because of major differences in affluence and technology. An undesirable 'factory-farm' approach could support a large human population at the lowest standards of living: certainly the maximum number of people is not the optimum. The higher the throughput of matter and energy, or the higher the consumption of environmental sources and sinks, the fewer the number of people that can enjoy it. Ehrlich and Holdren (1974) encapsulate this concept best: The impact (I) of any population or nation upon environmental sources and sinks[5] is a product of its population (P), its level of affluence (A), and the damage done by the particular technologies (T) that support that affluence.

$$I = P \cdot A \cdot T$$
$$[I = P \cdot Y/P \cdot I/Y]$$

Population (P) refers to human numbers.
Affluence (Y/P) is output (Y) per capita.
Technology (I/Y) refers to environmental impact per unit of output, i.e., a dollar's worth of solar heating stresses the environment less than a dollar's worth of heat from a lignite-fired thermal power plant.

In the $I = P \cdot A \cdot T$ identity, *affluence* means per capita consumption; *technology* refers to technological efficiency defined in terms of the number of units of environmental impact required to produce one dollar's worth of human well-being. Thus, where P is population, and Y is total production, then $I = P \cdot Y/P \cdot I/Y$, and EnvSus = I iff I < CC where *iff* means 'if and only if' and *CC* means carrying capacity. Where $P \cdot A \cdot T > CC$, the situation is environmentally unsustainable.

The elements of P, A, and T are not independent of each other. For example, increased A enables more money to be spent on existing T. The bind is that increased A + P can force the use of more powerful and damaging T. Also, increased A can increase or decrease P. The fact that the three factors are not independent over time does not alter the fact that it holds true at each and every instant. Increased income *will* impact on the environment, other things being equal. If one argues that other things are not equal, then one has to show that increased A causes a more than offsetting reduction in P.

There are only three means of reducing environmental impacts of human activities upon the environment. These are (1) limiting population growth; (2) limiting affluence; and (3) improving technology, thereby reducing throughput intensity of

[5] Early studies of environmental limits to human activities (e.g., Meadows et al. 1972) emphasised the limits to environmental resources (i.e., petroleum, copper, etc.). Experience has shown, however, that the sink constraints (i.e., waste assimilation such as air and water pollution, greenhouse gases and ozone depletion, etc.) are more stringent (Meadows et al. 1992).

production. There is much to be done to limit the impact of human activities upon the environment, although so far politically unpopular and difficult to achieve. The paths through which 'I' can be limited—population, affluence, technology—are each examined in more detail below.

Population

Population stability is fundamental to environmental sustainability. Today's 5.6 billion people are increasing by nearly 100 million a year, or a quarter of a million more people every day. Just the basic maintenance of 100 million extra people per year needs an irreducible minimum of throughput in the form of clothing, housing, food, and fuel. There is so much momentum in population growth that even under the United Nations' most optimistic scenario, the world's population may level off at 11.6 billion in 2150! Since under current inequitable patterns of production, consumption and distribution, a full one-fifth of humanity is not adequately provided for, even at today's relatively low population, the prospects for being either able or willing to provide better for double that number of people look grim indeed. Pimentel (1994) plausibly calculates that the world can support about one or two billion people sustainably in relative prosperity. He clarifies the choice: 'does human society want 10 to 15 billion humans living in poverty and malnourishment, or 1 to 2 billion living with abundant resources and a quality environment?' The political problem of unwillingness to share is not the biophysical problem of encountering limits to total product. Much greater sharing seems essential for environmental sustainability. It is important not to make the opposite error of suggesting that more equitable sharing will permit us completely to avoid the issue of biophysical limits to total production.

The human population is totally dependent on energy from the sun, fixed by green plants, for all food, practically all fibre (cotton, wool, paper), most building materials (wood), and practically all the cooking and heating fuels in developing countries (fuelwood). The human economic subsystem now appropriates 40 percent of all that energy, according to Vitousek et al. (1986). It is probably impossible, and certainly undesirable, to use the 100 percent that is implied in less than two doublings of the human population. This calculation suggests that we have less than 35–40 years left for the world to avoid using all of the world's fixed energy. Either that, or we reduce our appropriation of photosynthetic production markedly at a time when we are moving in the other direction (e.g., expanding cultivated area by 50 percent).

Several factors are all working in the same direction to reduce irreversibly the energy available globally through plants. Greenhouse warming, damage to the ozone shield, and less predictable, unstable climates seem inescapable and may have started. Depending on the models used, these will reduce agricultural, forest, range-land, fisheries, and other yields. The increases in UVb light reaching the earth through the damaged ozone shield may decrease the carbon-fixing rates of marine plankton, one of the biggest current carbon sinks. In addition, UVb light may damage young or germinating crops. According to some reports, tiny temperature elevations have already begun to increase the decomposition rates of the vast global deposits of peat, soil organic matter, and muskeg (the huge circumboreal mossy peat bog ecosystem with open stunted trees), thus releasing stored carbon. Only in mid-1992 did the circumboreal muskeg and tundra become net global carbon sources (instead of being net C-sinks). George Woodwell (pers. comm. 1993) goes as far as to claim that at least an immediate 50 percent reduction in global fossil fuel use is necessary to stabilise atmospheric composition. Whether one accepts this estimate or not, it dramatises the gravity of the situation. There is no ground for complacency.

Government policy should cherish their citizens by encouraging improvements in their well-being, not by encouraging increases in their numbers. People should not be treated as "easily replaceable materials" (Serageldin 1993a). One dimensional capitalists (so vividly described in 19th Century literature) love cheap labour brought on by overpopulation and poverty, and if there is not enough cheap labour available domestically they can move capital abroad or promote free immigration at home. Unfortunately, such socially irresponsible capitalists are not yet an extinct species.

Affluence

Overconsumption by the OECD countries contributes more to the lack of environmental sustainability today than does population growth in low income countries (Mies 1991; Parikh and Parikh 1991). Energy consumption is used as a crude surrogate for environmental impact on the earth's life support systems, since the type of energy used is not taken into account. "A baby born in the United States represents twice the impact on the Earth as one born in Sweden, three times one born in Italy, 13 times one born in Brazil, 35 times one in India, 140 times one born in Bangladesh or Kenya, and 280 times one born in Chad, Rwanda, Haiti or Nepal" (Ehrlich and Ehrlich 1989a, b). Switzerland, Japan and Scandinavia show that it is possible to make great progress in reducing the energy intensity of production. But the key question is: can humans lower their per capita impact (mainly in OECD countries) at a rate sufficiently high to counterbalance the explosive increases in population (mainly in low income countries)? The affluent are reluctant to acknowledge the concept of 'sufficiency,' preferring their current faith in consumption and the ideology that 'more is better'. The sooner OECD countries begin emphasising quality and nonmaterial satisfaction over volume of consumption, the sooner the world can reach sustainability. Redistribution from rich to poor on any significant scale is, at present, felt to be politically impossible.

Technology

There is much misplaced technological optimism. New technology is often adopted in order to improve labour productivity, which in turn can raise material standard of living. The impact of a particular technology depends on the nature of the technology, the size of the population deploying it, and the population's level of affluence. Certainly investment is needed in more sustainable technologies, such as solar energy.

In the $I = P \cdot A \cdot T$ identity, I equals population x affluence x technology. Environmental sustainability occurs when I is less than the carrying capacity. If it is accepted that world population will double in 40 years (as projected), and that rich country per capita income ($21,000) is roughly 21 times that of the poor and middle income countries ($1,000) (World Bank 1993), then to raise Southern affluence to today's level of the North (holding both impact and Northern incomes constant) means technology must improve 2 x 21 or 42 times by the year 2035. This of course is a very rough order of magnitude. Our point still would have force if the true factor were only 20 instead of 42. Since historical technological improvement rates (i.e., resource productivity improvement, or reduction in throughput intensity of GNP) never have exceeded a fraction of the needed 42 times, it will be exceedingly difficult for poor countries to catch up with rich countries in 40 years—even if the OECD countries maintain only current levels of income. Furthermore, this 42-fold increase must be in resource efficiency and not just in capital or labour efficiency. Historically, much of the increase in capital and labour efficiency has been at the expense of resource efficiency. In agriculture, for example, the increase in labour and capital productivity has required a significant increase in the complementary resource throughputs (energy, fertiliser, biocides, water) whose productivity has **fallen.** It will be that much more difficult if developing countries try to catch up with a moving target.[6]

---

[6] Throughput growth is a source of both income growth and of environmental damage. If the activities contributing to national income are disaggregated into two components, environmentally friendly (e.g., government services such as administration and justice), and environmentally burdening (e.g., industry, agriculture, utilities), about one quarter of the activities (measured in labour volume) generates about 65 percent of increases in national income in The Netherlands. "Unfortunately, that 25% is precisely the activities which impair the environment" (Tinbergen and Hueting 1991). Increases in productivity generated by a relatively small part of the economy spreads over the whole society via labour supply-demand linkages. For example, the labour volume and real output of individual barbers (or hairdressers) have not appreciably increased over the last 40 or 100 years, but their (deflated) income or value added has risen by a factor of four. The barber's increased real income has been generated by activities other than the barber's own, by activities that have used large amounts of energy and materials to produce more goods. This additional resource-based production is shared by the entire labour force through the competitive labour market. These other activities are much harder on the environment than the barber's own activities. Average Northerners now consume vastly more than they did 40 years ago all the way up the income scale: more than twice as much in the case of the US and Japan. For example, 88 percent of US households now own one car (up from 55 percent in 1935), and the average number of vehicles per household is two.

SOME COMMON MISCONCEPTIONS ABOUT ENVIRONMENTAL SUSTAINABILITY

1. *Is environmental sustainability the same as sustained yield?*

There is a lively debate, especially in forestry and fishery circles, whether EnvSus is 'sustained yield' (S–Y), in the form of timber removals from forest for example. Clearly EnvSus includes, but certainly is far from limited to, sustained yield. EnvSus is more akin to the simultaneous S–Y of many interrelated populations in an ecosystem. S–Y is often used in forestry and fisheries to determine the optimal—most profitable—extraction rate of trees or fish. EnvSus counts all the natural services of the sustained resource. S–Y counts only the service of the product extracted, and ignores all other natural services. S–Y forestry counts only the timber value extracted; EnvSus forestry counts all services. These include protecting vulnerable ethnic minority forest dwellers, biodiversity, genetic values, intrinsic as well as instrumental values, climatic, wildlife, carbon balance, water source and water moderation values, and, of course, timber extracted. The relation between the two is that if S–Y is actually achieved, then the stock resource (e.g., the forest) will be nearer sustainability than if S–Y is not achieved. S–Y in tropical forestry is doubtful now (Ludwig 1993), and will be more doubtful in the future, as human population pressures intensify. But even were S–Y to be achieved, that resource is unlikely to have also attained environmental sustainability. The optimal solution for a single variable, such as S–Y, usually (possibly inevitably) results in declining utility or declining natural capital sometime in the future, therefore is not sustainable.

2. *Is environmental sustainability a variable, or a constant?*

EnvSus is a variable, but it changes so slowly that it is probably best to assume it is constant as a first approximation. If humans evolve lungs that can use hitherto unbreathably polluted air, or if we carry cylinders of oxygen on our backs, then EnvSus is a variable. On the output side, in general, assimilative capacity cannot be substantially increased. Because 'waste is our fastest growing resource' this is significant. On the non-renewable input side, non-renewables can be used slower or more efficiently, or more ores and substitutes can be found, but the stock of non-renewables is fixed and cannot be increased. Technology and efficiency squeeze more utility out of inputs, but do not increase the stock. It is difficult to get renewables to regenerate faster! Even well fertilised and irrigated trees in the US, for example, grow slower than *laissez faire* trees in Southern Chile, which has a short winter. Light is often more limiting than water and nutrients. Human-made capital such as pond fish, intensive agriculture such as sugarcane or hydroponic laboratory greenhouse crops have reached 'maximum' levels of productivity. So again, EnvSus appears to be more constant than variable—i.e., a very slowly changing variable.

3. *Is environmental sustainability more of a concern for developing countries?*

EnvSus is even more relevant to industrial countries than to developing countries. The big difference is in burden sharing. For example the North can contribute more to decreasing the global warming risks by reducing greenhouse gas emissions, and the release of substances such as CFCs which damage the ozone shield. The North has to adapt to EnvSus more than the South, and arguably before the South. The North can afford to exert leadership on itself. But because developing economies depend to a much greater extent than OECD economies on natural resources, especially renewables, the South has much to gain from reaching EnvSus. In addition, because much tropical environmental damage is either impossible or more expensive to rehabilitate than temperate environments, the South will gain from a preventive approach, rather than emulating the curative approach and similar mistakes of the North.

4. *Does environmental sustainability imply reversion to autarky or the stone age?*

Certainly not; EnvSus is not sacrifice. On the contrary, EnvSus increases welfare. The message that affluence and overconsumption do not increase welfare is being acted on by a few people. It is important to recognise the over-riding importance of poverty alleviation first. As the diseases of overconsumption increase (heart attack, stroke), this message will spread. The concept of sufficiency (doing more with less) needs dissemination. Education is needed that love, pleasure, fulfilment, enjoyment and other rewards do not depend on overconsumption, rather, in fact, are decreased by it.

THE POSSIBILITY OF ENVIRONMENTAL SUSTAINABILITY

Experts in development are well aware that bringing the low income countries up to the affluence levels in OECD countries, in 40 or even 100 years, is an unrealistic goal. They may well accuse us of attacking a straw doll—who ever claimed that global equality at current OECD levels was possible? We acknowledge the force of that objection, but would suggest that most politicians and most citizens have not yet accepted the unrealistic nature of this goal. Most people would accept that it is desirable for low-income countries to be as rich as the North—and then leap to the false conclusion that it must therefore be possible! They are encouraged in this *non sequitur* by the realisation that if greater equality cannot be attained by growth alone, then sharing (i.e., redistribution) and population control will be necessary. Politicians find it easier to revert to wishful thinking than to face those two issues.

Once we wake up to reality, however, there is no further reason for dwelling on the impossible, and every reason to focus on what *is* possible. One can make a persuasive case (see Serageldin 1993b: 141-143) that achieving per capita income levels in low-income countries of $1,500 to $2,000 (rather than $21,000) is quite possible. Moreover, that level of income may provide 80 percent of the basic welfare provided by a $20,000 income—as measured by life expectancy, nutrition, education, and other measures of social welfare. But to accomplish the possible, we must stop idolising the impossible.

The contribution to approaching global environmental sustainability differs markedly between politico-geographic regions. OECD's main contribution to environmental sustainability should surely be to cease its long history of environmental damage from overconsumption and pollution (corollaries to affluence under today's technology), such as greenhouse warming and ozone shield damage. The contribution of low-income countries lies in stabilising the human population. The former centrally planned economies' contribution to environmental sustainability seems to be more in accelerating the modernisation of its technology, and reducing acid rain by removing subsidies on dirty coal, which together, will stop poisoning the land, reduce *waldesterben* or forest death, and will reduce nuclear risks. It is in the OECD's self-interest to accelerate technology transfer to the former centrally planned economies and to the low income countries. It is possible that with current types of technology and production systems, the global economy has already exceeded the sustainable limits of the global ecosystem and that manifold expansion of anything remotely resembling the present global economy would not alter the current trajectory towards unsustainability and collapse. We believe that in conflicts between political feasibility and biophysical realities, the former must eventually give way to the latter, although we cannot specify exactly how long 'eventually' will be.

PRIORITIES TO APPROACH GLOBAL ENVIRONMENTAL SUSTAINABILITY

The main means to accelerate the two crucial transitions to environmental sustainability—to population stability and to renewable energy—are by human capital formation, technology transfer, and by direct poverty alleviation. Human capital formation means more investment in education, training and employment creation particularly for girls equivalent to that now enjoyed by boys. Human capital needs constant renewal each generation, as well as during the lifetime of each individual. Vulnerable minorities and the poor need special support. Empowerment of women, and upholding women's reproductive and health rights are important on their own, with the added benefit that they lead to smaller families. Incentives to delay marriage, to postpone the first birth, and to space subsequent births also are tremendously effective demographically. They must be supported by meeting unmet demand for family planning. Population stability is more important in the overconsuming North at present. It is also important where 80 percent of future population growth will occur, namely in the South.

Technology transfer is needed to enable the South and East to leapfrog the North's environmentally damaging and inefficient 'smoke stack' stage of economic evolution. For developing countries, this requires creating an incentive framework conducive to efficient investment. For industrial countries, this requires adequate investment in renewable energy and clean technologies.

Direct poverty alleviation also is essential. Low-tech, labour-intensive job creation, with training, is the best start. Social safety nets and directly targeted aid (e.g., coarse grains) also are needed in most cases (World Bank 1992; Goodland and Daly 1993a,b).

The final transition needed for environmental sustainability, although not amplified in this chapter, is the crucial social and behavioural transition from overconsumption to sufficient consumption; from waste to efficiency and sufficiency. It is exceptionally important to educate mainly OECD citizens to understand that overconsumption does not increase their welfare. On the contrary, overconsumption decreases welfare. For example, limousines are slower than metro commuting; three steaks a day promote heart disease. Both examples postpone environmental sustainability. The challenge is to persuade overconsumers to change their behaviour.

PROGRESS TOWARDS ENVIRONMENTAL SUSTAINABILITY AND SUSTAINABLE DEVELOPMENT

The World Bank, although criticised for its environmental record in the past (Payer 1982; Rich 1994; Schwartzman 1986), has now adopted sustainability as an important platform for assessing future projects (Serageldin 1993a; World Bank 1994) and has directly invested in environmental management projects in developing countries. Lending for environmental projects totalled $2.4 billion in 1994; a fivefold increase over the last five years. In addition, funding for the social sector, such as population, education and human resource development, has been boosted as they are essential for sustainability. Furthermore, as of 1994, 47 countries had prepared national environmental action plans with assistance and encouragement from the World Bank.

The biggest failure in sustainability to date has been in improperly measuring income and investment, particularly failure to reflect deterioration of natural capital. To correct this failure, the National Income Accounts needs to be adjusted to take account of environmental and sustainability concerns (Ahmad et al. 1989; Lutz et al. 1993; El Serafy 1993). However, without an environmentally adjusted National Income Accounts, it is not possible to know whether an economy is genuinely growing, or merely living unsustainably on asset liquidation beyond its true income.

There are eight physical measures that could potentially be used in an index of environmental sustainability:

1. *Energy intensity.* The energy intensity of aggregate output (total national energy consumption/GNP), with the lower the energy intensity, the greater the sustainability.
2. *Renewable energy proportions.* The fraction of total energy from renewable sources, assuming the renewable sources are being utilised renewably.
3. *Material intensity.* The material intensity of output for various basic materials and for different sectors.

4. *Recycled proportions.* The recycled fraction of materials, and its complement, the dissipated fraction.
5. *Transport intensity.* The transport intensity of output for various basic materials and for different sectors.
6. *Water use.* The percent use of available water supply; aquifer withdrawal/recharge and contamination ratios.
7. *Cropland cultivated.* Percent of potential croplands in cultivations; topsoil erosion and depletion/regeneration ratios.
8. *Pastoral use.* Percent of rangelands and pastures being irreversibly overgrazed.

THE PRECAUTIONARY PRINCIPLE

An important and overriding principle that should dominate the environmental impact assessment process is the *precautionary principle*. Ecologists and economists disagree on the immediacy of global limits to throughput growth, on the rate of resource substitutions, and the rate of technological efficiency gains. The prudent view is that the costs of planning development with incorrect assumptions are much higher with overestimates of such rates than with underestimates. Exceeding carrying capacity reduces it for the future to below the level at which it was impaired. In many environmental capacities, the damage response curve is not linear. Overshoot of a carrying capacity can exceed a threshold and can lead to sudden crash. This emphasises the fundamental importance of the precautionary principle, a basic normative principle of international environmental law. "Rather than await certainty, regulators should act in anticipation of any potential environmental harm to prevent it" (Costanza and Cornwell 1992).

Application of the precautionary principle can be by a modified deposit-refund system which incorporates both risks and uncertain environmental costs[7] into the economic incentive system and promotes positive technological innovation. The 'flexible environmental assurance bonding system' charges an economic agent directly for known environmental damage, and levies an assurance bond equal to the best current estimate of the largest potential future environmental damages. The bond is held in an interest-bearing escrow account for a predetermined time. Portions of the bond are returned, plus interest, if and only when the agent demonstrates that the suspected worst-case damage has not occurred or would be less than originally assessed. Any damage that does occur is rehabilitated or

---

[7] Risk, or statistical uncertainty, is an event with known probability. Uncertainty, or indeterminacy is an event with unknown probability. For example, car driving risks are so well known that they are used to set automotive insurance premiums. Living near a toxic chemical dump imposes health uncertainties; no one knows the probability of health damage. Most environmental problems suffer from uncertainty, not merely risk (Costanza and Cornwell 1992).

compensated for from the bond account (Costanza and Cornwell 1992). Such a forced saving system shifts the costs of uncertainty and the burden of proof from the public onto the resource user. The resource user has a strong incentive to reduce the uncertainty of the environmental impacts as soon as possible.

CONCLUSION

This paper presents the current conceptualisation of environmental sustainability. The aim has been to contribute to and focus the debate on the essence of sustainability. The concept is a dynamic one and will be continually refined. The monumental challenge facing the world is of ensuring that possibly ten billion people are decently fed and housed within one human generation or so—but without damaging the environment on which we all depend. This means that it is imperative that the goal of environmental sustainability must be reached as soon as humanly possible.

REFERENCES

Ahmad, Y., S. El Serafy, and E. Lutz, eds. 1989. *Environmental accounting*. Washington DC: The World Bank.

Brown, L.B. et al. 1994. *State of the world: 1994*. Washington DC: Worldwatch Institute.

Cernea, M. 1993. "The sociologist's approach to sustainable development." *Finance and Development* 30(4): 11–13.

Costanza, R. and L. Cornwell. 1992. "The 4P approach to dealing with uncertainty." *Environment* 34(2): 12–20,42.

Daily, G.C. and P.R. Ehrlich. 1992. "Population, sustainability and the earth's carrying capacity." *BioScience* 42(10): 761–771.

Daly, H.E. and J. Cobb. 1989. *For the Common Good*. Boston: Beacon Press.

Ehrlich, P. and A. Ehrlich. 1989a. "Too many rich folks." *Populi* 16(3): 3–29.

Ehrlich, P. and A. Ehrlich. 1989b. "How the rich can save the poor and themselves." *Pacific and Asian Journal of Energy* 3(1): 53–63.

Ehrlich, P.R. and J.P. Holdren. 1974. "Impact of population growth." *Science* 171: 1212–1217.

El Serafy, S. 1991. "The environment as capital." In *Ecological Economics*. R. Costanza, ed. New York: Columbia University Press, 168–175.

El Serafy, S. 1993. "Country macroeconomic work and natural resources." Environment Working Paper No. 58. Washington DC: World Bank.

Goodland, R. and H.E. Daly. 1993a. "Why northern income growth is not the solution to southern poverty." *Ecological Economics* 8(1): 85–101.

Goodland, R. and H.E. Daly. 1993b. "Poverty alleviation is essential for environmental sustainability." Environment Working Paper No. 42. Washington DC: World Bank.

Goodland, R., H.E. Daly, and S. El Serafy. 1993. "The urgent need for rapid transition to global environmental sustainability." *Environmental Conservation* 20(4): 297–309.

Goodland, R., H.E. Daly, and J. Kellenberg. 1994. "Burden sharing in the transition to environmental sustainability." *Futures* 26(2): 146–155.

Hardin, G. 1993. *Living within Limits.* New York: Oxford University Press.

Hicks, J.R. 1946. *Value and Capital.* Oxford: Clarendon Press.

Ludwig, D. 1993. "Uncertainty, resource exploitation and conservation: Lessons from history." *Science* 260 (2 April): 17, 36.

Lutz, E., ed. 1993. Toward improved accounting for the environment. An UNSTAT-World Bank symposium. Washington DC: The World Bank.

Meadows, D. et al. 1972. *The Limits to Growth.* New York: Universe Books

Meadows, D., D. Meadows, and J. Randers. 1992. *Beyond the Limits.* Post Mills VT: Chelsea Green Publishing.

Mies, M. 1991. Consumption patterns of the North: The cause of environmental destruction and poverty in the South: Women and children first. Geneva, United Nations Conference on Environment and Development, United Nations Children's Emergency Fund, and United Nations Fund for Population Activities.

Orr, D.W. 1992. *Environmental literacy: Education and the transition to a postmodern world.* Albany: State University of New York Press.

Parikh, J. and K. Parikh. 1991. Consumption patterns: The driving force of environmental stress. Geneva: United Nations Conference on Environment and Development.

Payer, C. 1982. *The World Bank: A critical analysis.* New York: Monthly Review Press.

Pimentel, D. 1994. "Natural resources and an optimum human population." *Population and Environment* 15(5): 347–370.

Rich, B. 1994. *Mortgaging the Earth: The World Bank, Environmental Impoverishment, and the Crisis of Development.* Boston: Beacon Press.

Schwartzman, S. 1986. *Bankrolling Disasters: International Development Banks and the Global Environment.* San Francisco: Sierra Club.

Serageldin, I. 1993a. "Making development sustainable." *Finance and Development* 30(4): 6–10.

Serageldin, I. 1993b. *Development Partners: Aid and cooperation in the 1990s.* Stockholm: Swedish International Development Agency.

Simonis, U.E. 1990. *Beyond Growth: Elements of sustainable development.* Berlin: Edition Sigma.

Tietenberg, T. 1992. *Environmental and Natural Resource Economics.* New York: Harper Collins.

Tinbergen, J. and R. Hueting. 1991. "GNP and market prices: Wrong signals for sustainable economic success that mask environmental destruction." In *Environmentally Sustainable Economic Development: Building on Brundtland,* Environment Paper 36. R. Goodland, H. Daly, and S. El Serafy, eds. Washington DC: World Bank, 36–42.

Vitousek, P.M. et al. 1986. "Human appropriation of the products of photosynthesis." *BioScience* 36: 368–373.

World Bank, 1992. *World Development Report 1992: Development and the environment.* New York: Oxford University Press.

World Bank, 1993. *World Development Report 1993: Investing in health.* New York: Oxford University Press.

World Bank, 1994. *Making Development Sustainable: The World Bank Group and the Environment.* Washington DC: The World Bank.

Index